ROYAL COMMISSION ON ENVIRONMENTAL POLLUTIO
EIGHTEENTH REPORT
TRANSPORT AND THE ENVIRONMENT

ROYAL COMMISSION

ON

ENVIRONMENTAL POLLUTION

CHAIRMAN:
SIR JOHN HOUGHTON CBE FRS

EIGHTEENTH REPORT

TRANSPORT AND THE ENVIRONMENT

Presented to Parliament by Command of Her Majesty
October 1994

OXFORD UNIVERSITY PRESS 1995

Oxford University Press, Walton Street, Oxford OX2 6DP
Oxford New York
Athens Auckland Bangkok Bombay
Calcutta Cape Town Dar es Salaam Delhi
Florence Hong Kong Istanbul Karachi
Kuala Lumpur Madras Madrid Melbourne
Mexico City Nairobi Paris Singapore
Taipei Tokyo Toronto
and associated companies in
Berlin Ibadan

Oxford is a trade mark of Oxford University Press

Published in the United States
by Oxford University Press Inc., New York

British Library Cataloguing in Publication Data
Data available

Library of Congress Cataloging in Publication Data
Data available
ISBN 0-19-826065-2

1 3 5 7 9 10 8 6 4 2

Printed in Great Britain
on acid-free paper by
Biddles Ltd., Guildford and King's Lynn

Foreword to new edition by the Chairman of the Royal Commission

The Royal Commission's report on transport and the environment, published in October 1994, was the culmination of more than two years work. Its starting-point was the clear message in the evidence received from a large number of organisations and individuals that the projected rate of growth of transport, in particular road transport, is unsustainable in its long-term impact.

There are over a hundred recommendations in the report. Some of the most important deal with integrated planning, charging for environmental costs, switching investment from road-building to forms of public transport which are less damaging to the environment, and improved and safer facilities for walking and cycling. We believe these measures will produce not only a better environment and a healthier population, but also a more efficient transport system offering greater choice.

The report attracted considerable attention among transport specialists and among the public generally. Much of the attention focused on the annual increases in fuel duty we recommended, and to a lesser extent on the recommended reduction in the national road programme. To pick out only these two measures is misleading. One of our important conclusions is that the changed approach to transport needed in this country will require a co-ordinated strategy involving action by government and industry on a number of fronts over a long period.

The Royal Commission deliberately set its sights on the long term in order to outline the kind of transport system that will meet the needs of the 21st century and benefit our children and grandchildren. Achieving a sustainable transport system will not be easy. In addition to actions industry and government must take, there will have to be some fundamental changes in our attitudes towards transport.

The government has set up an interdepartmental task force to prepare a full response to our recommendations, and promoted a wide-ranging debate. The Secretary of State for Transport has called for a wide-ranging debate. Making our findings more accessible will facilitate public debate and help create a wider understanding of the issues and challenges involved. I am delighted therefore that Oxford University Press has agreed to make the whole report available to a larger audience.

John Houghton

Sir John Houghton, CBE FRS

ROYAL COMMISSION
ON
ENVIRONMENTAL POLLUTION

EIGHTEENTH REPORT

To the Queen's Most Excellent Majesty

MAY IT PLEASE YOUR MAJESTY

We, the undersigned Commissioners, having been appointed "to advise on matters, both national and international, concerning the pollution of the environment; on the adequacy of research in this field; and the future possibilities of danger to the environment";

And to enquire into any such matters referred to us by one of Your Majesty's Secretaries of State or by one of Your Majesty's Ministers, or any other such matters on which we ourselves shall deem it expedient to advise:

HUMBLY SUBMIT TO YOUR MAJESTY THE FOLLOWING REPORT.

The high roads may in one sense be said to bear more grass and corn than any other ground of equall bulk, as by facilitating carriage they cause all the other ground to be more improved and encourage cultivation, by which means a greater quantity of corn and grass is produced. But of themselves they produce nothing. Now if by any means you could contrive to employ less ground in them by straightening them or contracting their breadth without interrupting the communication, so as to be able to plow up 1/2 of them, you would have so much more ground in culture and consequently so much more would be produced.

Adam Smith, Lectures on Jurisprudence (1762–63), vi.128–9

CONTENTS

An environmentally sustainable transport system

Chapter 14

APPENDICES

ILLUSTRATIONS

The illustrations are located between pages 82 and 83.

INFORMATION BOXES

TABLES

FIGURES

PREFACE

An effective transport system is vital for economic well-being and the quality of life. There is widespread concern that continuing growth of transport (especially road and air traffic) will be damaging to the environment, to health and to the efficient functioning of the economy. The challenge we have faced is to propose ways in which the longer-term development of transport can be made environmentally sustainable. That means providing the access people want for continued economic growth, for their livelihoods and for leisure, but eliminating the many forms of damage which are already all too apparent.

The need for a change in direction, and a new strategy, is starkly highlighted by forecasts of road traffic growth and by the road building programme. Expenditure on the present programme is the largest ever, and many schemes in it are the subject of intense controversy. On the basis of past official forecasts even this programme would not prevent congestion worsening. We do not regard this cycle of continual road building facilitating continual growth of road traffic as environmentally sustainable.

We endorse the general framework put forward by the government for a sustainable transport policy. We have sought to extend this framework by identifying clear objectives for reducing the environmental impact of transport, which we set out below, and by proposing targets and measures through which these can be achieved.

The environmental issues we have investigated include air quality (especially its effects on health), carbon dioxide emissions (and their impact on climate change), noise, and the use of non-renewable materials. Of particular concern has been the use of land for transport infrastructure and the resulting loss of amenity and conservation value. A broader cause of concern is that land use policies and planning permissions have until recently assumed that people will travel increasing distances by car. A sustainable transport policy will require a thoroughgoing integration of transport (in all its modes) and land use policies, at national, regional and local levels.

Our economic analysis of the environmental costs of different aspects of transport has led us to recommend measures to increase over time the cost of private transport. We recommend a balancing package of investment to increase the capacity, convenience and reliability of public transport, which is environmentally less damaging. The result of these measures should be improved access and a better quality of life both in towns and cities (where cars and lorries need to be restricted in order to improve the quality of life) and outside urban areas (where minimising further landtake is paramount). Similar measures are recommended to encourage the transfer of freight to environmentally less damaging modes.

We have reviewed the ways in which innovative technology can assist in achieving sustainability. Technology is already helping to improve air quality; we advocate further advances here. We have also sought to stimulate the manufacture and sale of more fuel-efficient vehicles, and the development of appropriate technologies for improved traffic management.

Our aim has been to recommend measures which will reduce the environmental effects of all aspects of transport during the next decade and well into the next century. Although our emphasis is on the longer term, we wish to stress the importance of beginning now to implement a new transport strategy which will be genuinely sustainable. Much further work will be required to develop the strategy in detail. There are many areas where we have been able to indicate only in broad terms the direction for future policy. Our hope is that the overall vision we have presented in this report will generate the impetus and commitment that are necessary if the work is to be carried forward with urgency and vigour.

Objectives for a sustainable transport policy

A. To ensure that an effective transport policy at all levels of government is integrated with land use policy and gives priority to minimising the need for transport and increasing the proportions of trips made by environmentally less damaging modes.

B. To achieve standards of air quality that will prevent damage to human health and the environment.

C. To improve the quality of life, particularly in towns and cities, by reducing the dominance of cars and lorries and providing alternative means of access.

D. To increase the proportions of personal travel and freight transport by environmentally less damaging modes and to make the best use of existing infrastructure.

E. To halt any loss of land to transport infrastructure in areas of conservation, cultural, scenic or amenity value unless the use of the land for that purpose has been shown to be the best practicable environmental option.

F. To reduce carbon dioxide emissions from transport.

G. To reduce substantially the demands which transport infrastructure and the vehicle industry place on non-renewable materials.

H. To reduce noise nuisance from transport.

Chapter 1

SCOPE OF THE REPORT

1.1 In the general survey of the environment which formed its First Report in 1971, the Royal Commission drew attention to the possible deterioration in air quality as a result of a forecast doubling in the number of motor vehicles by 1995.[1] It also identified transport as 'the main menace' among sources of noise and discussed the effects of emissions of carbon dioxide and other substances on the global atmosphere. The Commission warned that it would be 'dangerously complacent' to ignore the potential implications of the increasing number of motor vehicles and commercial flights. The Fourth Report in 1974 returned to these subjects. By that time it was 'becoming increasingly apparent that it is not possible to cater for [the] unrestricted use [of vehicles] without engineering works on a scale that is socially unacceptable'.[2] The Commission concluded: 'We may therefore expect that limitations on their use in some urban areas will be imposed in order to safeguard the local environment. This will lead to a reduction not only in their exhaust gases but also of their noise, which many regard as a worse problem.'

1.2 A quarter of a century after the Commission was established the same issues remain. The increase in the number of vehicles has been broadly in line with the Ministry of Transport forecast which the Commission quoted in 1971.[3] The risk of complacency which the Commission identified has now been displaced by a deep and widespread concern about the prospect of further large increases in road traffic. Road construction and pollution from vehicles have become controversial issues, not only in densely populated urban areas, but in all parts of the country. The unrelenting growth of transport has become possibly the greatest environmental threat facing the UK, and one of the greatest obstacles to achieving sustainable development.

1.3 The development of modern transport systems has brought many benefits. People and goods can now be moved rapidly to or from every part of the world in a way never previously imagined. This has improved the quality of life by giving people an enormously wide choice of goods, activities and lifestyles. It has transformed the scale of the markets for firms supplying goods and services. It has provided the basis for great increases in trade and in wealth. Those developments in turn have vast repercussions for the environment. The subject of this report, however, is the transport systems themselves: the effects that construction and use of infrastructure and vehicles have on people and the environment at every level from local to global.

1.4 At local level proposals to build new roads have long been the subject of often bitter controversy. Objections have been based both on the foreseen effects on people (nuisance and disruption caused by construction, loss of homes, loss of public open space, the physical barrier created by the road, the noise of vehicles) and on the impact on the natural or built environment (visual intrusion, loss of habitats or species, damage to historic buildings). Similar controversies have been caused on occasion by proposals for new or upgraded railways, airports or ports.

1.5 Increasing levels of road traffic have given rise to more general concerns. The finite nature of oil resources has received less attention in recent years than in the 1970s. There has been progressively greater emphasis on the direct and indirect impact of the pollutants emitted by vehicles. This ranges from localised effects (for example, the toxicity of carbon monoxide) through regional effects (ground-level ozone episodes) and national and transnational effects (acid deposition) to the global significance of carbon dioxide as a greenhouse gas contributing to what could be irreversible climate change. A particular public concern, strongly voiced, is the increasing number of asthma cases and the possibility that vehicle emissions are one of the causes. Further increases in numbers of vehicles and their use would tend to erode the benefits of the more stringent limits now placed on emissions from new vehicles.

1.6 The environmental effects of transport systems are so many and various that it would not be practicable to discuss all of them in detail in a single report. Some aspects have been addressed in previous Commission reports. Some have recently been covered in reports by other official bodies, notably the Donaldson Inquiry into the prevention of pollution from merchant shipping and the committees advising the Department of the Environment and the Chief Medical Officer about air pollution. Our aim in preparing this report has been to identify, and give priority to, the issues which have the most pressing and lasting significance for human health or the environment and to take full account of the conclusions reached by other bodies about those issues.

1.7 Almost every member of the community contributes to the pollution caused by transport. Tens of millions of individuals in the UK take decisions about whether to buy a car or replace an existing car, which car to buy, what journeys to make, whether to make them by car or in some other way. The extent of pollution is also affected by the style of driving people adopt. This is a ground for hope because it means individual citizens have the capacity to contribute directly to environmental improvements. It is also a ground for caution. It may be difficult to change attitudes. People's use of transport is bound up with their livelihoods and lifestyles. Because the fleet of existing vehicles is very large, it will take time for improvements in the design of new vehicles to achieve their full effect. Changes in the present patterns of land use could reduce the need for travel, but will take even longer to have their full effect.

1.8 The present situation and the current trends are the result, not only of individual choices, but of policies pursued by successive governments. There is now widespread recognition of the need to analyse the environmental implications of government policies and adopt policies which give more weight to protecting the environment. The consequences of public expenditure on roads have been much discussed, and environmental assessments are now carried out routinely for particular schemes, but there has been no procedure for assessing the combined effects of the individual road schemes making up a major new route or for analysing systematically the overall impact of the national road programme or other transport policies.

1.9 Our focus in this report therefore is on the overall impact of transport policies. We set out the case for new policies which we have concluded are necessary or highly desirable from an environmental point of view but we do not attempt to specify them in the kind of detail that would be appropriate for a Transport White Paper. We discuss the future scale and nature of the national road programme for example, because this is a major environmental issue, and also the procedures that should apply to road schemes, but it is not our function to express a view about individual proposals for the construction of roads or other transport infrastructure. Many aspects of transport can be dealt with most effectively at the regional or local level. We hope this report will help make the environmental awareness being displayed by some local authorities more general throughout the UK, and also help ensure there will be a national framework within which local initiatives can be more effectively pursued.

1.10 The growth of transport and its environmental impact are major issues for all countries. Our role is to help form policy for the UK, but our perspective has necessarily been wider. Preventing global warming requires concerted international action. The UK is also subject to many obligations on environmental matters as a Member State of the European Union and under international conventions on civil aviation, shipping and air pollution. Many of the companies involved in transport are multinational and direct their activities and planning not only to Europe but to world markets. Last, but not least, much can be learned from the efforts of other countries to cope with the same basic problems. In arriving at our conclusions and recommendations, we have tried to take advantage of innovations and experiences in other countries, but also to make allowance for differences in circumstances.

1.11 We have recommended measures for reducing the environmental effects of all aspects of transport. The primary focus of this report is on the period from 2000 to 2020, which we regard as the medium term. This is the period in which international obligations to limit carbon dioxide emissions from all sources are likely to entail significant overall reductions and therefore become more difficult to fulfil. To arrive at realistic conclusions and recommendations about the period after 2000, it has of course been necessary to make assumptions about the course of events between now and then. In general our recommendations are for action to be taken immediately. In some cases this will produce almost immediate benefits. In other cases the need for action is urgent because of the long time-lag before the full benefits will be seen.

Sustainability as the aim

1.12 We have also had constantly in mind the position after 2020. We have published separately our views about the general approach to sustainable development. In all fields the need is to identify and adopt policies which are likely to be sustainable for as far ahead as we can foresee, and certainly to the middle of the next century and beyond. We accept the definition of 'sustainable development' which was put forward by the World Commission on Environment and Development in 1987[4]: 'development that meets the needs of the present without compromising the ability of future generations to meet their own needs'. Transport provides a crucial test of the commitment to sustainable development which the UK and other countries made at the Earth Summit in Rio in 1992.

1.13 In the Sustainable Development Strategy for the UK, published in January 1994, the government put forward the following framework for a sustainable transport policy:

> to strike the right balance between the ability of transport to serve economic development and the ability to protect the environment and sustain future quality of life.
>
> To provide for the economic and social needs for access with less need for travel.
>
> To take measures which reduce the environmental impact of transport and influence the rate of traffic growth.
>
> To ensure that users pay the full social and environmental cost of their transport decisions, so improving the overall efficiency of those decisions for the economy as a whole and bringing environmental benefits.[5]

1.14 Within this framework the government has acknowledged that, in addition to some significant measures already taken since the Rio conference, it will need to take action to:

> influence the rate of traffic growth and provide a framework for individual choice which enables environmental objectives to be met;
>
> improve understanding of the costs and benefits associated with transport, to ensure that transport decisions reflect the full costs they impose, and to ensure that measures affecting the transport sector are the most efficient ones for the economy as a whole;
>
> improve the environmental performance of vehicles;
>
> increase understanding of environmental impacts and pollutant emissions from transport. . .;
>
> explore the role for new technologies such as telecommunications.[6]

1.15 At the beginning of our study we identified a number of respects in which the present use of vehicles and the construction of vehicles and transport infrastructure might be in conflict with the aim of sustainability. These respects are:

a. using up finite resources of fuel;
b. using up finite resources of other critical materials;
c. contributing to irreversible changes in climate through emission of greenhouse gases;
d. producing other forms of pollution with widespread effects which are either irreversible or cause serious long-term damage;
e. causing serious damage to human health or quality of life;
f. materially eroding the stock of natural and semi-natural habitats and areas with amenity or cultural value;
g. promoting patterns of land use which depend for their viability on transport systems which will have one or more of the effects at a–f above.

In the course of our study we examined the extent to which the present transport system should be regarded as unsustainable in these respects and the actions that can be taken to make it more sustainable.

1.16 A serious dimension of transport is the pain, grief and loss of life and limb caused by accidents. The government has set the target of reducing deaths and injuries from road accidents to two-thirds of the 1981–85 level by 2000. As such an ambitious target already exists, and as road accidents are monitored and analysed extensively elsewhere, we have not made them a main focus of our study. Some measures to reduce accidents, for example, better enforcement of speed limits, also have environmental advantages. In framing our own recommendations, our aim has been that they should so far as possible have the effect of reducing the probability and severity of accidents. Our analysis of the economic aspects of transport includes the costs imposed by accidents.

1.17 We endorse the general framework for a sustainable transport policy which the government has put forward. Our aim has been to formulate a set of specific objectives to provide a solid basis for such a policy. We believe government action in pursuit of those objectives should be designed to achieve clear and, wherever possible, quantified targets. We have made proposals about what those targets should be. We have also recommended measures which will contribute to meeting those targets. As in the case of road accidents, the government has accepted that setting targets is a legitimate and powerful method for carrying policies forward. We believe the targets we have proposed are achievable. There are many uncertainties, especially in the longer term. The prudent way of dealing with such uncertainties is to establish explicit parameters for policy, so that these can be reviewed, and if necessary revised upwards or downwards, as further information becomes available.

1.18 The targets we have proposed are also deliberately challenging. Because transport is such an important dimension of national life they have far-reaching repercussions. Transport is a subject of keen, and often ambivalent, concern to almost everyone. The acceptance and consistent implementation of our recommendations will require a much broader consensus than was required in the case of previous Commission reports. We were impressed by the seriousness with which the environmental effects of transport are being addressed in policy statements from a very wide range of different organisations and interests. There has been a convergence towards a recognition that current trends are not sustainable. We judge it realistic to seek a new consensus about what will constitute a sustainable transport policy.

Structure of the report

1.19 The first part of this report (chapters 2–5) is devoted to analysis and assessment of the environmental problems created by transport. In **chapter 2** we describe briefly the present pattern of transport in the UK and its place in the economy. We review the social changes which have been associated with the massive growth in car travel. We then describe how the official forecasts of growth in road traffic made in 1989 were prepared and what they show.

1.20 The following two chapters assess the effects which have already been produced on health and on the environment. **Chapter 3** summarises the effects of vehicle emissions on air quality and on the Earth's atmosphere. **Chapter 4** looks at other major environmental issues associated with surface transport: the effects on people through accidents, noise and disruption to communities; the land requirements and wider impact of constructing new infrastructure; and the resource requirements for vehicle manufacture and road building. In both chapters we consider whether further constraints need to be imposed in future in order to safeguard health and the environment.

1.21 Air transport is the most rapidly growing form of transport. Regulatory action has to be taken globally to be effective. **Chapter 5** considers the environmental effects of aircraft and what steps can be taken to reduce such effects.

1.22 In the light of this analysis, the second part of the report (chapters 6–9) discusses various approaches to the future of transport. **Chapter 6** examines six possible perspectives. Letting congestion find its own level would not be acceptable. The 'predict and provide' approach which has been adopted in the UK up to now, and has shaped the present national road programme, does not appear to offer a coherent solution. Of the other four perspectives discussed, none can by itself achieve a satisfactory reconciliation between people's desires for mobility and the need to achieve sustainable development, but each of them has an important contribution to make. In the following chapters we develop these four perspectives in more detail.

1.23 **Chapter 7** uses economic analysis to clarify why the opportunities now available for mobility give rise to behaviour which is not sustainable in environmental terms. It considers the extent to which money values can be placed on the forms of environmental damage described in chapters 3 and 4.

Having identified the factors that affect decisions by transport users, we make recommendations about the most effective ways of ensuring that environmental costs are taken into account in such decisions. We also consider the likely effects of our recommendations on the wider economy.

1.24 **Chapter 8** considers how far changes in the technology and performance of road vehicles could reduce environmental damage. It looks first at emissions affecting air quality, then at fuel efficiency, then at noise. It considers the potential of alternative fuels and alternative methods of propulsion, both in the near future and in the longer term. We recommend a number of regulatory, fiscal and other measures to influence the design of vehicles and the way they are used.

1.25 **Chapter 9** emphasises, and explores, the closeness of the relationship between transport and land use. It assesses the likely effectiveness of planning policies intended to have the long-term effect of reducing the need to travel; and recommends improvements to the planning system as it applies to transport infrastructure.

1.26 The third part of the report (chapters 10–13) examines present transport policies to identify where modifications are needed in order to achieve sustainability, and looks at three aspects of surface transport: freight (**chapter 10**), local journeys (**chapter 11**), and the long-distance transport system (**chapter 12**). Under these headings we review the likely impact on demand of measures recommended elsewhere in the report, identify those modes of transport which are less damaging environmentally, assess the scope for increased use of such modes in preference to private road transport (or air transport), and consider how best use can be made of existing infrastructure.

1.27 **Chapter 13** considers whether the present and planned organisation of government and the main transport services in the UK is capable of devising and delivering a coherent strategy for a sustainable transport system, and what changes need to be made. We also consider the relevance of EC policies and legislation. In some contexts these are likely as they stand to have adverse effects; but there are important fields where the objective of a sustainable transport system can be most effectively pursued through action at European level.

1.28 The final part of our report (**chapter 14**) brings together our conclusions and recommendations. It relates the specific measures to our proposed targets for achieving essential environmental objectives; and considers what the overall effect would be on traffic levels, and on public expenditure.

Acknowledgements

1.29 We requested written evidence, and in some cases oral evidence, from many organisations. We also published a general invitation to submit evidence, the text of which is at **appendix G**. The Commission held most useful discussions with government and other bodies in the Netherlands and several other countries and with the German Council of Environmental Advisors. We are grateful to all the individuals and organisations listed in **appendix H** for providing us with help, information and views for the purposes of this study. To anyone we may have overlooked, we offer our apologies as well as our thanks.

1.30 The Commission had as its special advisers for this study Dr Claire Holman and Dr Susan Owens. We are grateful to them, and to our Secretariat, for the enormous amount of assistance they gave us.

1.31 We take this opportunity of paying tribute to a Member of the Commission, Mr D A D Reeve, who died suddenly after a short illness, just before this report was sent to the printers. Don Reeve had played a major part in this study, chairing one of the groups in which we carried out much of our preparatory work, and taking part in visits to France and Japan. We are much the poorer for the loss of his wisdom and experience. He is deeply missed.

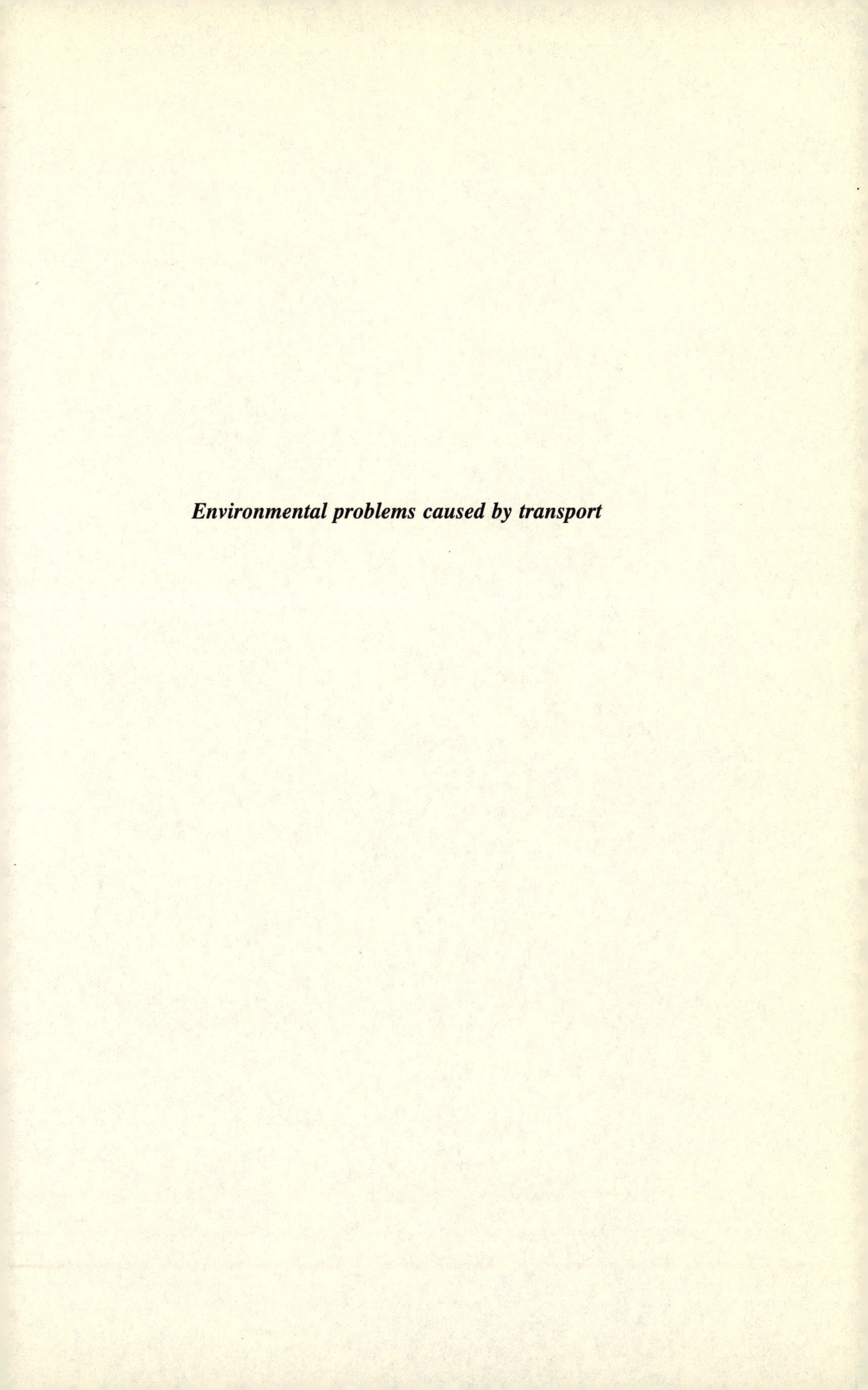

Environmental problems caused by transport

Chapter 2

GROWTH OF MOBILITY

2.1 The search for a sustainable transport policy has to start from an understanding of the present situation and how it has come about. This chapter emphasises the enormous growth in transport which has occurred, and summarises how people and goods move about in this country, and why. It looks at the place of transport in the economy, the relationship between transport and economic growth, and the changes in lifestyles which have been associated with the growth in road traffic. It then examines the forecasts of future road traffic which the Department of Transport (DOT) produced in 1989 by extrapolating past trends.

The present pattern of transport

2.2 There are various ways of measuring movements of people and goods (box 2A[1]). Over the last 25 years the average distance travelled per person in Britain each day has risen by almost three-quarters, to nearly 18 miles.[2] People now make more journeys and the average length of their journeys has increased.[3]

2.3 This growth in travel has taken the form of an enormous increase in the distances travelled by car (figure 2-I[4]), a tenfold increase over the last 40 years. Seven out of ten journeys of a mile or more are now by car[5] and these account for over 86% of total distance travelled.[6] As a consequence travel by bus and coach, which in the early 1950s was more important than travel by car, has roughly halved and now accounts for only 6% of distance travelled. Although the estimates for cycling may be less reliable,

Figure 2-I
Growth in surface transport: movement of people by mode 1952-93

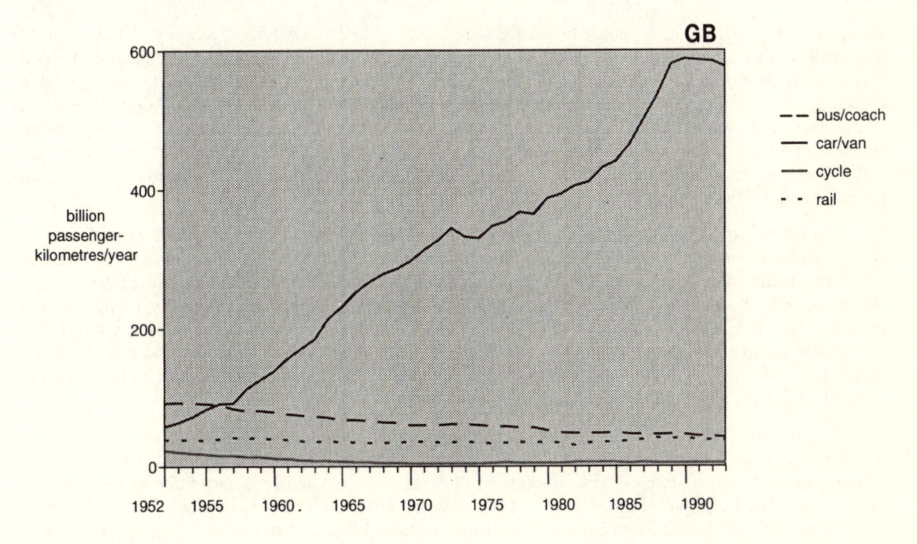

Figure 2-II
Growth in personal travel: number of journeys 1965-91

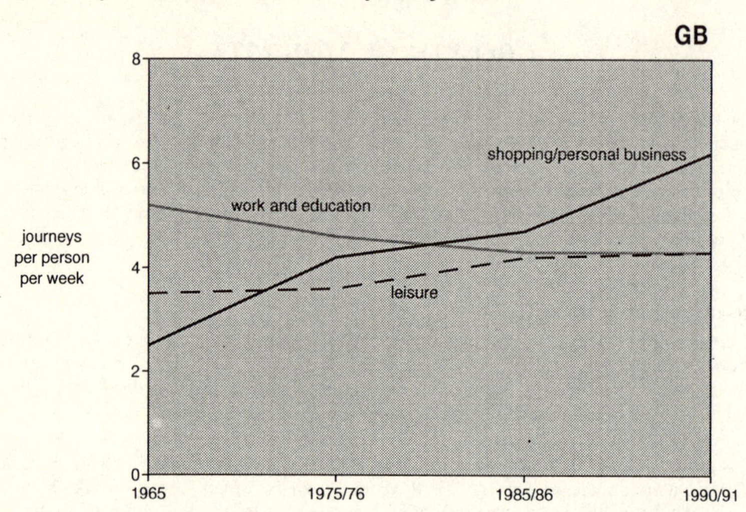

it has certainly declined even more sharply: it now accounts for only about 1% of distance travelled, the majority of journeys being made by children.[7] Rail travel is at much the same level as in 1952 but is now less than 6% of the total. Domestic air travel has grown rapidly in recent years, with passenger numbers increasing by 62% between 1982 and 1992[8], but accounts for only about 1% of passenger-kilometres.[9]

2.4 The purposes for which people travel are analysed in table 2.1.[10] A fifth of journeys[11] and a similar proportion of distance travelled are to and from work. As figure 2-II[12] shows, work and education-related journeys are now a smaller proportion of the total than 30 years ago. However, such journeys have shown the most striking increase in average length: 83% between 1965 and 1989/91.[13] Over two-thirds of journeys, and almost as high a proportion of the distance travelled, are for shopping, other personal business or leisure. Table 2.1 also shows the modes people use to make journeys of a mile or more. Car is the most important mode for all types of journey; and, for all purposes except education, cars are used for well over half of journeys.

2.5 Freight transport has also expanded, although less rapidly. The weight of goods carried rose by two-thirds between 1952 and 1968, to 2 billion tonnes a year; but has fluctuated subsequently and in 1993 was at the same level as in 1968. The trend in tonne-kilometres has been rather different; after slower growth during the 1950s, this indicator has subsequently risen more rapidly to 212 billion tonne-kilometres in 1993, an increase of 141% on 1952.[14] The primary cause of this growth was a 128% increase in the average length of freight trips by road, from about 37 km in 1952 to 84 km in 1993.[15] As the non-UK portion of international trips is not included in the statistics, the increase in the distances over which goods are carried must have been even greater.

2.6 Road is now the predominant mode of freight transport, accounting for 63% of tonne-kilometres in 1993. Substantial quantities of freight (mostly oil, petroleum products and coal) are carried by coastal shipping and inland waterways (25% of tonne-kilometres). The remainder is carried by rail (6.5%) and on-shore pipelines (5.5%).[16] Figure 2-III[17] shows how the different modes have changed in importance over the last 40 years. Road transport of freight has increased fourfold since 1952. Transport by water has more than doubled, largely because of the growth of the North Sea oil industry, and there has been a big increase in the use of pipelines for petroleum products. Rail on the other hand, which was the most important mode in 1952, has declined substantially.

Table 2.1
Why people travel

percentages (rounded to nearest whole number)

proportion of journeys	proportion of distance travelled		proportion of journeys for each purpose made wholly or mainly							
			on foot	by car: driver	by car	by car: passenger	by local bus	by rail	in other ways	
29	40	leisure	12	37	74	37	6	2	6	100
23	15	other personal business	7	52	83	31	6	1	3	100
20	20	to and from work	5	58	72	14	10	5	8	100
19	11	shopping	10	43	70	27	16	1	3	100
5	11	business	3	81	92	11	3	3	3	100
4	2	education	18	6	38	32	21	3	10	100
100	100	all purposes	9	46	74	28	9	2	6	100

This analysis excludes journeys of less than a mile and distances travelled outside Great Britain.

Figure 2-III
Growth in surface transport: movement of goods by mode 1952-93

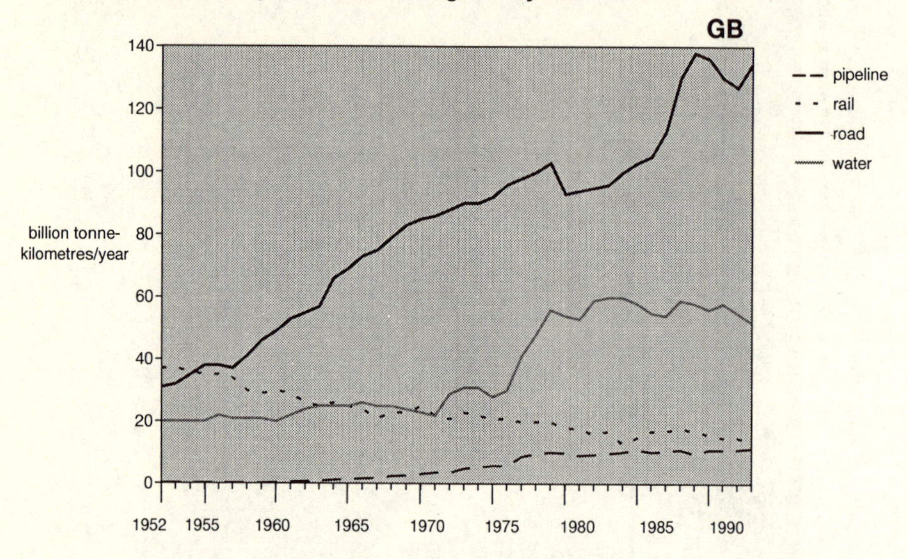

The coverage of statistics of waterborne freight was extended in 1972 and now includes North Sea oil traffic and other one-port freight movements.

Transport in the economy

2.7 Since 1952, the growth of freight transport, measured in tonne-kilometres, has broadly parallelled the growth in gross domestic product (GDP) (figure 2-IV[18]). The fluctuations in the weight of goods carried have been an amplified version of fluctuations in the economy, probably because the transport-intensive construction industry is so cyclical. For example, after a sharp fall in the early 1980s, tonnes lifted rose by 29% between 1984 and 1989 but dropped by 13% between 1989 and 1992.[19] The absence of consistent long-term growth in the weight carried reflects changes in the type of load. Coal has become much less important.[20] Other categories of goods have tended to become lighter because different materials have been used in their manufacture and packaging.

2.8 Movements of people within Britain have tended to grow more rapidly than the economy. People also travel widely outside the country, predominantly by air. When the statistics are adjusted to allow for that, figure 2-IV shows that the total movements of people have grown even more rapidly.[21] Acquisition of a car is the primary mechanism through which growth in personal incomes has affected personal travel. Car ownership has been closely related to the growth in personal incomes. Between 1971 and 1992 real household disposable income per head in the UK increased by nearly 80%[22] and the number of private cars rose by 93%, to 20.1 million.[23] The proportion of households with two or more cars increased more rapidly, from 7% in 1971 to 24% in 1993. It remains the case that almost a third of households do not own a car. Of the 20% of households with the lowest incomes, 74% do not own a car and people in these households travel less than half the average distance. Of the 20% of households with the highest incomes only 7% do not own a car.[24]

2.9 Transport is itself an important sector of the economy. In 1992, 15% of UK consumer expenditure was for personal travel.[25] Out of that, 68% was for the purchase and running of private vehicles, 13%

BOX 2A	MEASURING MOVEMENTS OF PEOPLE AND GOODS

Movements of people

The most common measurement is *passenger-kilometres*, the aggregate of the distances travelled by car drivers, passengers in cars, public transport passengers and cyclists.

Personal travel is defined as travel for which (irrespective of the purpose) the main reason is for the traveller himself or herself to reach the destination. Personal travel can be measured either as distance or as number of journeys.

A *journey* is any course of travel undertaken by a person for a single main purpose, unless it is of less than 200 metres and undertaken on foot. The outward and return halves of a return journey are treated as two separate journeys. Analyses sometimes exclude journeys of less than a mile.

Where a journey involves using more than one *mode* of transport it is categorised according to the mode used for the longest stage.

Movements of goods

The usual measurement is *tonne-kilometres*, the weighted total of the distances travelled by each vehicle-load of goods.

An alternative measurement is *tonnes lifted*, the total weight of goods carried.

Movements of vehicles

The most common measurement is *vehicle-kilometres travelled.* This takes no account of the number of people in each vehicle or the weight of goods carried by each vehicle.

A *vehicle movement* or *trip* is undertaken by a vehicle in one direction to convey either people or goods.

An overall measure of movements

To provide an approximate measure of overall *net mass movement*, covering both people and goods, passenger-kilometres can be divided by 20 (on the assumption that people weigh 50 kg on average) and added to tonne-kilometres for goods. *Gross mass movement* is the sum of the net mass movement and the movements of the vehicles used to carry people and goods (for example, cars, goods vehicles and trains), including movements by empty public transport and goods vehicles.

Geographical coverage

Unless otherwise indicated, statistics in this report relate to Great Britain, because that is the basis on which the government prepares transport statistics. In most respects, trends in Northern Ireland have been broadly in line with those in Great Britain.

Movements of people or goods to places outside Great Britain are normally included in government statistics up to the point at which the boat or plane was boarded.

was for air travel, and 6% each for bus and coach travel and for rail travel. In the case of firms the Treasury estimates that, of that part of GDP generated by industry and business, 10–15% is spent on the transport of goods and the use made of transport by the service sector.[26] Calculations of this kind are subject to considerable uncertainties. We have not found it practicable to arrive at an estimate of the overall proportion of GDP which represents expenditure on travel and transport by individuals, firms and government.[27]

2.10 It is a matter of definition whether the purchase of cars by or for private individuals is treated as consumption or investment. If all expenditure on the purchase of cars is categorised as investment, it represents over three-fifths of transport investment (figure 2-V[28]). On that definition transport investment was £17 billion in 1985/86 (26% of total investment), rose to £31 billion (33%) in 1989/90 and fell to £28 billion in 1992/93 (which still represented over 33% of total investment). Over this period, road-related investment represented 90% or more of transport investment.

2.11 The growth of transport should not be regarded as a simple and necessary consequence of economic growth. The nature of the economy has undergone far-reaching changes. Coal has largely

Figure 2-IV
Transport growth in relation to growth of economy 1952-92

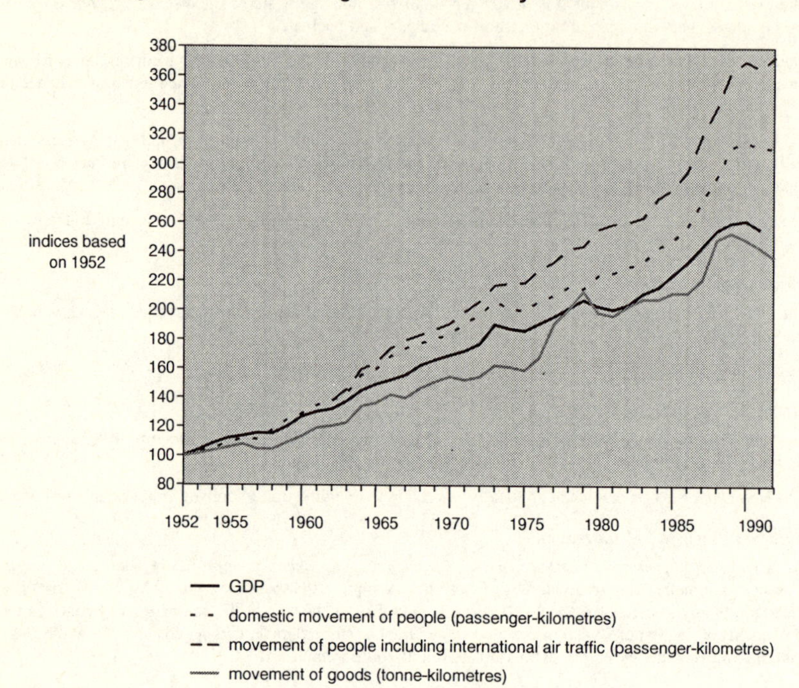

been superseded as an industrial and domestic fuel by gas and electricity, which are moved by pipeline and the national grid respectively and do not feature in transport statistics. Similarly the value of trade carried into and out of the UK by ships and aircraft is now rivalled by the value of invisible trade, conducted largely through telecommunications.[29] The development of the information superhighway is likely to loosen still further the links between economic growth and the growth of transport.

2.12 The existence of an efficient transport system has been essential for the development of the UK economy, and remains a necessary part of the UK's competitive position. Over and above that, transport has often been regarded as providing specific levers for governments to use in promoting economic growth and competitiveness. Vehicle manufacture has been seen as a leading sector in the economy. The car industry is an important source of demand for other industries and a significant source of innovation in manufacturing methods.[30] UK production of vehicles declined sharply in the early 1980s, however; although it has shown some recovery, the number of vehicles produced in the 5-year period 1988–92 was only 72% (cars) and 63% (commercial vehicles) of the level 20 years earlier.[31] The rapid increase in the number of vehicles in use has been largely brought about by imports.

2.13 The construction industry has often advocated a large programme of road building or, more recently, an early start to major rail projects, as a way of providing additional employment and boosting the economy. While investment for such purposes leads to an increase in overall economic activity if it utilises resources which would not otherwise have been used, a comparison also has to be made with alternative forms of capital investment. It has been argued for example that road building is less cost-effective in creating employment than investment in the upgrading of railway lines or in housing.[32]

Figure 2-V
Investment in transport 1992/93

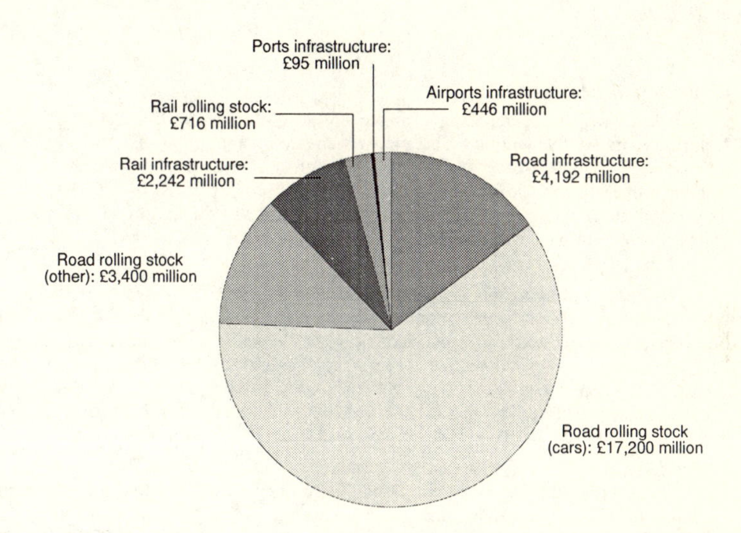

2.14 It has also been an assumption of both UK and European Community transport policy that further investment in transport infrastructure will stimulate the growth of other sectors of the economy by facilitating the movement of goods between areas with different comparative advantages. In the Treasury's view it is not possible to generalise about the importance of transport infrastructure as a factor in bringing about economic growth in depressed or deprived regions.[33] We consider that it would be desirable to carry out further research on the subject. A recent study concluded that road building is not the key to economic growth in the regions.[34] It is clear that the existence of a network of uncongested motorways in an area such as Merseyside has not been sufficient to overcome the influence of other factors which inhibit economic growth. Indeed, it seems that good roads can sometimes speed the decline of less prosperous areas by allowing their needs to be met conveniently from sources outside the area.

2.15 An industry can organise its activities in various ways. The techniques of advanced logistics have been developed to help industries and freight operators solve their particular transport problems within given constraints. The solution adopted will be the one which gives the highest profit to the firm making the decision. If a road network is improved, firms are likely to use it to make longer, faster or more frequent journeys, which may not represent the best overall solution if environmental costs are taken into account.[35] Other methods of reducing transport costs in order to maintain competitiveness, such as bringing together different activities on the same site or adjoining sites, may be more attractive in economic and environmental terms than constructing or improving roads.

2.16 Improvements to transport systems can contribute to economic growth in some circumstances, but they can contribute to growth which is environmentally sustainable only if decisions take into account the external costs which the construction and use of such systems impose on the environment and on society generally. In the past, the government has tended to place a strong emphasis on the contribution of roads to economic development and to pay little attention to the potential contribution from alternative modes of transport, which may be less damaging to the environment. The UK already has a high-quality road network.[36] This could be adequate, broadly speaking, for present and foreseeable industrial and commercial needs. However, it is used predominantly for personal travel. This is also the fastest growing component of demand (2.8), in particular journeys for shopping, other personal business and leisure (2.4). We now go on to examine the factors which have been associated with that growth.

Changes in lifestyles

2.17 The growth in personal travel has been closely linked with social changes. Some of the key features can be briefly identified. There have been demographic changes. Although the population of the UK increased by only 14.5% between 1951 and 1991[37], there was a more rapid increase in the number of people over 16 and an increase of 44% in the number aged 65–79.[38] People are remaining active longer and making more use of post-retirement leisure. In doing so they are even more likely than younger people to welcome the comfort and convenience of a car.

2.18 Women use buses more frequently than men. In 1989/91 women aged 16–29 made 133 trips by bus a year on average compared to only 84 by men, and women aged 30–59 made 83 trips by bus a year on average compared to only 38 by men. However, women now increasingly have a car or access to a car, and the gap between women's and men's travel patterns is narrowing. The proportion of car mileage driven by women rose from 16% in 1975/76 to 25% in 1989/91.[39] Over the same period the ratio of average annual distance travelled in a car by women to the average for men rose from 0.53 to 0.67 in the 17–29 age group and from 0.43 to 0.56 in the 30–59 age group.[40]

2.19 The organisation of work is becoming more flexible and decentralised. For a number of reasons, including the shift from manufacturing to services[41] and improvements in labour productivity, employment is now spread more thinly over a larger number of establishments. There have been increases in the numbers of self-employed[42], part-time workers[43], women workers[44] and workers with more than one job.[45] An increase in the number of workers without a fixed place of employment may help to account for the increase by a third in the late 1980s in the average number of journeys made on business[46], which now represent over a tenth of distance travelled (table 2.1).

2.20 The travel patterns of parents and children have also changed. Partly because of the increase in female employment, the number of places in day nurseries and playgroups and with registered child-minders has grown by 400,000 in the last 15 years.[47] Many more children now start school before the age of 5. There has been a marked change in the way older children travel to school. The proportion of children aged 7–11 taken to school by car increased to 30% in a 1990 sample from 1% a generation earlier; the proportion walking dropped from 87% to 67%. One factor has been the closure of many smaller schools. Despite that, 80% of the 1990 sample lived within a kilometre of their school; but parents now have a heightened perception of the risks to children, their main concerns being danger from traffic (43%), unreliability of the child (21%) and fear of molestation (21%).[48] The trend towards use of cars has been partly offset as a cause of traffic growth by a reduction of 13% in the late 1980s in the average number of journeys made primarily for education, as a result of a fall in the school age population.[49]

2.21 There have been major changes in the way people shop. Between 1982 and 1992 more than half of new retail floorspace opened was at out-of-town sites, compared with only one-seventh between 1960 and 1981.[50] Between 1988 and 1992, the number of superstores (defined as having an area of more than 25,000 square feet, and mostly on out-of-town sites) increased by more than half.[51] There are also estimated to be over 2,000 retail warehouses. Half of do-it-yourself sales are now at out-of-town locations, which are also becoming increasingly important for electrical goods, clothing, sports equipment, footwear and office supplies. The last decade has seen the creation of a small number of massive out-of-town shopping centres, each containing hundreds of shops: the Gateshead MetroCentre, Lakeside at Thurrock (Essex), Meadowhall at Sheffield and Merry Hill at Dudley. In all, 37% of retail sales in 1992 were in out-of-town locations, compared to only 5% in 1980.

2.22 The growth of out-of-town stores and centres has increased dependence on cars. Customers travelling by car account for 97% of the sales from Marks and Spencer edge-of-town stores compared with about 70% from high street stores.[52] An early survey of people shopping at the Gateshead MetroCentre found 80% had travelled by car, compared with 27% of those shopping in Newcastle city centre.[53] At a site adjacent to the M40, 98% of customers at an Asda food store came by car.[54] These customers tend to come from a wide area. About a quarter of customers at the Asda store had a journey time of 20 minutes or more and most used the motorway. Marks and Spencer define the catchment area of an edge-of-town store as 40 minutes driving time (which can be equivalent to a distance of more than 40 miles), compared with 20 minutes for minor high street stores. Lakeside in Essex attracts visitors along the A12 and M25 from as far away as Ipswich and Norwich.

2.23 The government has recently made important changes in planning policies on out-of-town stores, which are discussed in chapter 9. The success of out-of-town stores and centres reflects a fundamental modification in the way people view shopping, the result ultimately of increasing affluence. It is now for many people a form of leisure, and stores and centres are designed and operated on that basis. In other spheres of life as well, the increased opportunities for mobility have highlighted the scope for choice and variety. People are now less likely to use a particular shop or school or other facility simply because it is the nearest. Many of them choose to travel further in order to obtain a quality or kind of service which they see as more desirable. Between 1985/86 and 1989/91 the number of journeys made for leisure increased by 10% on average, but the average distance travelled for leisure increased by 23%. The corresponding figures for shopping journeys were 16% and 24%.[55] Another strong trend has been towards the 'chaining' of journeys to serve a number of purposes in succession. For example the journey to work may be combined with delivering one or more children to school or playgroup; the journey home from work may be combined with collecting children and with shopping. In order to make such chains of journeys a car is almost essential.

2.24 These new lifestyles are not shared by everyone. Two-thirds of people do not shop out of town. Some of them dislike the wide ranges of often unfamiliar products stocked by out-of-town stores and the time taken up by a visit to one.[56] Others are among the third of households which do not have access to a car. For those households opportunities have often been reduced: the growth of superstores and out-of-town centres has reduced the profitability of many shops which could be easily reached on foot or by public transport, and in some cases brought about their closure.[57] Many households, especially in rural areas, have difficulty in affording a car but find they now need one in order to buy food and obtain employment.

2.25 Although only limited statistics are available, it is clear that increased use of cars, especially for short journeys, has greatly reduced the distances many people walk or cycle. This is of considerable concern in a health context. Physical activity contributes to the prevention and management of weight problems and obesity and protects against coronary heart disease. Present levels of fitness are very low. About one-third of men aged 55–64, almost two-fifths of women aged 45–54 and over half of women aged 55–64 are not fit enough to walk on level ground at 3 mph.[58] In the light of findings like this the government is developing strategies for promoting physical exercise.[59]

2.26 The changes that have taken place in lifestyles can be regarded as a response to the greater opportunities for mobility. Their general effect, however, is to provide a powerful further stimulus to the use of cars and, as a result of changes in the location of population and the pattern of land use (discussed in chapter 9), they have now developed a dynamism of their own. The car has also developed deep and powerful symbolic meanings, which add to its attractiveness.[60] If our lifestyles are not sustainable, however, they will have to be modified sooner or later.

2.27 Travel by car has shown a rapid growth in all developed countries, but there is no simple relationship between car use and ownership and GDP per head. Britain has the lowest GDP per head of the ten developed countries shown in figure 2-VI[61], and comes seventh in terms of car ownership per thousand people, but third in the average distance travelled by car annually. The USA has the most car-intensive lifestyle; but people in the other countries listed have a less car-intensive lifestyle at present than the British either in absolute terms (Belgium, France, Germany, Italy, Japan, Netherlands) or in

Figure 2-VI
Car ownership and use in developed countries in relation to wealth (1991)

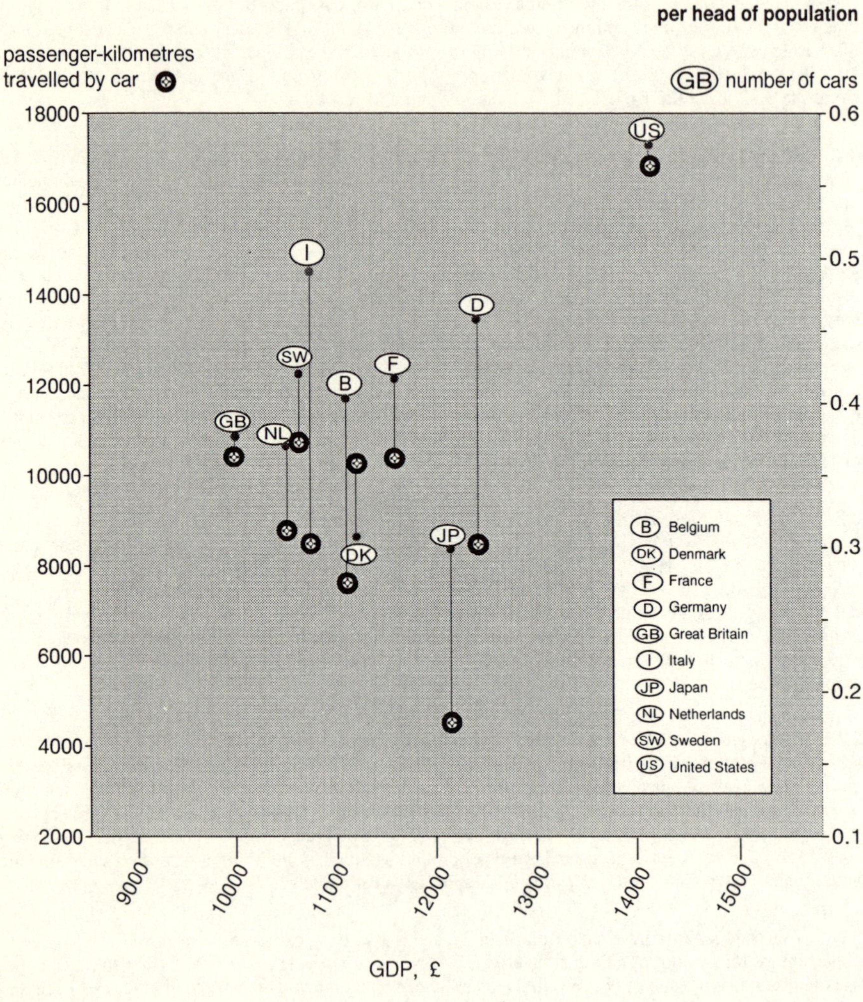

relation to their wealth (Denmark, Sweden). We do not believe that the extrapolation of past trends in the UK represents the only possible future.

Forecasts of road traffic

2.28 DOT's practice has been to produce long-term forecasts of road traffic (that is, vehicle-kilometres), rather than overall forecasts of movements of people and goods. These forecasts have covered Great Britain for a period of at least 30 years ahead. At one time they were given the status of government policy statements. More recently their primary purpose has been described as facilitating economic appraisal of proposals to build or improve trunk roads or motorways. The most recent

Figure 2-VII
Road traffic growth and 1989 forecasts

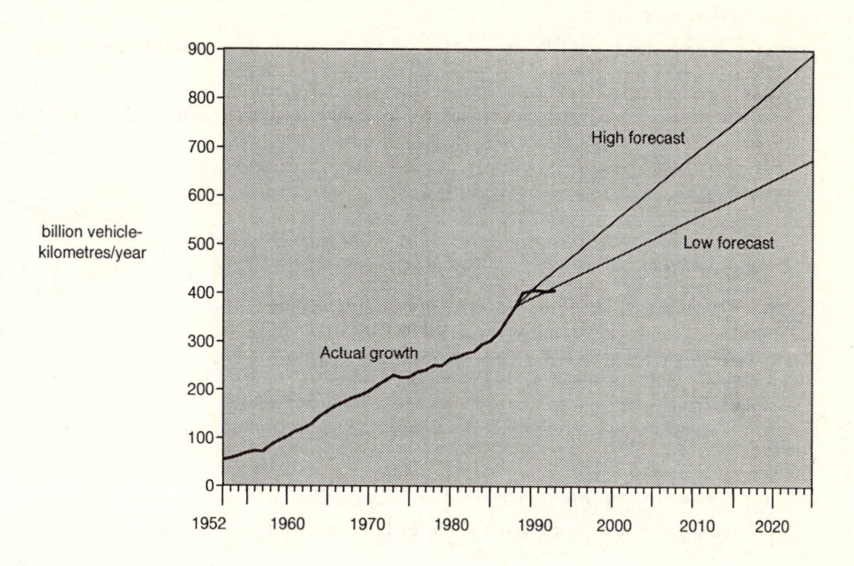

forecasts are the National Road Traffic Forecasts made in 1989.[62] Box 2B describes how they were produced.

2.29 Two forecasts were published: a high forecast of a 142% increase in traffic between 1988 and 2025 and a low forecast of an 83% increase (figure 2-VII[63]). It was assumed that rising real incomes will spread the lifestyles at present characteristic of higher income groups to a larger and larger proportion of the population. The forecast levels of traffic were therefore largely determined by the assumptions made about economic growth. In fact the growth forecast between 1988 and 2000 was virtually the same as the assumed growth in GDP, although traffic was forecast to grow significantly less rapidly than GDP thereafter. The 1989 forecasts considerably exceeded those made in 1984[64], partly because higher rates of economic growth were assumed, and partly because the baseline incorporated the rapid growth in traffic in the mid-1980s (also shown in figure 2-VII).[65] DOT subsequently reached the conclusion that traffic levels in 1988 had been higher than originally estimated[66]: the levels of traffic in 2025 implied by the 1989 forecasts are therefore now 13–14% higher than when they were first published.

2.30 The high forecast made in 1989 assumed an average growth in GDP of 3% a year and the low forecast 2%. The Treasury regards an appropriate range of assumptions about long-term economic growth as 1–3½% a year, and the most probable range as 1½–3% a year.[67] In view of that, and the government's decision to raise the price of fuel above the levels assumed in the forecasts (7.52), we consider that the low forecast made in 1989 has a higher probability than the high forecast. Figure 2-VII allows a comparison to be made with what actually happened in the period 1988–93, when economic growth was considerably less than had been assumed. In the first year traffic exceeded the high forecast; but by 1993 it was below the low forecast.

2.31 There is general recognition that further growth in road traffic on anything like the scale shown in the 1989 forecasts would have enormous repercussions for the environment, and widespread doubt about whether it would be practicable to accommodate such growth. In chapters 3 and 4 we analyse the environmental impact of the present system of surface transport. In chapter 6 we discuss the implications of road traffic forecasts for the environment and for transport policy.

BOX 2B **GROWTH IN ROAD TRAFFIC: HOW THE 1989 FORECASTS WERE MADE**

Alternative assumptions about the annual rate of economic growth (3% and 2%) gave rise to high and low forecasts of road traffic.

Separate forecasts of future traffic levels were made for each class of vehicle, on the following bases:

Cars
The previous relationship between increases in income and increases in *car ownership* will continue to apply until 90% of the population of driving age (17–74) own a car.

The *distance travelled* by each car will increase one-fifth as fast as GDP per head; and will be reduced by 1.5% for every 10% increase in the price of fuel in real terms.

Light goods vehicles (those with a gross vehicle weight of up to 3.5 tonnes)
Vehicle–kilometres will increase in direct proportion to GDP.

Heavy goods vehicles (those with a gross vehicle weight over 3.5 tonnes)
Vehicle–kilometres will be largely determined by GDP. The forecast also took into account:

— the extent to which growth in GDP will increase the tonnages carried
— factors such as changes in the nature of goods carried
— an assumption that road's share of freight will continue to grow
— an assumption that most of the additional freight will be carried by the heaviest HGVs.

Buses and coaches
Vehicle–kilometres will remain at the 1989 level.

No forecasts were made for travel on foot, by cycle, or by moped or motorcycle.

It was assumed that the price of fuel in real terms will increase by:

— 8% by 2000 and 21% by 2025 in the case of the high forecast
— 45% by 2000 and 61% by 2025 in the case of the low forecast.

Although different assumptions were made for the high and low forecasts, fuel price and the rate of economic growth were treated as independent variables.

Chapter 3

EFFECTS OF VEHICLE EMISSIONS

Use of petroleum as a fuel

3.1 This chapter describes the emissions from road vehicles, places them in the context of emissions from other forms of transport, and assesses their significance for human health and for the environment. It looks first at airborne pollutants and their effects on air quality and human health; then at their effects on the natural and built environment; then at the contribution transport is making to the increasing concentrations of greenhouse gases, especially carbon dioxide, in the Earth's atmosphere. The outcome of this analysis is to identify objectives which we believe should form the basis for future policy and clear targets for moving towards those objectives.

3.2 Modern transport is powered predominantly by combustion of fossil fuels in the form of petroleum products. In 1992 the UK transport sector used more than 43 million tonnes of such products, 54% of the UK total (table 3.1[1]). There has been a steady increase in the relative importance of transport use, which accounted for only 33% of the UK total in 1962.[2] Road transport accounted for 80% of transport use of petroleum products in 1992 and over 40% of total use. Most of the other transport use was by aircraft.

3.3 Crude oil is a mixture of substances, mainly hydrocarbons, which is distilled to produce transport fuels and industrial raw materials. Gasoline (petrol) is made from lighter fractions than kerosene (aviation fuel), which is in turn lighter than diesel fuel. Less energy is required to produce diesel. Figure 3-I shows the principal ways in which petroleum-powered vehicles give rise to pollution; pollutants are represented in the figure by black arrows.

Table 3.1
Petroleum products: direct use in transport by mode in 1992

	petroleum products used (million tonnes)	% transport use	% total UK use
road	34.99	80.3	43.4
air	6.69	15.3	8.3
water	1.26	2.9	1.6
rail	0.64	1.5	0.8
total	43.58	100.0	54.1

Figure 3-I
Pollutants emitted by petroleum-powered vehicles

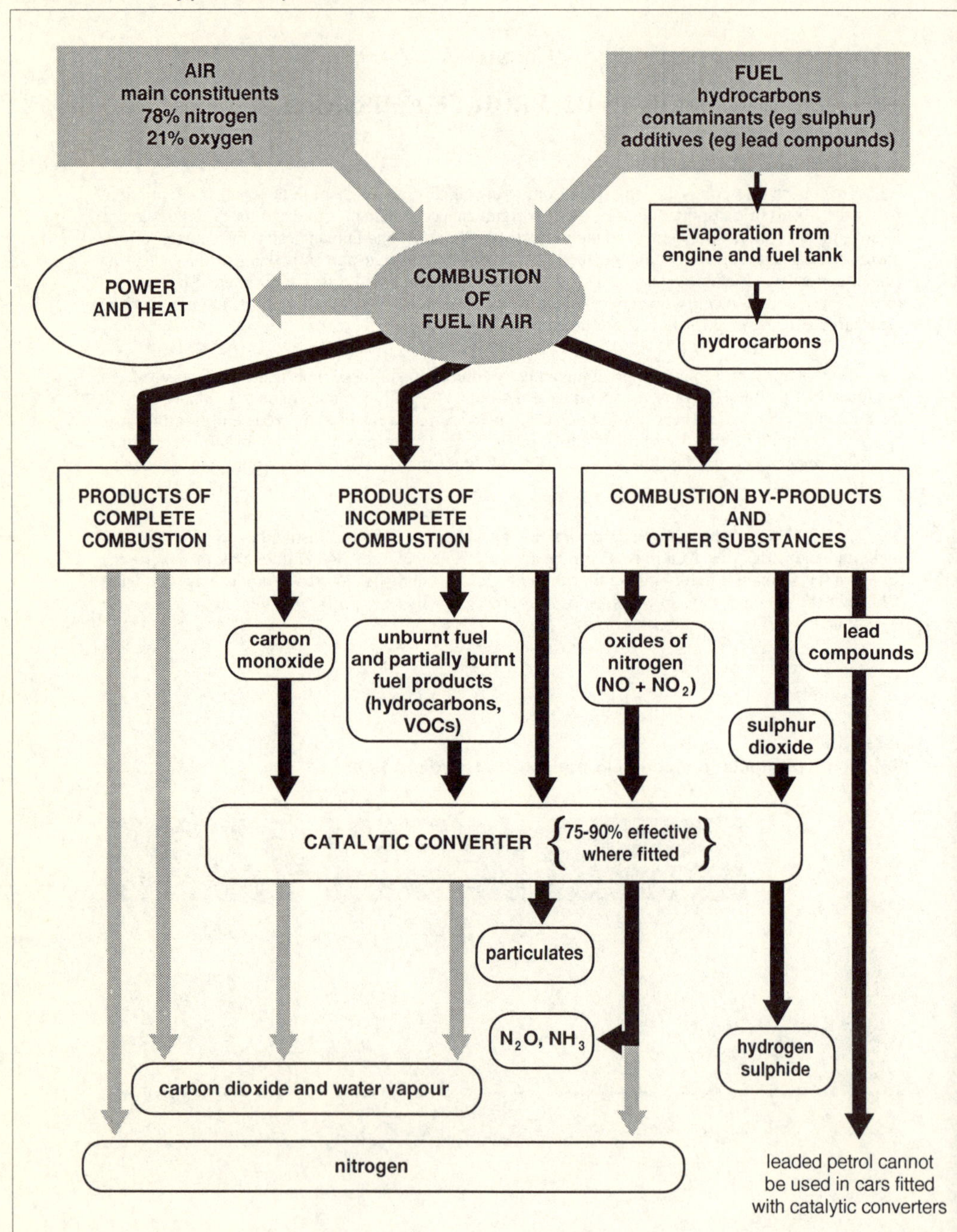

Table 3.2
Airborne pollutants from transport: estimated emissions in the UK (1992)

thousand tonnes

	road	rail	air	shipping	total emissions from transport	transport as % of UK emissions from all sources[a]
carbon monoxide	6029	12	11	19	6071	90 (91)
nitrogen oxides	1398	32	14	130	1574	57 (61)
volatile organic compounds	949	8	4	14	975	38 (48)
particulates	215	not reported	1	3	219	48 (48)
sulphur dioxide	62	3	3	60	128	4 (7)

a Figures in parentheses include transport-related emissions from oil refineries and power stations.

3.4 The amounts and proportions of substances in the exhaust gases from an engine depend on a number of factors, including:

— the design of the engine

— its size

— the characteristics of the fuel

— the conditions in which the vehicle is being used

— how it is being driven

— its age and state of maintenance.

The legislation controlling vehicle emissions and fuel quality is summarised in chapter 8, which discusses the scope for reducing emissions.

3.5 In a petrol engine, a mixture of air and fuel is drawn into the cylinder, compressed and ignited by a spark. In a diesel engine, fuel is injected into air which is already very hot as a result of compression and thus no spark is needed. Because a diesel engine works with a surplus of air, the combustion process is more efficient; the amount of fuel consumed is therefore at least 25% less by volume and 15% less in terms of mass than in a petrol engine of comparable power.

3.6 A diesel engine produces much less carbon monoxide, and less nitrogen oxides and unburnt or partially burnt hydrocarbons, than a petrol engine. On the other hand it produces far more particulates than a petrol engine. New petrol-engined cars are now fitted with three-way catalytic converters designed to reduce emissions of carbon monoxide, nitrogen oxides and hydrocarbons by 75–90%; they emit a smaller amount of nitrogen oxides than a diesel-engined car of comparable power. The overall desirability of diesel engines in environmental terms has therefore become a matter of controversy, even though they are preferable in terms of fuel efficiency. We discuss this issue later in the report (8.31–8.34).

Airborne pollutants

3.7 The airborne pollutants produced by transport and their effects are described in appendix A. Estimated emissions of the main pollutants are given in table 3.2.[3] Transport is the most important source of most of them, with almost all the emissions coming from road transport. The exception is sulphur dioxide, most of which comes from power stations, and some from oil refineries and other industrial sources. Transport is the most important source of *volatile organic compounds* (VOCs) if transport-related emissions from oil refining and distribution are included.

3.8 The estimates in table 3.2 do not include emissions from aircraft, except those below 300 feet during take-off and landing, or emissions from shipping outside coastal waters (the 12 mile limit). The effects of emissions from cruising aircraft are a major issue and are discussed in chapter 5. Emissions from ships, which usually use diesel with a relatively high sulphur content, can produce high concentrations of sulphur dioxide near large ports. Sweden and Finland have local regulations to limit emissions from ships; and the US Environmental Protection Agency has proposed reductions in emissions of nitrogen oxides from ships and a limit on the sulphur content of diesel in an area extending 100 miles from the Californian coast.[4] Emissions from shipping are not discussed further in this report, but we welcome the indications that the next significant regulation under the International Convention for the Prevention of Pollution from Ships (MARPOL) will cover emissions to air from marine diesel engines.

3.9 Most types of emissions from road vehicles have shown large overall increases, reflecting the growth in traffic. Between 1982 and 1992 the estimated increases were 85% for black smoke, 61% for nitrogen oxides, 31% for carbon monoxide and 8% for VOCs.[5] It will be apparent from table 3.2 that emissions from road vehicles are the main influence on air quality over the large areas of the UK in which there are no significant industrial emissions. Estimates of the amounts of the main pollutants emitted by different classes of road vehicle are given in table 3.3.[6] As well as the primary pollutants listed in tables 3.2 and 3.3, vehicle emissions give rise to the *secondary pollutants* nitrogen dioxide and ozone as a result of chemical reactions in the atmosphere which are described in box 3A.

3.10 The *particulates* in exhaust gases consist mainly of carbon and unburnt or partially burnt organic compounds from the fuel or lubricating oil. In addition, secondary particulates may be formed in the atmosphere; these include nitrates and sulphates formed from nitrogen oxides and sulphur dioxide. The

Table 3.3
Airborne pollutants from road transport, by class of vehicle (1990)

percentage of total road transport emissions

	cars	light goods vehicles	heavy goods vehicles	public service vehicles	motor cycles
carbon monoxide	88	7	3	1	1
nitrogen oxides	72	7	19	3	-
volatile organic compounds	84	7	6	1	2
particulates	6	7	77	10	-
sulphur dioxide	37	9	47	6	-

BOX 3A **SECONDARY POLLUTANTS**

Secondary pollutants are formed as a result of chemical changes which primary pollutants undergo in the atmosphere. The principal secondary pollutants attributable to transport are *nitrogen dioxide* and *ozone*.

Although small amounts of *nitrogen dioxide* (NO_2) are released directly in vehicle emissions, most of the NO_2 attributable to transport comes from the reaction of nitric oxide (NO) emitted by vehicles with oxidants in the atmosphere. Most NO is oxidised rapidly by ozone or the hydroperoxy radical (HO_2):

$$NO + O_3 \rightarrow NO_2 + O_2$$
$$NO + HO_2\cdot \rightarrow NO_2 + \cdot OH$$

In polluted atmospheres, other oxidation reactions may take place, involving molecular oxygen, hydrocarbons, aldehydes or other compounds. NO also combines with hydroxyl (OH) radicals to produce nitrous acid (HNO_2), which is in turn converted to nitric acid (HNO_3) and mostly removed from the atmosphere by dry or wet deposition. HNO_3 may also be formed from NO_2.

NO_2 is also the main precursor of *ozone* (O_3), which is produced through the following reactions (where $h\nu$ is energy from sunlight and M is a molecule of oxygen, nitrogen or other naturally occurring gas which, by absorbing excess energy from the newly formed ozone molecule, prevents the reaction from being immediately reversed):

$$NO_2 + h\nu \rightarrow NO + O\cdot$$
$$O\cdot + O_2\ (+M) \rightarrow O_3\ (+M)$$

The net amount of ozone produced is greater in the presence of volatile organic compounds from vehicle emissions and other sources, which are broken down by sunlight into complex mixtures of substances.

Where ozone is present NO removes it from the atmosphere in the course of forming NO_2. High levels of NO_2 tend to be associated with low levels of O_3 and *vice versa*.

atmosphere also contains particulate matter emitted from power stations or industrial processes and dust from a variety of natural and other sources. Small particles of solid or liquid form *aerosols*. Individual particles vary considerably in size and are variously categorised: *PM10* is a term applied to particles with a diameter of less than 10 μm, *inhalable particles* are those which are small enough to breathe in, *thoracic particles* those which penetrate beyond the larynx, and *respirable particles* (a term frequently applied to particles with a diameter of less than 2.5 μm) are small enough to penetrate to the deep lung and remain there.[7] Concentrations of particulates in air are estimated in various ways: by calibrating the darkness of a white filter after the air has been drawn through it (to measure *black smoke*), by assessing the total mass of particulates (gravimetric method) or by controlling the size of particulates collected (to measure PM10).

3.11 Those pollutants which are products of incomplete combustion (especially carbon monoxide and hydrocarbons) are produced in larger amounts when an engine is not operating efficiently. When an engine is first started, the air/fuel mixture is rich in fuel; there is not enough oxygen available to achieve complete combustion in all parts of the combustion chamber. This is also the period in which present designs of three-way catalytic converter are not operating effectively (8.5–8.6). Petrol cars may have to be driven for several minutes before the engine has warmed up sufficiently for efficient combustion to take place; diesel cars will take slightly less time to warm up. Both types of engine take longer to warm up in cold weather. Products of incomplete combustion are also produced in greater quantities when a vehicle is accelerating, when it is moving slowly, and in the stop-start conditions created by congestion. In contrast, formation of nitrogen oxides from the nitrogen and oxygen in air increases as the combustion temperature increases. Thus the amounts of nitrogen oxides in exhaust gases are highest at high speeds and during rapid acceleration. In general, the amounts of pollutants produced are lowest, and fuel efficiency is highest, when a vehicle is driven at a steady speed in the range 35–55 mph with a warm engine.

Behaviour of pollutants in the atmosphere

3.12 The highest concentrations of primary pollutants in the atmosphere occur near the point of emission. At the kerbside of major roads, concentrations are two to three times the urban background level. In samples taken in vehicles travelling on major roads, concentrations are on average five times the urban background level. Concentrations up to ten times the urban background level have been recorded in enclosed spaces such as road tunnels.[8] There is ground for concern that young children are exposed to high concentrations of pollutants both as passengers in cars and when they are on pavements not far above the height of exhaust pipes. Of the 20 models of car which sell in largest numbers in the UK, 16 have their exhaust pipe on the pavement side, apparently because they were originally designed for left-hand drive.[9] The secondary pollutant nitrogen dioxide is also found at high concentrations inside vehicles and at sites on major roads.

3.13 The weather gives rise to wide variations from day to day in concentrations of pollutants. The atmospheric boundary layer, extending typically to a height of about 1 km above the Earth's surface, is usually turbulent enough to disperse pollutants rapidly. In anticyclonic weather, which can persist for several days or even weeks, a stable layer frequently forms at the top of the boundary layer, reducing vertical movements of air and acting as a lid to confine pollutants near the ground. In winter, such conditions lead to 'anticyclonic gloom'. Anticyclonic weather is associated with light winds, and horizontal movement of pollutants is thus also reduced. In cold weather, when emissions of pollutants from vehicles are higher, anticyclonic conditions can result in severe urban pollution episodes, such as the one in London in December 1991, described in box 3B.[10] Plate I shows the poor visibility in London during the episode.

3.14 Strong sunlight promotes the formation of ozone, as described in box 3A. High ozone concentrations may give rise to 'photochemical smog'. In the UK, ozone concentrations are highest between April and September. In areas where there is heavy traffic, nitric oxide emitted by vehicles removes ozone from the air almost as rapidly as it is formed. In consequence, although the precursors of ozone (nitrogen dioxide and VOCs) come to a large extent from urban areas, the highest ozone concentrations are usually found in rural or suburban areas, in an air mass which has drifted away from an urban area. During anticyclonic weather it may be several days before an inflow of cleaner air dilutes a polluted air mass.

3.15 Ozone differs from other transport-related pollutants in being a regional rather than a local problem. High concentrations are often reported simultaneously from several parts of the UK, or even across a large area of Europe. Depending on the direction of air flow, a substantial proportion of the ozone present during an episode in southern England may be the result of vehicle emissions on the continent, or *vice versa*. Box 3C describes the high ozone concentrations recorded in England in July 1994.

3.16 The UK rarely experiences the conditions that regularly give rise to very high ozone concentrations in some other countries. In the Los Angeles area, for example, there were 20–30 days in 1987 on which the maximum ozone concentration was higher than 400 $\mu g/m^3$; and the hourly mean can reach 600–800 $\mu g/m^3$.[11] No hourly mean above 360 $\mu g/m^3$ was recorded in the UK in the five-year period 1988–92.[12]

3.17 Experience in the USA, where three-way catalytic converters have been fitted to cars for some years, shows that reducing emissions of primary pollutants does not always lead to a corresponding reduction in concentrations of secondary pollutants.[13] There could even be some local increases in urban ozone levels in the UK in future, as the increased use of catalytic converters reduces emissions of substances which scavenge ozone.[14] There is also a complex relationship between emissions of nitrogen oxides and concentrations of nitrogen dioxide. The atmosphere in urban areas has only limited capacity to oxidise nitric oxide (NO) and form nitrogen dioxide (NO_2). If the concentration of nitrogen oxides is relatively low, the level of nitrogen dioxide present quickly reaches a plateau and thereafter is not much affected by variations in the concentration of nitrogen oxides. At high concentrations of nitrogen oxides, however, variations in the concentrations of nitrogen oxides are reflected in the concentration of nitrogen dioxide. This is probably because in such circumstances oxidation of nitric oxide by molecular oxygen (O_2) becomes more important than the oxidation mechanisms described in box 3A.[15] The practical implication is that reducing emissions of nitrogen oxides from vehicles during a pollution episode could have a significant benefit in terms of reducing the concentration of nitrogen dioxide.[16]

BOX 3B THE LONDON POLLUTION EPISODE OF DECEMBER 1991

Over the period 12–15 December 1991, unusually high concentrations of nitrogen dioxide were recorded in London. An anticyclone centred over the Alps was affecting south-east England and producing low wind speeds, low temperatures, mist and high stability.

As figure 3–II shows, nitrogen dioxide pollution reached its peak in the early hours of 13 December. This contrasted with previous episodes because there was no obvious correlation with peak traffic flow. Nevertheless, the correlation of concentrations of nitrogen oxides with concentrations of carbon monoxide, and the absence of correlation with sulphur dioxide, strongly suggests that vehicle emissions were the dominant source of pollution.

The highest concentrations of nitrogen dioxide were recorded at the urban background monitoring sites at Bridge Place and Earls Court, rather than at the kerbside site at Cromwell Road. The WHO guideline of 150 $\mu g/m^3$ averaged over 24 hours was exceeded, and at Bridge Place the guideline of 400 $\mu g/m^3$ averaged over an hour was also exceeded. There was no breach of the EC Directive because, although hourly means exceeding 200 $\mu g/m^3$ were recorded at the urban background sites at which compliance is measured, the requirement in the Directive is for 98% compliance over a year.

Figure 3-II
Nitrogen dioxide concentrations in London 11-17 December 1991

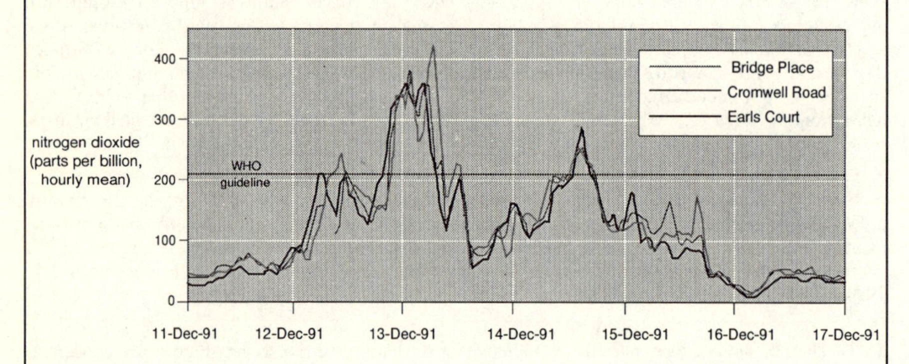

The Department of Health (DH) funded a study of the health effects of this episode. In giving oral evidence to the House of Commons Select Committee on Transport in July 1994, DH referred to unpublished findings that the death rate in London rose by 10% during the four days of the episode; 160 additional deaths were recorded in that week.

Health effects of transport-related pollutants

3.18 Concern that pollution caused by vehicle emissions may be adversely affecting human health, especially in urban areas, has been emphasised in evidence to the Commission from a wide range of sources, including local authorities, environmental pressure groups and individuals. The majority of these submissions have focused on the respiratory effects of exposure to nitrogen dioxide, ground-level ozone and particulates; the risk of cancer from exposure to benzene and other VOCs has also been mentioned.[17] The subject has received much attention from the press, television and radio, particularly

BOX 3C **HIGH OZONE CONCENTRATIONS IN JULY 1994**

During July 1994, periods of hot weather gave rise to elevated levels of ozone across Europe. These elevated levels were prolonged in some areas as a result of the calm conditions.

In addition to its normal air quality bulletins, DOE issued public pollution alerts during these episodes, which, as well as giving health advice, urged drivers to leave their cars at home.

On a number of days the levels recorded at national monitoring sites exceeded the lower bound, and often the upper bound, of the WHO health-based guideline, which is 150–200 $\mu g/m^3$ averaged over an hour. In many cases, they entered the 'poor' air quality band defined by DOE (180–358 $\mu g/m^3$). The peak hourly levels recorded on those days are shown below.

Day	Site	Peak level
Friday 1 July	— Lullington Heath (Sussex coast)	— 190 $\mu g/m^3$
Saturday 2 July	— London (Bridge Place)	— 190 $\mu g/m^3$
Monday 11 July	— Lullington Heath	— 204 $\mu g/m^3$
	— Harwell (Oxfordshire)	— 196 $\mu g/m^3$
	— London (Bridge Place)	— 192 $\mu g/m^3$
Tuesday 12 July	— Sibton (Suffolk)	— 238 $\mu g/m^3$
	— Bexley (south-east London)	— 218 $\mu g/m^3$
	— Ladybower (Peak District)	— 214 $\mu g/m^3$
	— Harwell	— 204 $\mu g/m^3$
	— Lullington Heath	— 202 $\mu g/m^3$
Friday 22 July	— Lullington Heath	— 206 $\mu g/m^3$
	— Southampton	— 152 $\mu g/m^3$
Saturday 23 July	— Harwell	— 204 $\mu g/m^3$
Sunday 24 July	— Lullington Heath	— 218 $\mu g/m^3$
	— Harwell	— 212 $\mu g/m^3$
	— Southampton	— 188 $\mu g/m^3$
	— London (Bridge Place)	— 174 $\mu g/m^3$
	— Bexley	— 168 $\mu g/m^3$

Data supplied by the National Environmental Technology Centre, Harwell; peak levels were converted from the ppb values supplied using a factor of 1 ppb = 2 $\mu g/m^3$.

in the context of the possible link between air pollution and asthma. The evidence has been reviewed in several published reports by government advisory bodies.[18,19,20,21,22,23]

3.19 One of the basic difficulties in investigating the health effects of air pollution is that a number of pollutants are usually present together in the atmosphere. Most epidemiological studies have been concerned with the effects of mixtures of pollutants. Laboratory studies on the other hand have generally been concerned with the effects of high concentrations of individual substances. They have provided evidence that exposure to very high levels of nitrogen dioxide may increase the probability that respiratory infections and bronchitis will develop. Nitrogen dioxide at concentrations of 950–1,910 $\mu g/m^3$ has been shown in animal studies to reduce resistance to infection.[24] Concentrations of 760 $\mu g/m^3$ may lead to inflammation in cells from the lining of the bronchial tubes in humans.[25] In the light of studies of this kind the Advisory Group on the Medical Aspects of Air Pollution Episodes (MAAPE) concluded in 1992 that, although statistical associations with nitrogen dioxide concentrations had been found in a small number of epidemiological studies, attribution of a causal role to nitrogen dioxide did not seem biologically plausible; and that a general health warning would not be necessary until the nitrogen dioxide concentration exceeded 1,128 $\mu g/m^3$ (roughly twice the starting-point of the 'very poor' band used in the Department of the Environment's air quality bulletins, and a level never recorded in the UK).[26]

3.20 Individuals vary considerably in their response to ozone. Those who are sensitive to it can experience temporary breathing difficulties if they take vigorous outdoor exercise when ozone concentrations are at or above about 160 $\mu g/m^3$; this level is often exceeded during hot summers, especially in southern England. In terms of lung function, people who suffer from asthma or other respiratory disorders are no more likely to be sensitive to ozone than other members of the population, although laboratory studies have shown that ozone may produce an enhanced inflammatory response in the airways of asthmatics.[27]

3.21 A number of known or suspected carcinogens are detectable in vehicle emissions. It is difficult to assess what risk this represents. The individual substances which have attracted most attention are benzene, which has a direct effect on genetic material, and the suspected carcinogen 1,3-butadiene. Vehicle emissions are the most important source of exposure to benzene for people in urban areas who do not smoke, are not heavily exposed to other people smoking and do not encounter high concentrations of benzene in their workplace.[28]

3.22 Attempts have recently been made to quantify the cancer risks to the general public resulting from vehicle emissions. The German Council of Environmental Advisors has made a comparative analysis of the emissions from petrol and diesel engines. It did not consider it possible to quantify the absolute risk because comprehensive data are not available about the substances present in emissions and their contribution to atmospheric concentrations. The Council emphasises the significance of particulates, which may have polyaromatic hydrocarbons and possibly other carcinogens adsorbed to them.[29] The California state authorities have published a draft health risk assessment for diesel exhaust and sought views on the proposal that diesel exhaust should be regarded as a human carcinogen.[30] They have drawn attention to the large margins of uncertainty involved.

3.23 There has been a general increase throughout the developed world in the *prevalence* of respiratory problems among children, that is, the proportion of children who have such problems. For example there has been a fivefold increase since 1983 in the number of families in the UK applying for disability grants because their children have severe asthma.[31] Some epidemiological studies indicate that other allergic diseases, such as hayfever, are becoming more common in industrialised societies[32] and that both asthma and hayfever seem to be more common in urban areas.[33] There is evidence that environmental influences are important in the initial development of asthma[34] but exposure to outdoor pollutants is only one possible cause. Other possible explanations fall into two main categories: increased exposure to other types of substance, such as indoor pollutants, tobacco smoke[35] or allergens (particularly from house dust mites) and changes in lifestyle over recent decades which may have led to reduced resistance.[36] The latter include changes in diet (especially a lower intake of anti-oxidants, which might increase susceptibility to infections and airway inflammation)[37]; decline in family size (with the result that children are exposed to fewer viral infections in early life)[38]; and the possibility that children are now so completely protected from infections that their immune systems fail to develop fully.[39] Smoking by the mother during pregnancy has been linked to impaired lung development in children and the persistence of asthma into adulthood.[40] Allergic diseases are more common among infants whose mothers smoke[41] and there is evidence from studies of occupational exposure that adults who smoke are more likely to become sensitised to new allergens.[42] A combination of some or all of these factors may be involved.

3.24 There has been a marked increase in the *incidence* of respiratory problems in recent years.[43] Between 1976 and 1987, acute attacks of asthma more than doubled in England and Wales (from 10.7 to 27.1 per 100,000 patients a week). The greatest increase was in children (from 13.5 to 74.4 per 100,000 in the 0–4 age group and from 17.4 to 58.9 in the 5–14 age group).[44] It seems unlikely that these increases, which are parallelled elsewhere in the developed world, can be dismissed as a consequence of changes in reporting or diagnosis.[45] There is some evidence that hayfever symptoms are exacerbated when concentrations of pollutants are high.[46] Vehicle emissions may enhance sensitivity to pollen allergens.[47] Some recent epidemiological studies have indicated that individuals suffering from respiratory disorders, including asthma, may experience a worsening of their symptoms when there are elevated ambient levels of nitrogen dioxide and associated pollutants, especially particulates.[48] The factors which lead to an increase in the number of attacks of asthma in already susceptible individuals are not, however, necessarily the same as the factors which cause individuals to have respiratory problems in the first place.

3.25 Because of the concern expressed by so many who submitted evidence, we commissioned Dr Jon Ayres (a specialist in respiratory medicine at Birmingham Heartlands Hospital) to review work on the health implications of transport pollutants in the UK, USA and elsewhere and draw conclusions as to the possible links between respiratory problems and air pollutants from traffic; he reported his findings to the Commission in June 1993. This study confirmed that the amount of UK work in this field was small compared with the USA, some other European countries and Japan. Dr Ayres concluded that the lack of quantitative data for the UK at that time made it difficult to assess the overall importance of transport-related air pollution in public health terms but that any effects would be significant in terms of the large number of people likely to be affected.

3.26 Although the increasing prevalence of asthma has coincided with large increases in vehicle emissions, no causal relationship with levels of air pollution has been demonstrated. The Committee on the Medical Effects of Air Pollution (COMEAP), which advises the Chief Medical Officer, has set up a Working Group on Asthma and Air Pollution to advise on the possible link; its report is due before the end of 1994.

3.27 A study of six urban areas in the USA (published in December 1993) found a statistical relationship between air pollutants and mortality, although the factor most strongly associated with mortality rates was cigarette smoking. After adjusting for smoking and other risk factors, the mortality rate was 1.26 times higher in the most polluted than in the least polluted city. The statistical relationship was stronger for fine respirable particles than for total suspended particulates, sulphur dioxide, nitrogen dioxide or the acidity of the aerosol. The authors concluded that fine particles, including sulphates, either alone or in a more complex mixture of air pollutants, contributed to excess mortality in some US cities.[49] It should be noted that transport was not the only major source of the particulates present in the atmosphere in these areas.

3.28 The concentrations of PM10 in large UK cities are within the range recorded in this US study (although, as in US cities, transport is not necessarily the most important source). By making the assumption that the particulate material in urban air is similar in the two countries, and extrapolating from the results of the US study, it has been suggested that PM10 may be responsible for as many as 10,000 extra deaths a year in England and Wales.[50] Although there are not sufficient data about the origin, nature and ambient concentrations of particulates in UK cities to allow the validity of this extrapolation to be tested, it seems to us that the health implications of particulates need to be taken very seriously.

3.29 Another study looked at the relationship between mortality and air pollution in London for the winters 1958–1972, following the coming into effect of the Clean Air Act 1956, and examined data for black smoke (which includes inhalable particles and larger material), sulphur dioxide, temperature and humidity. A significant relationship was found between mortality and concentrations of both particulate matter and sulphur dioxide, but was stronger in the case of black smoke. The annual mean level of black smoke fell sharply over the period, from 536 $\mu g/m^3$ in 1958 to 59 $\mu g/m^3$ in 1972. The authors suggested that, even at the lower levels of black smoke now prevailing, a 10% reduction in particulates would result in several hundred fewer early deaths a year in London.[51]

3.30 Although a 1992 report by MAAPE concluded that particulates did not pose a significant threat to health[52], COMEAP has advised that a precautionary approach should be taken in the light of recently published epidemiological studies, and would welcome any reduction in levels of particulates. It has set up a Working Group on Fine Particles to investigate the relationship between particulates and health and advise on whether an estimate of deaths due to PM10 can be made.[53] We welcome this. The Royal Commission's Fifteenth Report in 1991 (paragraph 7.6) recommended that a precautionary approach should be adopted which would seek to reduce emissions of particulates as far as is practicable.

Monitoring and research

3.31 Concentrations of various airborne pollutants or combinations of pollutants are measured at 48 automated monitoring sites in the UK. The national monitoring programme also includes 1,200 sites measuring nitrogen dioxide by means of diffusion tubes, 13 sites measuring lead and other elements and 5 sites measuring toxic organic micropollutants, as well as sites monitoring acid deposition or smoke and sulphur dioxide.[54] In addition, many local authorities regularly measure at least one pollutant: about twenty have sophisticated monitoring programmes and three monitor a wide range of pollutants.[55]

3.32 There are few data to show how concentrations of the main transport-related pollutants have varied in the past. The absence of long-term records, except at a few sites, reflects the past emphasis on smoke and sulphur dioxide. The present national monitoring programme is of high technical quality, but its scale has been criticised. For example, the UK has fewer monitoring sites than most other member states for the purposes of the EC Nitrogen Dioxide Directive. There is a much less comprehensive coverage of the main transport-related pollutants than programmes in California and Japan provide.[56] Considerable uncertainty also remains about the spatial variation in pollutant concentrations, and thus about the extent to which data from fixed monitoring points reflect individual exposure to certain pollutants.[57]

3.33 In its first report[58] the Quality of Urban Air Review Group (QUARG) established by the Department of the Environment (DOE) recommended:

i. the extension of the urban monitoring network to cover at least 24 major towns and cities;

ii. the central co-ordination of monitoring carried out by local authorities and others so that the best use can be made of the data collected;

iii. additional monitoring and modelling to quantify the exposure of individuals inside cars, next to busy roads and in suburban locations;

iv. limited monitoring of some additional pollutants which pose a risk to health (for example, acid aerosols, dioxins, polyaromatic hydrocarbons (PAHs) and toxic metals such as mercury).

We endorse the measures recommended by QUARG for improving the monitoring of concentrations of air pollutants and developing modelling techniques, so as to gain a more accurate picture of exposure.

3.34 In addition to enhanced monitoring, **we recommend that further research be carried out into the health effects both of individual transport-related pollutants and of substances in combination.** This research should include:

a. using personal monitoring devices to make accurate measurements of the total doses of particular pollutants received by typical individuals in their daily lives, so that models of personal exposure can be constructed and tested;

b. further epidemiological work;

c. further study of the effects of pollutants, especially in combination;

d. examining how pollutants interact with other factors such as allergens, diet and housing.

We are glad to note that the effect of air pollution on human health is a priority topic for the new Institute for Environment and Health established by the Medical Research Council.

Present standards for air quality

3.35 The present standards for concentrations of transport-related pollutants are shown in table 3.4. Because these pollutants differ in the nature of their physiological effects, a period of time is specified as well as a concentration. For lead, a cumulative poison, standards take the form of an annual average concentration. For pollutants which act rapidly, such as carbon monoxide, the concentration over periods as short as 15 minutes is important.

3.36 European Community Directives adopted in 1980 and 1985 set legal limits for sulphur dioxide and suspended particulates[59] and for nitrogen dioxide.[60] They have been implemented in the UK by regulations[61] which require the Secretary of State to ensure that concentrations are monitored and kept below specified levels (shown in bold type in table 3.4). The means by which he should do this are not

Table 3.4
Air quality standards: WHO guidelines, **EC standards (bold type) and** *EC guide values (italics)*

	less than 1 hour	1 hour	8 hours	24 hours	1 year
					mg/m³
carbon monoxide	100 (15 mins) 60 (30 mins)	30	10		
					µg/m³
nitrogen dioxide		400 **200**[a] *135*[a] *50*[b]		150	
sulphur dioxide	500 (10 mins)	350		*100-150* *40-60*[c]	
combined exposure to sulphur dioxide and suspended particulates (black smoke)				125 SO_2 plus one of the following: 125 black smoke 120 total suspended particulates 70 thoracic particles **120 SO_2 if smoke ≤40**[d] **80 SO_2 if smoke > 40**[d] **180 SO_2 if smoke ≤60**[e] **130 SO_2 if smoke > 60**[e] **350 SO_2 if smoke ≤150**[f] **250 SO_2 if smoke > 150**[f]	50 SO_2 and 50 black smoke
suspended particulates (black smoke)				**80**[d] **130**[e] **250**[f] *100-150* *40-60*[c]	

	less than 1 hour	1 hour	8 hours	24 hours	1 year
ozone (health)		150-200	100-120 **110[g]**		
ozone (vegetation)		200 **200**		65 **65**	60 (April to September, over the growing season)
lead					0.5-1.0 **2.0**
formaldehyde	10 (30 mins)				

Concentrations of gaseous pollutants can be expressed in two ways:
as the mass of pollutant in a given volume of air, usually expressed as milligrammes per cubic metre (mg/m^3) or as microgrammes per cubic metre ($\mu g/m^3$) or
as the ratio of the volume of the gaseous pollutant to the volume of the air in which the pollutant is contained, usually expressed as a ratio in parts per million (ppm) or parts per billion (ppb).

WHO Guideline Values and EC Limit and Guide Values are specified as mass/volume concentrations.The following conversion factors can be used to convert from one set of units to the other at standard atmospheric pressure and 20°C:

carbon monoxide - 1 mg/m^3 = 0.86 ppm
nitrogen dioxide - 1 $\mu g/m^3$ = 0.52 ppb
sulphur dioxide - 1 $\mu g/m^3$ = 0.38 ppb
ozone - 1 $\mu g/m^3$ = 0.50 ppb

For black smoke, measurements made in the UK using the BSI (British Standards Institution) method have to be divided by 0.85 for comparison with EC standards, which assume use of an OECD method of measurement specified in the relevant Directive.

a - limit or guide value expressed as 98th percentile of hourly means over a year (in other words, if higher values are recorded for 175 hours in a year, it does not count as an exceedance of the Directive).
b - guide value expressed as 50th percentile of hourly means over a year.
c - guide value expressed as mean of daily means over a year.
d - limit expressed as median of daily means over a year.
e - limit expressed as median of daily means over the winter, October to March.
f - limit expressed as 98th percentile of daily means over a year (in other words, if higher values are recorded for 7 days in a year, it does not count as an exceedance of the Directive); in addition, the limit values should not be exceeded on more than 3 consecutive days.
g - expressed as an 8 hour mean (calculated as a non-overlapping moving average).

specified. These Directives also contain more stringent guide values (shown in italics in table 3.4) to provide the basis on which member states can develop longer-term policies for improving air quality. There is also a Directive setting a limit for lead in air[62] but concentrations now being recorded are well below this limit.[63]

3.37 A Directive on ground-level ozone[64] was adopted in 1992 and has been implemented by regulations.[65] Member states are required to monitor ozone concentrations, exchange information, and provide information and guidance to the general public if one of the threshold values shown in bold type in table 3.4 is exceeded. This last aspect of the Directive is implemented through DOE's Air Quality Bulletins (13.29). The thresholds cover both vegetation and health protection.

3.38 The European Commission has recently published a draft Directive on Ambient Air Quality Assessment and Management. This would be a framework Directive and supersede the existing Directives: quality objectives would be established by the end of 1996 for the substances covered by existing Directives and by the end of 1999 for a further list of substances. The additional transport-related pollutants listed in the draft are carbon monoxide, benzene and PAHs.

3.39 The guidelines for air pollutants published by the World Health Organization (WHO) Regional Office for Europe in 1987[66] cover a larger number of pollutants and are also shown in table 3.4. They are intended to provide background information and guidance to governments making risk management decisions, particularly in setting standards. They allow a margin of protection below the minimum concentrations associated with adverse effects on the health of the general population and are derived from the lowest concentration of a pollutant at which effects have been observed in humans, animals and plants. For pollutants that have irritant or sensory effects on humans, the no-observed-effect level is adopted where the necessary information is available. A revised version of this document will include a commentary on the consequences of exceeding the guideline values. The WHO guidelines are expected to play an important role in the formulation of further EC standards.[67]

3.40 DOE has set up an Expert Panel on Air Quality Standards (EPAQS), with a secretariat provided jointly by DOE and the Department of Health, to recommend standards for individual pollutants. It has produced reports on benzene and ozone, and plans to publish reports on 1,3-butadiene, carbon monoxide and sulphur dioxide before the end of 1994.

The need to improve air quality

3.41 The monitoring data show that many people in the UK are exposed from time to time to concentrations of pollutants which exceed WHO guidelines.[68] This is mainly the result of emissions from road vehicles. In particular:

> the health-based guidelines for **ozone** are exceeded on occasions in both urban and rural areas, particularly in southern Britain. In 1992 the lower bound of the guidelines was exceeded for a total of 41 hours at 7 urban monitoring sites, for 35 hours at the sole suburban site and for 551 hours at 15 rural sites; the upper bound was exceeded for 38 hours in the rural network. Examples of exceedances during July 1994 are summarised in box 3C above.
>
> The guidelines for **nitrogen dioxide** are exceeded on occasions at urban sites. The highest urban background levels ever recorded in the UK were measured in 1991 and 1992. The London pollution episode of December 1991 is described in box 3B above. In 1992 the daily guideline was exceeded for 8 days at the sole kerbside site and for 15 days in total at the 13 urban sites; the hourly guideline was exceeded for 10 hours in total at the urban sites.
>
> The results of the first year's monitoring show that, although the daily guideline for thoracic particles has not been exceeded, there have been shorter periods when levels of **PM10** exceeded the concentration specified in that guideline.

3.42 Although the monitoring data are from a small number of sites, there is no reason to suppose they are not representative of conditions in many areas of the UK. Modelling studies submitted in evidence by the Meteorological Office indicate that the hourly mean concentration of nitrogen dioxide would frequently exceed the EC standard of 200 $\mu g/m^3$ (expressed in the Directive as the 98th percentile of hourly means over a year) near any road carrying more than 10,000 fast-moving vehicles an hour if these are not in most cases fitted with three-way catalytic converters.

3.43 We are concerned that the present use of road vehicles may be causing serious damage to human health by triggering or exacerbating respiratory symptoms and by exposing people to carcinogens from vehicle emissions. The situation should therefore be regarded as unsustainable (1.15). Despite the many uncertainties about the effects of transport pollutants on human health and the environment, there is a clear case, on the basis of what is already known, for increasing the precautionary action taken to improve air quality. It is especially important to reduce concentrations of particulates and nitrogen oxides. The overall policy objective should be:

TO ACHIEVE STANDARDS OF AIR QUALITY THAT WILL PREVENT DAMAGE TO HUMAN HEALTH AND THE ENVIRONMENT.

3.44 WHO has left it to national governments to set a date for bringing pollution levels below its health-related air quality guidelines.[69] We propose as the target:

To achieve full compliance by 2005 with World Health Organization health-based air quality guidelines for transport-related pollutants.

In planning how best to achieve this target it will of course be necessary to take into account other sources of such pollutants. The primary responsibility for achieving the target must lie with central government. We discuss in chapter 13 the respective roles of government departments and local authorities, and in chapter 8 the contribution that lower levels of emissions from road vehicles can make. It will be necessary to keep this target under review as the work of EPAQS and the revision of the WHO guidelines proceed.

3.45 WHO does not recommend a guideline for benzene because no safe level of exposure is known. Measurements made at national monitoring sites show wide variations in the hourly means.[70] **We endorse the recommendation of EPAQS that for practical purposes the standard for benzene should be 5 ppb as an annual running average, and should be reduced to 1 ppb at a later date.** At 5 ppb the risks from exposure are extremely small.[71] The effect of reducing concentrations to less than 1 ppb would be that ambient air will no longer be the main source of an individual's exposure to benzene.

Other effects of transport-related pollutants

3.46 In addition to their effects on human health, nitrogen dioxide and ozone also have effects on the natural environment. High concentrations of nitrogen dioxide retard plant growth and may cause visible damage to plants, but lower concentrations may promote plant growth, especially on nitrogen-deficient soils. Nitrogen dioxide is also involved in the formation of acid rain and the acidification of soils and aquatic ecosystems. Emissions of nitrogen oxides are thought to be responsible for about one-third of the acidity of rainfall, and the proportion appears to be increasing. The nitrate to sulphate ratio in acid aerosol in the UK has increased steadily since 1954 and measurements in Arctic lakes and ice cores also show an increase.[72] The effects of acid rain on vegetation and natural habitats were discussed in the Commission's Sixteenth Report on freshwater quality. Nitrogen oxides are additionally involved in eutrophication, which was discussed in the Sixteenth Report in relation to freshwater, but also affects soil and the marine environment.

3.47 Exposure to high concentrations of ozone can damage plants. In experimental studies UK ozone episodes have significantly reduced growth in several crop species[73] and caused visible injury to sensitive annual crop species.[74] The evidence from these and other studies suggests that, although it is not possible to quantify the effects, present ozone levels in the UK affect crop yield in some years and may be affecting the species composition of natural vegetation.[75]

3.48 Trees can be damaged by a wide variety of agents, such as wind, frost, drought and pests. Atmospheric pollutants may alter the sensitivity of trees to some of these agents, either indirectly through changes in soil chemistry or by their direct effects on leaf tissue. A recent review of air pollution and tree health in the UK concluded that pollution is contributing to tree damage.[76]

3.49 Transport-related pollutants also damage buildings. Carbon and other particulate material soils buildings and can damage their fabric, especially if acid particles are deposited.[77] The interactions of particulate material with the surface of a building are complex, but it is known that carbon particles act as a catalyst for reactions in which calcium carbonate (in the form of limestone, for example) is converted to gypsum or calcium nitrate.[78]

3.50 The Convention on Long-Range Transboundary Air Pollution of the United Nations Economic Commission for Europe (UNECE) originated in concern about the effects of acid rain. It was adopted in 1979, came into force in 1983, and commits governments to endeavour to limit and, as far as possible, gradually reduce and prevent air pollution, including long-range transboundary air pollution. The action taken to limit emissions of sulphur dioxide is not a main concern in the present context, as transport is only a minor source (table 3.2). The Sofia Protocol of 1988 requires that total UK emissions of nitrogen oxides be reduced to the 1987 level (2.6 million tonnes) by 1994, with further reductions thereafter. Under the terms of the Geneva Protocol of 1991, UK emissions of VOCs must be reduced by 30% of the 1988 level by 1999.

3.51 Transboundary effects are significant in determining ground-level ozone concentrations. It has been estimated that, by the turn of the century, the effect over Europe as a whole of general compliance with the Sofia and Geneva Protocols will be:

a reduction in peak hourly ozone concentrations of about 20–40 $\mu g/m^3$ (10–20%);

about 10–20 fewer days a year on which the running average 8-hour concentration exceeds 100 $\mu g/m^3$.

The same study suggested that, in order to prevent ozone episodes with a running average 8-hour concentration greater than 100 $\mu g/m^3$ (the air quality standard for ozone recommended by EPAQS), it would be necessary either to reduce present European emissions of VOCs by 75–85% or to reduce present European emissions of nitrogen oxides by more than 95%.[79]

3.52 The WHO air quality guidelines incorporate separate guidelines for the protection of vegetation from ozone and the EC Ozone Directive incorporates separate thresholds for vegetation (table 3.4). Both have adopted a concentration of 65 $\mu g/m^3$ over 24 hours for this purpose. Full compliance with this guideline is likely to be difficult to achieve.

3.53 *Critical levels* for nitrogen oxides and ozone were set by UNECE following a workshop in 1988 and have been modified and refined at subsequent international meetings. Critical levels are defined as 'the concentration in the atmosphere above which direct adverse effects on receptors such as plants, ecosystems or materials may occur according to present knowledge'.[80] The critical level specified for nitrogen dioxide is an annual mean of 30 $\mu g/m^3$. This concentration is much lower than that specified in the EC guide value, which is in turn much lower than those specified in the WHO health-related guidelines (table 3.4). It is exceeded regularly in many areas of the UK, especially in central and southern England and close to large cities.[81] The critical level for ozone now takes the form of an index representing both the length of time and the extent to which an hourly mean concentration of 80 $\mu g/m^3$ is exceeded. Separate critical levels have been recommended for forests and for crops, calculated over different periods during the summer months.[82]

3.54 We believe that, to protect particularly sensitive habitats, it is desirable to introduce additional pollution control measures in certain localities, aimed at reducing concentrations of pollutants to below the critical levels. In a discussion paper on air quality issued in March 1994 (13.30), UK Environment Departments have suggested that health-based national standards for air quality should be supplemented by local standards based on critical levels. We endorse this concept. We propose as an additional target:

To establish in appropriate areas by 2005 local air quality standards based on the critical levels required to protect sensitive ecosystems.

The measures required to achieve compliance with such local standards will not relate solely, or even in some areas primarily, to pollution from transport. Transport is probably not the main source of nitrogen oxides in many rural areas. In framing programmes of action to meet the standards, however, the contribution vehicle emissions make to the prevailing concentrations will have to be taken fully into account in the light of increasing knowledge of the processes involved.

Figure 3-III
Greenhouse effect: contribution of principal greenhouse gases 1765-1990

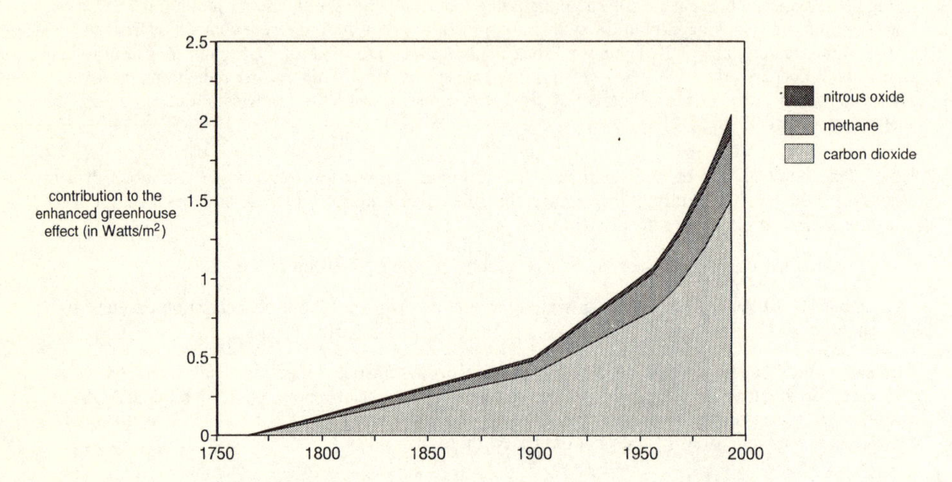

Changes in the Earth's atmosphere

3.55 A *greenhouse gas* absorbs heat from the Earth's surface which would otherwise have been radiated into space. Water vapour is the most abundant gas which has this effect but the total amount of water vapour in the atmosphere is not directly affected to any significant extent by human activity. The other principal greenhouse gases are carbon dioxide (CO_2), methane (CH_4) and nitrous oxide (N_2O). All occur naturally in the atmosphere, but at levels which have been significantly affected by human activity. Ozone, which is present both in the *troposphere* and in the *stratosphere* (5.19), also acts as a greenhouse gas, as do some synthetic organic compounds containing chlorine and related elements, including chlorofluorocarbons (CFCs).

3.56 Greenhouse gases differ in the extent to which they absorb radiated heat. Table 3.5[83] shows the concentrations of the principal greenhouse gases in the Earth's atmosphere and their relative contributions to the global greenhouse effect during the 1980s. Figure 3-III[84] shows the extent to which the greenhouse effect has been enhanced by increases in concentrations of the principal greenhouse gases.

3.57 Particulates present in the troposphere have various effects:

a. they scatter some solar radiation back out of the atmosphere, thus having a cooling effect;

b. they can act as nuclei for condensation of cloud droplets. When there are large numbers of nuclei, the droplets in clouds tend to be smaller. This enhances the scattering of solar radiation, and thus the cooling effect;

c. they can absorb some solar radiation, and this tends to warm the atmosphere;

d. they can act like a greenhouse gas by absorbing heat radiated from the Earth's surface.

Carbon dioxide emissions

3.58 Carbon dioxide is the dominant greenhouse gas affected by human activity. Although its concentration in the atmosphere has been increasing since the 18th century, there has been an acceleration

Table 3.5
Principal greenhouse gases

	atmospheric concentration 1750-1800	atmospheric concentration 1990	present annual rate of change	estimated contribution to enhanced greenhouse effect in the 1980s[a]
carbon dioxide	280 ppm	353 ppm	+ 0.5%	65%
methane	0.8 ppm	1.72 ppm	+ 0.9%	20%
nitrous oxide	288 ppb	310 ppb	+ 0.25%	7%
tropospheric ozone	17-23 ppb[b]	30-34 ppb[c]	+ 1-2%[d]	5-10%[e]

Units - ppm = parts per million by volume; ppb = parts per billion (thousand million) by volume.
a - any contributions to the enhanced greenhouse effect from CFCs are ignored.
b - atmospheric concentration ca. 1850s-1900.
c - current atmospheric concentration, mean for uplands of north and west Britain.
d - present annual rate of change in the Northern Hemisphere.
e - estimated current contribution to the enhanced greenhouse effect.

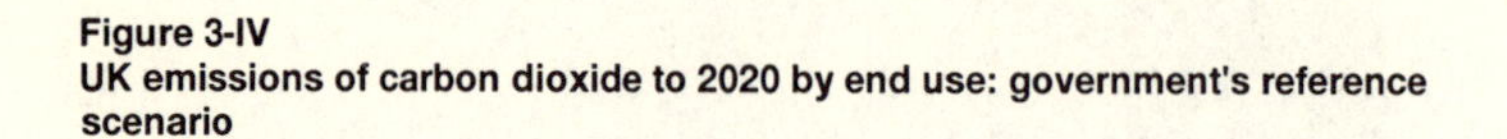

Figure 3-IV
UK emissions of carbon dioxide to 2020 by end use: government's reference scenario

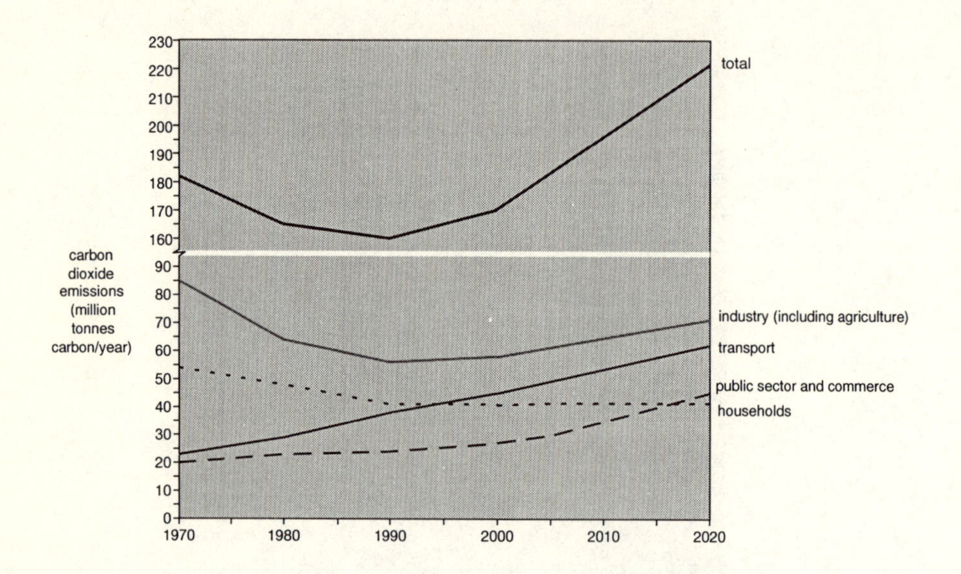

since the beginning of the 20th century, and markedly so since about 1950. Of UK carbon dioxide emissions, 21% come from surface transport, or about 24% if electricity generation for transport and the production of transport fuel are included. Road transport accounts for 87% of transport-related emissions. It is estimated that carbon dioxide emissions from road transport almost doubled between 1970 and 1990, from 16 million tonnes of carbon (mtC) to 30 mtC. Emissions from rail and coastal shipping have not increased significantly over the last ten years.[85]

3.59 On the basis of DOT's 1989 forecasts of road traffic and present trends in fuel consumption, carbon dioxide emissions from UK road transport will show further substantial increases over the next 25 years. The catalytic converters now fitted to new petrol cars do not remove carbon dioxide from exhaust gases, and in fact lead to a small loss in fuel efficiency. Figure 3-IV shows a reference scenario for UK emissions by end use[86], before taking into account the programme of measures announced by the government earlier this year (3.74).

3.60 Statistics of UK carbon dioxide emissions do not include those from climbing and cruising aircraft. Carbon dioxide emissions from aircraft are substantial (5.20) and growing (5.30): air transport accounts for 15% of the petroleum products used by the UK transport sector (table 3.1) and a similar proportion of the total energy used in transport.[87]

Other effects of transport on the atmosphere

3.61 Emissions from road vehicles are only a minor source of methane (insignificant in the case of the UK), but a significant source of nitrous oxide (estimated to account for 7% of UK emissions).[88] Catalytic converters increase emissions of nitrous oxide from cars by an order of magnitude.[89]

3.62 As a major source of both nitrogen oxides and VOCs, transport has made a significant contribution to the major increase in the total amount of ozone in the troposphere which has occurred over the last 50 years.[90] Emissions of nitrogen oxides from transport have increased by about 30% globally since 1970.[91] As in the case of carbon dioxide, there are substantial emissions from climbing and cruising aircraft (5.21), which are growing (5.31) and are not reflected in present UK statistics.

3.63 CFCs have been widely used in road vehicles and other forms of transport, particularly in refrigeration and air conditioning units; and may be released to the atmosphere when such units are serviced or scrapped. Following the adoption of the Montreal Protocol and its amendments[92], international action is in hand to phase out the use of CFCs because of their role in the destruction of stratospheric ozone, which protects the earth from ultraviolet radiation. Although their direct contribution to the greenhouse effect is significant, their net contribution is less because stratospheric ozone is also a greenhouse gas. The contribution of CFCs will in due course diminish as their concentration in the atmosphere is reduced as a result of the Montreal Protocol.

3.64 Transport is only a minor source for the sulphate aerosols which may be significant in the context of climate change (3.67). Of other types of particulates, those from diesel emissions, which contain free carbon, may be particularly effective in absorbing solar radiation, with a consequent warming effect. Overall, particulates from transport emissions probably have smaller effects on the atmosphere than those from other sources such as the burning of biomass, for example in tropical forests. It is important that work continues to quantify the effects on the atmosphere from all the significant sources of particulates.

3.65 Emissions of nitrogen oxides, VOCs and particulates from road transport will be substantially reduced by the fitting of catalytic converters to cars and further measures to improve air quality (8.23, 8.28). For the other substances mentioned above, either the proportion of emissions attributable to transport is small or there is no clear evidence that they make a significant net contribution to global warming. We therefore focus here on carbon dioxide.

The possible extent of climate change

3.66 A major programme of research, is forecasting future atmospheric concentrations of greenhouse gases, observing climatic variations, and investigating the possible extent and consequences of climate change resulting from the enhanced greenhouse effect. Important tools in this research are complex computer models of the climate, known as general circulation models, which bring together the physics and dynamics of atmosphere, ocean, ice sheets, land and biosphere. Assessment of the research findings is co-ordinated by the Intergovernmental Panel on Climate Change (IPCC), established by the World Meteorological Organization and the United Nations Environment Programme. The main conclusions in IPCC's 1990 report were that:

emissions resulting from human activities are substantially increasing atmospheric concentrations of carbon dioxide, methane, CFCs and nitrous oxide;

global mean surface temperatures have increased by between 0.3 and 0.6°C over the last hundred years, a change which is consistent with the predictions from models but of the same magnitude as natural climatic variability;

doubling the amount of carbon dioxide in the atmosphere is likely to increase global mean surface temperatures by between 1.5 and 4.5°C, but there are uncertainties about the timing, magnitude and regional effects of such a change;

it is likely to be a decade or more before unequivocal results are available from observations of climate change.[93]

3.67 In a more recent assessment[94], IPCC has reaffirmed these findings but pointed out that other factors may complicate the picture. For example, the effects on climate of sulphate aerosols created by sulphur dioxide emissions, largely from power stations, have been estimated to be comparable, in industrial areas of the Northern Hemisphere, to the enhanced greenhouse effect resulting from present concentrations of greenhouse gases.

3.68 While there is potential for reducing energy demand in developed countries, demand will continue to grow in developing countries as a result of population growth and the need for economic development.

Figure 3-V
Projected global emissions of carbon dioxide from energy generation to 2100: four scenarios (3-68)

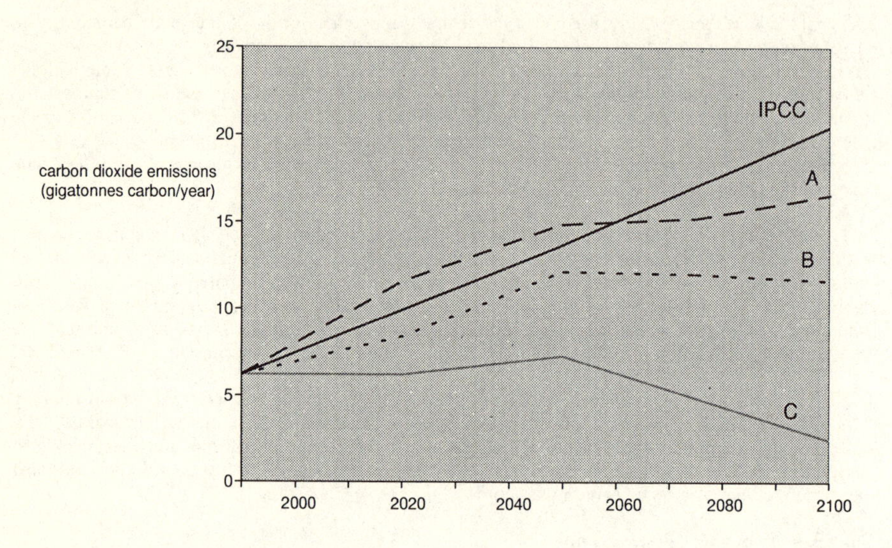

Several bodies have made estimates of global emissions of carbon dioxide over the next century. The World Energy Council considered the following scenarios[95]:

A assumes a higher rate of economic growth in developing countries

B is based on a moderate rate of economic growth

C assumes that strong environmental pressures will reduce energy demand.

All three scenarios assume economic and environmental pressures to achieve improvements in energy efficiency and reduce emissions of carbon dioxide. For scenario C these are described as 'strong pressures'. An IPCC scenario assumes moderate economic growth, a world population a little over double its present size by 2100 and no significant action to reduce energy demand on environmental grounds.[96] Figure 3-V[97] shows projected global emissions of carbon dioxide from energy generation for these four scenarios.

3.69 Because it is removed only slowly by natural processes, atmospheric concentrations of carbon dioxide will take decades to reflect reductions in emissions from human activities. Figure 3-VI[98] shows the atmospheric concentrations which would result from the four scenarios and compares these with the concentrations that would result if global carbon dioxide emissions from fossil fuel use were held constant at the 1990 level.

3.70 If global emissions were stabilised at the 1990 level, atmospheric concentrations of carbon dioxide would continue to increase for several centuries. Even with the large improvements in energy efficiency and major contributions from renewable energy entailed by scenario C, atmospheric concentrations of carbon dioxide would not level off until near the end of the next century.

International obligations

3.71 The Framework Convention on Climate Change, signed by some 150 nations at Rio de Janeiro in June 1992, came into force in March 1994. It recognises that emissions of greenhouse gases, in

Figure 3-VI
Carbon dioxide concentrations resulting from the emission scenarios in figure 3-V

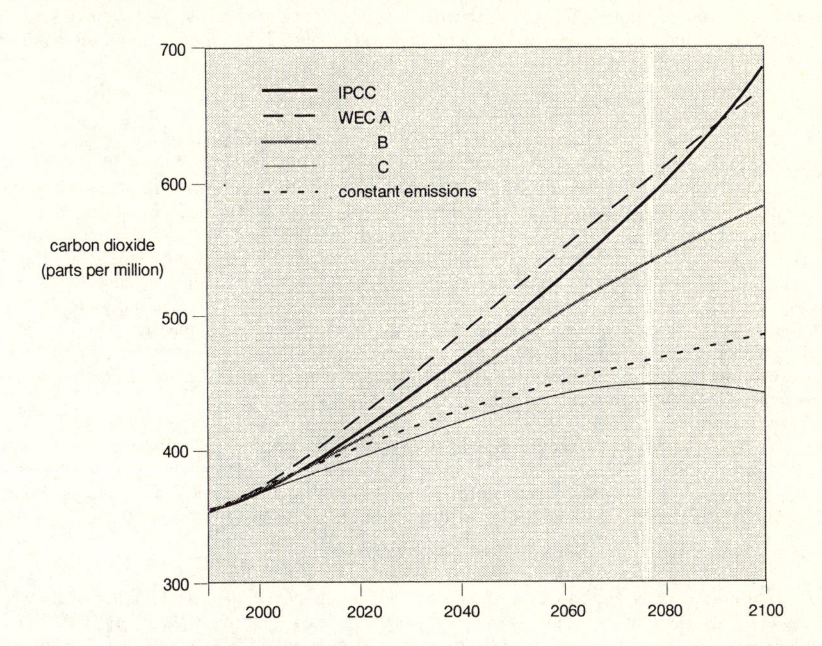

particular carbon dioxide, pose a serious threat to the Earth's environment. The 'ultimate objective' is the 'stabilization of greenhouse gas concentrations in the atmosphere at a level that would prevent dangerous anthropogenic interference with the climate system' and 'within a time frame sufficient to allow ecosystems to adapt naturally to climate change, to ensure that food production is not threatened and to make economic development proceed in a sustainable manner'.[99] Parties to the Convention are required to publish national inventories of emissions of greenhouse gases (other than CFCs and halons, which are covered by the Montreal Protocol) and to draw up, and regularly update, programmes of measures to mitigate climate change by 'addressing' emissions.

3.72 Developed countries are to take the lead 'in modifying longer-term trends in anthropogenic emissions . . ., recognizing that the return by the end of the present decade to earlier levels of anthropogenic emissions of carbon dioxide and other greenhouse gases not controlled by the Montreal Protocol would contribute to such modification'. They must submit a report on their policies within six months of the Convention coming into force 'with the aim of returning [emissions of greenhouse gases] individually or jointly to their 1990 levels by the end of the decade'.

3.73 The UK has only about 1% of the Earth's population but produces about 3% of the carbon dioxide emissions from human activities. In December 1992, DOE published a discussion document on the UK programme for limiting carbon dioxide emissions. In its response to this document (reproduced as appendix F), the Commission emphasised that, while total carbon dioxide emissions from UK sources had fallen by 10% in the previous 20 years, those from the transport sector had increased (figure 3-IV); that the projected increase in transport emissions between 1970 and 2020 was equivalent to the whole of the projected net increase in total UK emissions over that period; and that two-thirds of this increase was accounted for by private cars.

3.74 In fulfilment of its Convention commitments, the government has published a programme of measures aimed at reducing emissions of each greenhouse gas to the 1990 level or below by 2000.[100] For carbon dioxide, that involves reducing the projected emissions in 2000 by 10 mtC. The main measure affecting transport was an increase in duty on road fuel of 10% in 1993 and at least 5% in real terms in each subsequent year (7.52). Although this was expected to reduce emissions in 2000 by 2.5 mtC, the government has not committed itself to any specific reduction in transport emissions. Nor has it set further target dates or announced measures to deal with the position after 2000 (although the programme sets out some of the options).

The need to reduce carbon dioxide emissions

3.75 The objective of the Climate Change Convention is the stabilisation of the concentrations of greenhouse gases in the atmosphere, of which carbon dioxide is the most important. The carbon dioxide concentration to be aimed at has not yet been defined, although the Convention spells out the criteria. It is already clear however from figure 3-VI and other similar work on which IPCC will report later this year[101] that stabilisation at any level likely to be acceptable will require global carbon dioxide emissions in the next century to be well below the levels shown in the IPCC scenario and scenarios A and B in figure 3-V. Eventually they will have to be brought well below 1990 levels. That will involve a dramatic change in previous trends, and has very large implications for all countries, especially the developed countries. The World Energy Council's scenario C, for example, recognises that by 2020 there will be a substantial increase in primary energy use in developing countries, but shows that by that date a reduction of about 20% is realisable in OECD countries. It emphasises that extreme commitment will be required within those countries to achieve such a reduction.

3.76 We believe a substantial reduction in emissions from the transport sector is an essential element if stabilisation of carbon dioxide concentrations in the atmosphere is to be achieved in the longer term. Because transport is contributing to what could be irreversible changes in the Earth's climate, an important objective of a sustainable transport policy must be:

TO REDUCE CARBON DIOXIDE EMISSIONS FROM TRANSPORT.

3.77 Transport is not the largest source of carbon dioxide emissions. Some reductions in emissions from other sectors could be made relatively cheaply, for example by improving the energy efficiency of buildings or industrial processes. It has been argued that reducing carbon dioxide emissions from transport may not be the most cost-effective way of reducing total emissions, and even that some further increase in emissions from the transport sector might be accepted and compensated for by larger reductions elsewhere. Although it is clearly important to take decisions on how to reduce carbon dioxide emissions in a wider context than transport, we consider that transport must take its share of the reductions, for three reasons. Because of the large overall scale of the reductions which are likely to be required, all sectors will need to achieve reductions. The policies followed by government in relation to a particular sector cannot be regarded as sustainable unless they provide for this. Second, there is a large potential for increased efficiency in the use of energy by transport at relatively small cost. Third, as the analysis in the first part of this chapter and in the next chapter shows, transport has other undesirable effects on the environment, which are likely to increase markedly in scale if present policies remain unchanged. Policies designed to achieve major reductions in carbon dioxide emissions from transport can therefore bring other important benefits for the environment.

3.78 We propose that the UK government adopt the following target:

To reduce emissions of carbon dioxide from surface transport in 2020 to no more than 80% of the 1990 level.

This target implies a reduction of less than 1% a year between 2000 and 2020. Nevertheless, in view of the forecast growth in traffic, radical measures will be required in order to achieve it. In addition to the considerations set out above, the choice of a target which is near the bounds of what is achievable is justifiable because the Climate Change Convention implies that the reductions made by developed countries ought to be large enough to allow scope for some growth in the energy demands of developing countries. Although no country has yet produced a firm plan to achieve reductions in emissions on the scale likely to be required in the longer term, the proposed target is broadly consistent with the position being adopted in this respect by other northern European countries such as Germany and the Netherlands.

3.79 Action to meet the target for 2020 will need to start immediately. As it will span a period of 25 years, it is important to have an intermediate target in order to provide a clear framework and performance measure for action taken over the next few years. For this purpose it is reasonable to apply to the transport sector the target already set by the government for the economy as a whole. We propose as the intermediate target:

To limit emissions of carbon dioxide from surface transport in 2000 to the 1990 level.

3.80 Our recommended targets apply to surface transport. Aircraft contribute significant proportions of the global emissions of carbon dioxide and nitrogen oxides from human activities, and air transport is the most rapidly growing form of transport. It is therefore highly desirable that action be taken to limit such emissions. However, aircraft emissions cannot be effectively controlled by the UK government acting independently or even by the European Community; appropriate international negotiation and agreement will be required before any action can be taken. We discuss aircraft emissions as a separate issue in chapter 5.

3.81 The corollary of setting targets for carbon dioxide emissions from the transport sector is that we must exploit the potential for increased energy efficiency in transport, and also find ways of satisfying the needs of present and future generations in less transport-intensive ways. These are two of the central issues discussed in later chapters.

Chapter 4

OTHER MAJOR IMPACTS OF SURFACE TRANSPORT

4.1 Transport systems have many environmental effects besides those associated with emissions of airborne pollutants and greenhouse gases. Our study of transport and the environment has involved a general review of those effects, but this chapter is deliberately restricted to those other issues which we have concluded are of major significance in the context of formulating a sustainable transport policy for the UK. It considers first the effects on people in the form of noise, disruption to communities and accidents. It examines the effects on the natural environment when roads and other forms of infrastructure are constructed. It then discusses the requirements for materials to construct such infrastructure and manufacture vehicles.

4.2 A comprehensive life-cycle analysis of a transport system, tracing all the material and energy flows involved for each component of the system, would be an enormous task, which we have not attempted in this report. We have not for example made a further study[1] of the environmental impact of exploration for oil and its extraction, processing and distribution (except to the extent that emissions from oil refineries were taken into account in chapter 3).

4.3 The operation of transport systems gives rise to many solid and liquid pollutants, including waste oils, de-icing agents, particles of rubber and metal shed by vehicles and herbicides used in road and track maintenance. These pollutants may find their way into watercourses or groundwater and eventually into the sea. In addition there is widespread contamination of soil and groundwater caused by leaks from petrol stations.[2] Pollution of fresh water and groundwater as a result of transport operations was considered in the Commission's Sixteenth Report (paragraphs 7.76–7.81 and box 7.11). These forms of pollution from transport are not different in kind or in scale from those found in other sectors of the economy, and have not therefore been discussed here.

4.4 Over 95% of the goods carried into and out of the UK are transported by sea. The state of the seas, especially those around Europe, remains an important environmental issue.[3] Over three-quarters of marine pollution comes directly from sources on land and a third of it is airborne, including some pollutants from vehicle emissions. It is estimated that 12% comes from ships and boats, as a result of operational discharges, accidents or general litter.[4] Most aspects of marine pollution were considered either in the Commission's Eighth Report[5] or in its Eleventh Report.[6] The report published earlier this year of an inquiry by Lord Donaldson of Lymington[7] made a number of recommendations to provide more effective protection for the UK coastline against pollution from merchant shipping. These are based on the polluter pays principle and recognise that maritime transport is international, that most marine pollution incidents are the result of human failure, and that ships under way are difficult to identify. Three main areas are covered in the recommendations: discouraging substandard management and maintenance of shipping; providing adequate emergency assistance, especially salvage tugs; and establishing Marine Environmental High Risk Areas, where there is both a heavy concentration of shipping and a high risk of environmental damage if an accident were to occur. **We endorse the recommendations of the Donaldson Inquiry, particularly those relating to port state inspection**.

Noise

4.5 For the majority of people in the UK, transport is the most pervasive source of noise in the environment. In a survey carried out over a 24-hour period in 1990, noise from roads was recorded outside 92% of a sample of dwellings in England and Wales and noise from motorways outside 2%. The next most common source of noise was aircraft (at 62% of the sites). Noise from railways (recorded at 15% of sites) was much less common. A separate survey provided evidence that road traffic is the main source of noise outside more than 60% of dwellings.[8]

BOX 4A	**HOW NOISE IS MEASURED**

Sound is a form of energy which is transmitted through air (or any other medium, such as water) by vibrations associated with rapid cyclical pressure changes. What is measured is the size of the fluctuations in pressure caused by the sound wave.

The sound-pressure level (or magnitude of the sound) is expressed in decibels (dB) on a logarithmic scale related to a standard reference pressure. At any point on this scale an increase of 3 dB represents a doubling in the sound intensity. A doubling in the perceived 'loudness' of sound is represented by an increase of about 10 dB.

In assessing human exposure to noise, the sensitivity of sound meters is adjusted to correspond to the frequency range detectable by the normal human ear (the dB(A) scale). On this scale, commonly experienced noises fall in the range 30 to 100 dB(A).

Measurements of exposure to noise also need to take account of variation in noise levels over time. This can be expressed in a number of different ways, for example:

dBL_{Aeq}	the mean level of the sound
dBL_{A90}	the level of sound exceeded for 90% of the time (the background noise)
dBL_{A10}	the level of sound exceeded for 10% of the time.

Other suffixes are used to represent the level of exposure over all or part of a 24-hour period, for example:

$dBL_{Aeq.8h}$	typical night-time exposure (for example, 10 pm to 6 am)
$dBL_{Aeq.16h}$	typical exposure during the day (for example, 6 am to 10 pm)
$dBL_{A10.18h}$	the mean of the 18 consecutive hourly L_{A10} measurements between 6 am and midnight on a normal working day (used to determine whether dwellings qualify for insulation from road traffic noise).

4.6 Measurements of noise levels take into account the pressure of sound waves, the duration of sounds and the way people perceive noise. The units of measurement employed are described in box 4A. In the survey of England and Wales referred to above, over half the sites were exposed to more than 55 dBL_{Aeq} from all sources and 7% to more than 68 dBL_{Aeq}. An estimate made for OECD countries as a whole is that more than half the population is exposed to noise levels greater than 55 dBL_{Aeq} from road transport, and about 14% to more than 65 dBL_{Aeq}.[9]

4.7 People's reactions to noise vary, but in general depend, not only on the level of sound, but on characteristics such as pitch, and whether a sound is continuous, regularly repeated or intermittent. In a 1978 survey of reactions to outside noises heard in the home, road traffic bothered more people (23%) than any other form of noise and was most often named as the biggest nuisance (by 16%). Aircraft bothered 13% of people and were the biggest nuisance for 8%. Few people mentioned trains as bothering them, even though they could be heard by 35% of people in this survey.[10] In a more recent survey road traffic noise was the most common type of outside noise heard in the home (47%), followed by aircraft noise (41%), and annoyed more people in total than other external noises.[11] It seems that the public increasingly regard transport noise as intrusive: since 1979, complaints to Environmental Health Officers in England and Wales about the noise from road traffic have grown by more than a quarter, and complaints about aircraft noise have more than doubled.[12]

4.8 Transport also causes vibration. It has been suggested that prolonged or repeated exposure to vibration produced by traffic may be a contributory factor in stress-related diseases.[13] Many people living very close to major roads or railways are worried about the possibility of damage to their homes, but studies by the Transport Research Laboratory found no firm evidence that vibration from traffic causes structural damage to buildings.[14]

4.9 Conclusive evidence of general health effects from noise is limited to cases of hearing loss and tinnitus (ringing in the ears) caused by long periods of exposure to more than 75–80 dB(A). Although this level of noise can be produced by heavy traffic at 60 metres, it is unlikely that members of the general public are exposed to traffic noise at this level over a sufficiently long time to cause hearing loss. On the other hand traffic noise almost certainly contributes to, or aggravates, stress-related health problems, including raised blood pressure, and minor psychiatric illness.[15,16] Disturbance of sleep patterns by noise may be an important effect.[17,18]

4.10 Epidemiological studies of populations exposed to aircraft noise have produced conflicting or uncertain findings. However, the results of studies in Germany, Japan and the Netherlands suggest that growth of the foetus may be inhibited, or birth weight reduced, by exposure to high levels of aircraft noise during pregnancy.[19] In animal experiments, exposure to excessive noise has caused still births, birth defects and reduced birth weights.

4.11 Exposure to noise can be reduced by cutting down the amount of sound produced, by deflecting or absorbing it or by insulating buildings. European Community (EC) legislation sets limits on the noise produced by road vehicles (table 8.8). There have been significant reductions in railway noise as a result of electrification, improvements in the design of diesel locomotives and the replacement of rim brakes by disc brakes. There may also be ways of reducing track noise, such as staggering the intervals between sleepers.[20]

4.12 As a significant proportion of noise from road traffic, especially at high speeds, is produced by tyres, it can be influenced by the nature of the road surface. Traffic on a porous asphalt surface is up to 4 dB quieter under dry conditions than traffic on a conventional asphalt surface, and up to 8 dB quieter in wet conditions. Porous asphalt also reduces spray by up to 95% when first laid, although it becomes considerably less effective after a few years because the pores become clogged.[21] It also reduces glare and the risk of aquaplaning. However, it is more expensive than conventional asphalt, is more liable to damage from heavy traffic turning and braking, and requires more winter maintenance. The policy of the Department of Transport (DOT) is now to use porous asphalt on urban trunk roads and in other noise-sensitive areas 'where the benefits outweigh the higher cost'. Concrete is more durable than asphalt. The traditional transverse brushed texture of concrete is particularly noisy, especially if laid with deep surface ridges, and will not be used in future on heavily trafficked motorways and trunk roads. DOT is carrying out trials of 'whisper concrete', which has a random texture.

4.13 Noise from roads can be reduced by using earth banks or barriers made of timber, metal, plastics or other materials. Tall barriers can reduce noise levels by up to 25 dB(A), provided they are faced with absorbent material to counteract reverberation. They have been much less widely used in the UK than in France, Germany or Japan, partly on grounds of cost and partly because they have been considered to be unsightly.

4.14 Householders within 300 metres of a new or substantially upgraded road are eligible for a grant for soundproofing if the external noise level exceeds 68 $dBL_{A10,18h}$. Barriers have been constructed as part of UK road schemes where noise levels at the external walls of housing would otherwise have exceeded that figure. The government is reassessing the qualifying level, following the recommendation of a working party.[22] Secondary glazing with mechanical ventilation and the wide air gap necessary to provide effective sound insulation has a clumsy appearance, and is not popular with householders. Expenditure on grants is modest (£0.8 million in 1990/91). Much larger amounts are being paid as compensation under part I of the Land Compensation Act 1973, without a qualifying noise level, to householders who have been able to establish that their property has lost value as a result of road works nearby.

4.15 Although noise from railways causes much less annoyance overall, high-speed trains produce high noise levels (which have been a particular cause of complaint and opposition in Japan). It is fair and logical that the safeguards should be as effective as for noise from roads. The Department of Transport has proposed that there should be qualifying levels applied to noise from new rail lines, above which the owner of the rail line will have to provide insulation or a grant. The proposed levels are 68 dBL_{Aeq} over the 18 hours from 6 am to midnight and 63 dBL_{Aeq} over the 6 hours from midnight to 6 am.[23] Barriers may also be valuable in the case of railways: Railtrack and Kent County Council are sharing the cost of 12.5 km of barriers along existing lines in Kent used by Channel Tunnel freight services, in order to prevent night-time noise levels outside some 2,000 houses exceeding 65 dBL_{Aeq}.[24]

4.16 The World Health Organization recognises noise as a health hazard; and suggests that it is desirable for daytime outdoor levels to be less than 55 dBL_{Aeq} (or 57 $dBL_{Aeq,16h}$ for aircraft noise), and that they should not exceed 65 dBL_{Aeq}.[25] The European Community's Fifth Environment Action Programme[26] included targets for night-time noise from all sources which can be summarised as:

noise levels of 85 dBL_{Aeq} should not be exceeded;

exposure to noise levels greater than 65 dBL_{Aeq} should be phased out;

there should be no increase in existing noise levels in areas where the present noise level is below 65 dBL_{Aeq}.

There has not been any proposal for European Community legislation to achieve these targets. The Second Transport Structure Plan for the Netherlands has adopted the aim of not increasing exposure to noise from local traffic above 1986 levels; and has set targets of reducing the number of dwellings with external noise levels of above 55 dBL_{Aeq} by 5% between 1986 and 1995, and by 50% between 1986 and 2010.[27]

4.17 We have concluded that present exposure to noise from transport is environmentally unsustainable in that, in addition to the health implications, it causes serious damage to the quality of life (1.15). One of the objectives of a sustainable transport policy must therefore be:

TO REDUCE NOISE NUISANCE FROM TRANSPORT.

4.18 In pursuit of this objective we propose the following targets:

To reduce daytime exposure to road and rail noise to not more than 65 $dBL_{Aeq.16h}$ at the external walls of housing;

To reduce night-time exposure to road and rail noise to not more than 59 $dBL_{Aeq.8h}$ at the external walls of housing.

Compliance with these target levels should be an important factor in the environmental assessment of proposals to construct new roads and rail lines or make more intensive use of existing infrastructure. They should also apply to existing roads and railways. DOT should accept responsibility for making early and substantial progress towards the target levels of noise and should place corresponding obligations on the Highways Agency, on Railtrack and on local authorities.

4.19 An important part will be played by reductions at source in the sound produced. The EC legislation on sound levels from road vehicles is discussed with other aspects of vehicle technology in chapter 8. In future there should be a higher priority on laying quieter road surfaces. The Dutch government decided in 1990, on the basis of a cost-benefit analysis, to lay porous asphalt by 2010 on all national roads carrying more than 35,000 vehicles a day. **We recommend that, both in new road construction and resurfacing, porous asphalt or whisper concrete should be used at all appropriate sites; and that research and development continue in order to identify surfacing materials with an even better combination of characteristics.**

4.20 Reductions of noise at source will not be sufficient in themselves. In addition, **we recommend more extensive and innovative use of barriers to absorb and deflect sound from roads and railways, as the most cost-effective way to achieve our targets in some cases.** Even well-designed barriers may look ugly or out of place in some locations however. Before a decision is taken to use barriers in a particular case, considerable weight should be given to the views of the people who would benefit from them, especially if the noise levels without barriers are close to the recommended targets.

4.21 In cases in which, after the scope for using barriers has been fully explored, traffic noise is likely to remain above the target levels, the solution may lie in policies for traffic restraint, which we discuss in chapter 11. In framing traffic management policies local authorities should also have in mind the desirability of avoiding increases in noise levels in areas where the existing daytime noise level is 55–65 dBL_{Aeq}.

4.22 Reductions in external noise levels are much preferable to insulating buildings against sound because they improve the general environment as well as living conditions indoors. We recognise that it may be impracticable to achieve our targets for external noise levels in all cases in the foreseeable future, even when the effect of other recommendations in this report is taken into account. Moreover, a working party of the UK Environmental Law Association has concluded that a total phasing out of exposure to external noise levels above 65 dBL_{Aeq} would be unduly restrictive.[28] There will therefore continue to be a role for insulation grants. **We recommend that the qualifying level for insulation grants for both road and rail noise should be reduced to 65 $dBL_{Aeq.16h}$ to match our target for daytime noise.**

4.23 Under present legislation a reduction in the qualifying level will help only households living near a new or substantially upgraded road or, in future, a new rail line. These are also the cases in which it should be easiest to achieve the target levels for external noise because suitable provision can be made at the design stage of schemes. The continuing existence of housing exposed to higher levels of noise will in our view leave an unfair burden on the people affected. It is sometimes argued that, as noise pushes down property values, such people are compensated for higher noise levels by being able to obtain cheaper housing. For houseowners that argument is valid only to the extent that noise levels, and the sensitivity of property values to noise levels, have not increased significantly since they purchased their house. Compensation for loss in the value of property is available only where traffic has increased as a result of road works carried out in the vicinity of the property. Compensation cannot be obtained if increased traffic, and an increased level of noise, are the result of a road improvement scheme (for example, a motorway extension or a new link road) carried out at some distance from the property.

4.24 As 7% of dwellings in England and Wales may be exposed to noise levels (from all sources) above the existing qualifying level (4.6), and perhaps another 10% have external noise levels of 65–68 $dBL_{A10.18h}$, the cost of extending insulation grants to those affected by noise from existing roads and railways would be considerable. In 1983, the Department of the Environment (DOE) estimated the cost of extending grants to those affected by existing roads, with a qualifying level of 68 $dBL_{A10.18h}$, as £1.84 billion.[29] However, we do not regard insulation grants as the primary instrument for reducing exposure to noise, but as a fallback for use in a limited proportion of cases. Usually reductions in noise at source, construction of barriers or traffic restraint will provide a more cost-effective, as well as an environmentally more satisfactory, solution. Even in cases where other solutions are not practicable, the take-up of insulation grants will not be complete. **We recommend that the government study:**

- **i. how eligibility for insulation grants might be extended to householders affected by noise from existing roads or railways in cases in which it will be impracticable in the foreseeable future to achieve our targets for external noise levels;**
- **ii. whether eligibility under the Land Compensation Act 1973 can be extended to householders whose property has lost value since they purchased it as a result of increased traffic caused by a road improvement scheme at some distance from the property.**

4.25 Because of the high peak noise levels from aircraft, it would not be appropriate to apply to them the targets we have proposed for road and rail. The scope for reducing the noise made by aircraft is considered in the next chapter.

4.26 The land use planning system can help to ensure that new development does not lead to exposure to high noise levels from transport sources, including aircraft, and we welcome DOE's intention to issue guidance on the relevance of noise exposure for development control. **We recommend that guidance on development control should be based on the principle of preventing the exposure of new residential development or schools to noise levels which exceed our proposed targets.**

Impact on communities

4.27 Heavy road traffic disrupts the life of communities. A US study showed that, as traffic volumes increased, social contacts within streets declined. There were also other differences in behaviour. Where traffic was heavier, people no longer lingered on the pavement, they did not use their front gardens, and those living in the busiest of the three streets studied spent less time in the front rooms of their house. Families who could afford to do so moved away from the area.[30]

4.28 The most obvious obstacle to social contacts is a motorway or other fenced road (or a railway line) which pedestrians are not allowed to cross. A road which pedestrians can cross may also be a formidable obstacle, especially for the old, young children and their mothers. These groups are less likely to have access to a car and are therefore more dependent on contacts within walking distance.

Even if a pedestrian crossing or subway is provided, the detour involved may well discourage these less mobile groups. In the case of subways, anxiety about the risk of criminal attacks may be an important additional psychological obstacle.

4.29 Noise was discussed above primarily in the context of annoyance caused to people in their homes but has an even more marked effect on people in the streets. Traffic noise may well prevent conversation, or at least make it difficult and uncivilised. Where traffic is heavy, people on the pavements will be exposed to relatively high concentrations of pollutants (3.12). Quite apart from the implications for health many people notice the unpleasant smell and taste of exhaust fumes. Last but not least, there is the risk of accidents, which we discuss in its own right below.

4.30 The combined effect of these factors is to make streets unpleasant places in which to spend time. The use made of the street declines. Space which was an important possession of local people has been taken from them over the years, and made the preserve of people in cars who happen to be passing through. Children no longer play on the pavement or make unaccompanied journeys along the street to school. A reduction in the numbers of people walking through the streets creates the conditions for an increase in the crime rate, and a vicious circle may develop which drives down still further the number of journeys made on foot.

4.31 Local authorities are now making considerable efforts to reverse this process. In particular they are applying traffic calming measures with the aim of reducing speeds and discouraging traffic from using residential streets when there are other routes available. Often this is an essential element in revitalising town centres and inner city areas. Another approach to reducing the effects of road traffic, emphasised by DOT, is to build bypasses for villages and small towns.

4.32 While quality of life can be a subjective concept, there seems to be very widespread agreement that heavy car and lorry traffic is causing serious damage and conflict (plate II). We consider it should be a basic objective of a sustainable transport policy:

> **TO IMPROVE THE QUALITY OF LIFE, PARTICULARLY IN TOWNS AND CITIES, BY REDUCING THE DOMINANCE OF CARS AND LORRIES AND PROVIDING ALTERNATIVE MEANS OF ACCESS.**

In the next section, and in chapter 11, we propose some specific targets for moving towards this objective.

4.33 The construction of bypasses often improves conditions in the towns and villages bypassed and leads to a freer flow of traffic, which reduces emissions from vehicles. On the other hand some evidence suggested that the problems may reappear somewhere else (in effect being pushed along the road) and the bypassed centre may wither because of the loss of trade. It was also suggested that the extent to which traffic is transferred to bypasses is overestimated; and that in some circumstances it would be environmentally preferable, and more cost-effective, to achieve the benefit to the local community through traffic restraint and traffic calming, rather than by building a new road. At a minimum, traffic management could be adopted much more quickly as an interim solution.

4.34 Bypasses form an increasingly important component of the trunk road programme.[31] It is disturbing therefore that there is a lack of consensus about their overall effect. DOT's Bypass Demonstration Project is intended to establish best practice in this field by maximising the benefits which six towns obtain from bypasses. However, there is also a need for research to throw light on the overall effectiveness of bypass schemes. This should take the form of studying a representative selection of towns and villages three or four years after a bypass has been completed to ascertain whether the environmental, traffic and safety objectives have been achieved in practice.[32] **We recommend that DOT:**

- **i. make comparative studies of representative towns and villages before and after the completion of bypasses in order to improve understanding of their environmental and other effects;**
- **ii. investigate whether some towns and villages could obtain most of the benefits of a bypass, more cost-effectively and with less environmental damage, through traffic management measures.**

Figure 4-I
Road accident deaths 1960-93

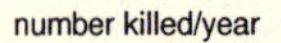
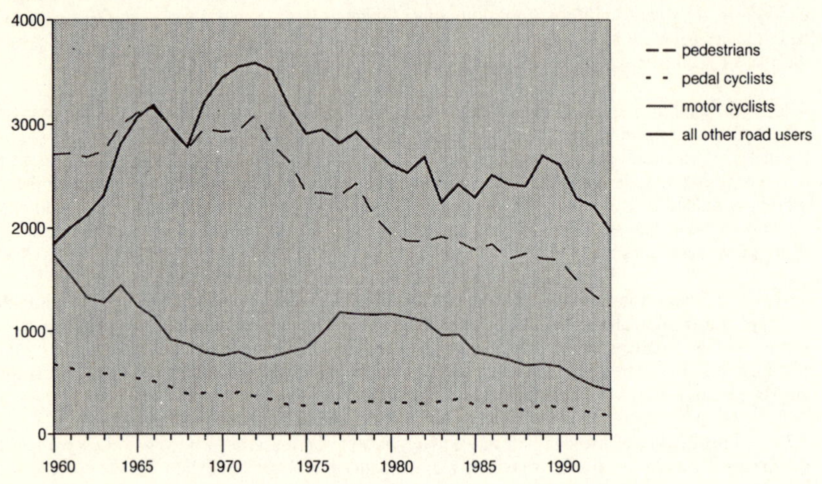

Accidents

4.35 Transport is responsible for nearly two-fifths of the accidental deaths in Britain: in 1992, 39% of accidental deaths took place on the roads, and a further 2% in connection with other forms of transport.[33] In 1993, road accidents killed 3,820 people, including 1,250 pedestrians and 190 cyclists; caused 44,890 people serious injuries, including 12,680 pedestrians and 3,740 cyclists; and involved over 0.25 million other casualties.[34]

4.36 The deaths and injuries caused by other modes of transport are not only far fewer in number but fewer in relation to the distances travelled. In the period 1982–92, 73 passengers and 94 other people were killed in rail accidents. In the financial year 1992/93, 5 people (none of them a passenger) died in rail accidents, 13 were seriously injured and 140 received minor injuries.[35] However, rail is not necessarily perceived as safer; a single rail accident can involve tens or even hundreds of casualties, and attracts a great deal of public attention. Air travel carries an even lower risk than rail travel; but there too a single accident may be on a very large scale.

4.37 Despite the great increase in traffic, deaths from road accidents have roughly halved over the last 30 years and injuries have fallen by over a third. As figure 4-I[36] shows, deaths have fallen among all classes of road user. The overall death rate in the UK is lower than in a number of other EC countries (figure 4-II[37]). The comparison is less favourable in the case of pedestrian deaths, particularly among children. A big reduction in the pedestrian death rate between 1990 and 1993 (from 3.1 to 2.4 per 100,000 population) has brought the UK near to or below the EC average, but the rate is still twice as high as in the Netherlands (1.0 per 100,000 in 1993).[38]

4.38 Denmark and Sweden have sought to reduce deaths among child pedestrians through traffic restraint policies which involve reducing speed limits in urban areas and designating zones in which pedestrians have priority. In contrast, the UK, the USA and New Zealand have placed the emphasis on educating children in road safety. A comparison of trends in child pedestrian mortality over the last 20 years (figure 4-III[39]) suggests that traffic restraint policies have been a more effective approach. The

Figure 4-II
Rates of road deaths in EC countries (1991)

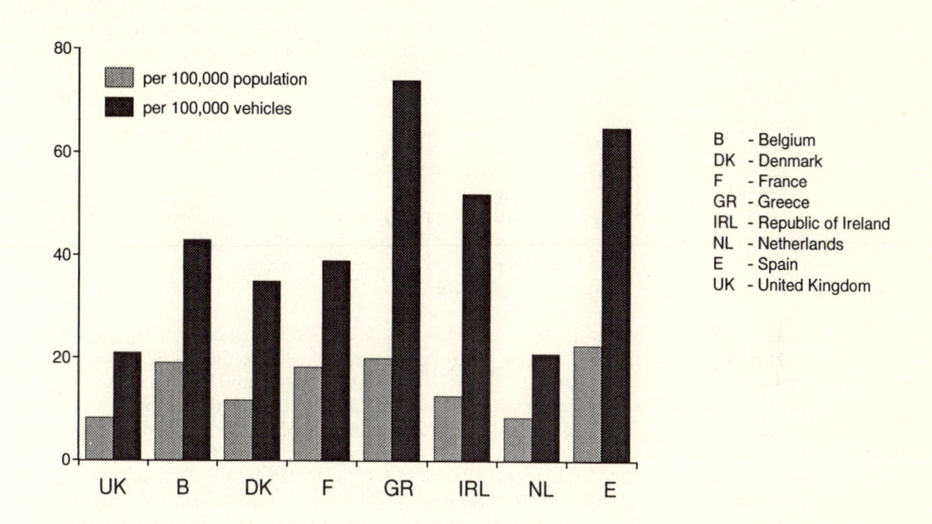

Figure 4-III
Deaths of child pedestrians: trends in selected countries since 1968

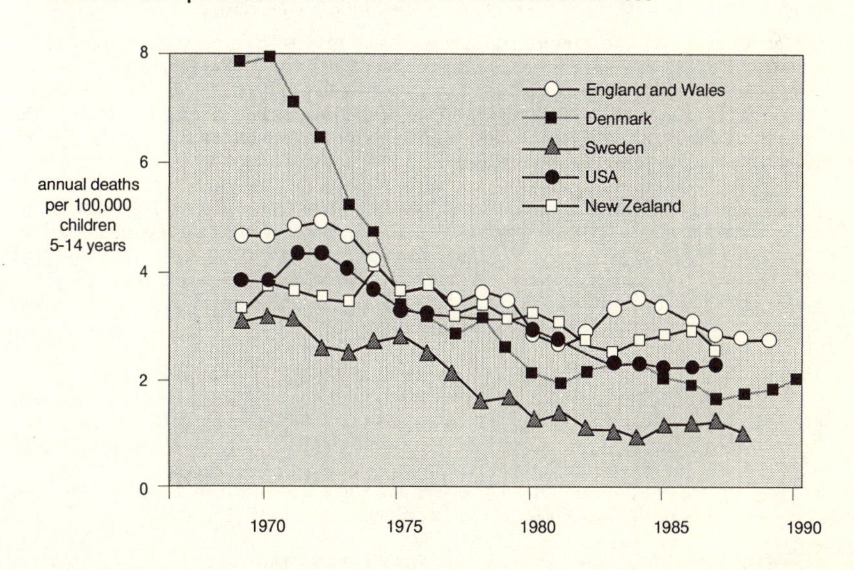

welcome reduction in pedestrian deaths over the last couple of years in Britain has brought about a substantial reduction in the death rate of children, and may reflect a wider adoption of traffic restraint policies in this country.[40]

4.39 There is some cause for concern, however, that pedestrian deaths may have fallen for a different reason, namely that there are fewer pedestrians. In London there was a 26% reduction in pedestrian casualties between 1981–85 and 1993, but the number of trips made on foot also seems to have fallen.[41] The implications are that the accident risk for pedestrians may have remained much the same, and that the disruptive effect of road traffic on social contacts and community life may have intensified. The decline already noted (2.20) in walking to school appears to have been accompanied by a much wider decline in the numbers of unaccompanied journeys made by children.

4.40 Casualties to cyclists can also be analysed in terms of risk. Despite a reduction in the last three years, the death rate per 100 million kilometres cycled is nearly 10 times that for car drivers and passengers, and two to three times higher than in the Netherlands, Sweden or Denmark. The overall casualty rate is up to 10 times higher than in those countries. The evidence from the continent is that increased cycle use can lead to lower risk because cyclists are less easily overlooked by motorists, and road layouts are increasingly designed with the safety of cyclists in mind.[42]

4.41 We have not made road accidents a main focus of our study because they are monitored and analysed extensively elsewhere. **We endorse the government's target of reducing deaths and injuries from road accidents to two-thirds of the 1981–85 level by 2000 and welcome the progress already made towards meeting that target.** For reasons discussed later in this report, we believe a sustainable transport policy must involve a major increase in the numbers of journeys made on foot and by cycle; we make recommendations in chapter 11 about the measures needed to bring that about. Accompanying measures are needed to reduce the risks of accidents for both pedestrians and cyclists. To supplement the government's overall target for road casualties, and as a contribution to the objective of improving the quality of life by reducing the dominance of motor traffic, we believe there should be specific targets for reductions in the accident rates for pedestrians and cyclists. We propose that these targets should be:

> **To reduce pedestrian deaths from 2.2 per 100,000 population to not more than 1.5 per 100,000 population by 2000, and cyclist deaths from 4.1 per 100 million kilometres cycled to not more than 2 per 100 million kilometres cycled by the same date.**

The target for deaths of cyclists has been related to the distance travelled by cycle in order to emphasise that the aim is a genuine improvement in the safety of cycling and remove the possibility that a target for reducing casualties could be met by policies which merely led to a further fall in the already low level of cycling in the UK. The same point applies equally to pedestrian casualties but the limitations of the present statistics make it inadvisable at the moment to try to relate that target to distances travelled on foot and we have therefore related it to population.

4.42 Another type of accident which is significant in environmental terms is the spillage of hazardous substances. Of 2,324 incidents involving hazardous substances in 1991, 674 (29%) occurred in transit. Table 4.1 analyses the circumstances in which incidents in transit occurred.[43] The substances most frequently involved in incidents were ammonia, LPG-propane 263-butane 88, acetylene, dichlorodifluoromethane, hydrochloric acid, natural gas and sulphuric acid; there is no separate analysis of the substances involved in incidents occurring in transit.

4.43 There are regulations governing the carriage of dangerous goods by each mode of transport.[44] In the case of road transport the regulations cover packaging, labelling, marking of vehicles, design and construction of vehicles, precautions against fire and explosion, and driver training. They are supported by approved codes of practice and guidance notes. The Health and Safety Executive has issued lists of substances classified as dangerous in transport. Another outcome of transport accidents can be the spillage of large quantities of fuel or refrigerants. Substances which are not hazardous to people, such as detergents or milk, can also be environmentally damaging if spilt in large quantities.

4.44 Action taken immediately after a spillage to protect life and property can cause damage to natural habitats and water resources, for example, if water is used to wash the spilt substance (possibly mixed

Table 4.1
Incidents involving hazardous substances attended by fire brigades (1991)

total number of incidents involving hazardous substances, including:	2,324
static incidents	1,650
in transit incidents	674
in transit incidents	
during loading/unloading	34
during road transport (not including motorways)	353
incidents on motorways	67
during rail transport	48
incidents at docks	52
incidents at airports	16
other transport-related incidents (not defined further)	104

Incidents involving fire, and thus presenting an increased risk of environmental contamination through the potentially large volumes of fire-fighting chemicals and/or water used, numbered 433; it was not stated how many of these were in transit incidents.

Incidents due to hazardous substances being washed ashore from cargo vessels at sea numbered 47.

with fire-fighting chemicals) into drains and thence into watercourses. The emergency action code linked to the hazard warning panels displayed by vehicles carrying dangerous goods takes account of possible pollution. The National Rivers Authority (NRA) maintains close liaison with the emergency services in order to limit the effects of transport accidents by preventing pollution from spreading or closing down abstraction points for public water supply. Some fire service teams carry equipment to prevent spillages of hazardous substances from entering drains.[45]

Construction of new infrastructure

4.45 Transport infrastructure occupies substantial areas of land. In broad terms roads take up a fifth of the surface of urban areas in the UK. In the large cities, railways take up about 4% of the surface.[46] In London, with its network of underground railways, the total proportion of the surface used for transport is rather less (17%).[47] The 1990 Countryside Survey indicated that roads occupy 3.3% of the land area of Britain: 1.4% in the form of roads in built-up areas and 1.9% (2.4% in England, 1.9% in Wales and 1% in Scotland) in the form of roads outside built-up areas.[48] This is a substantially higher figure than the 1.5% quoted by DOT[49], apparently in part because DOT includes only the paved areas of roads and not the verges. Railways occupy about 0.2% of the land area of Britain. The length of the rail network has been reduced by about half since the beginning of the century (to some 16,500 km).[50] In contrast, the total length of roads in Britain increased by over 28% between 1909 and 1992 (to some 360,000 km), including growth of 5% between 1982 and 1992.[51]

4.46 New infrastructure is one aspect of the environmental impact of the transport system which has aroused most concern and controversy. We referred at the beginning of this report to the wide range of effects such projects can have, both as a direct consequence of land acquisition and construction and as a result of their subsequent use (1.4). Effects such as pollution and disruption of communities have been discussed already. There is now a strong presumption against the building of new roads in urban areas, both in the UK and in most other developed countries. The few new roads that are built in urban areas tend to be across derelict land as part of urban regeneration schemes. Where there are proposals for investment in railways in urban areas, they usually involve either upgrading existing lines or constructing tunnels (which can be cost-effective as well as environmentally preferable because of the high price of urban land). We focus here on the use of open land for building new roads or other infrastructure or for widening existing roads.

4.47 Few new rail lines have been constructed this century, although plans are well advanced for the line linking London to the Channel Tunnel. Much larger areas of land have been used for new or expanded ports and airports. The total area occupied by Heathrow, Gatwick and Stansted airports is just under 3,000 hectares (equivalent to the area occupied by 720 km of three-lane motorway) and studies have taken place of possible major expansion at Heathrow or Gatwick. Both ports and airports generate large volumes of surface traffic, thus creating an additional demand for land for new road or rail links. Proposals for new or expanded ports often raise difficult environmental issues because of the relatively small proportion of suitable areas of coast which remains undeveloped and because of the importance of estuaries for conservation.[52] The major expansion at Felixstowe, for example, has seriously affected protected sites of international importance for birds.[53] In addition to on-shore development, dredging to create or restore deep water channels can affect marine life and littoral habitats.

4.48 In rural areas the routes proposed for new roads have tended to be through open country. Not only does this avoid the opposition aroused if communities and housing are affected, the lower land prices produce a more favourable cost-benefit ratio. In arable areas good quality land has the attraction from an engineering viewpoint that it is usually relatively low-lying, with low relief. Until recently it was government policy to protect good agricultural land from development and to use land of lower quality for road schemes wherever possible; because of the surplus of agricultural land that presumption no longer exists. However, poorer agricultural land often has a high conservation value, and at the same time a lower market price. These factors may explain why, as the Joint Nature Conservation Committee suggested in its evidence to us, land of high conservation value tends to be favoured for road construction. The difference in land prices is even more marked in the case of open land of conservation importance within urban areas.[54] It is an unsatisfactory feature of the present system of cost-benefit analysis (COBA) that use of low cost land of high conservation value gives a scheme a more favourable cost-benefit ratio. COBA does not in any case attempt to cover the value of land for the community, which is not reflected in its market price (7.12), and which may be considerable; this aspect is dealt with in a parallel environmental appraisal. Within the context of the cost-benefit analysis, however, it is a paradoxical feature that construction of a road effectively breaches the restrictions which were previously placed on development of such land and which have contributed to lowering its market price.[55]

4.49 The amount of land occupied by transport infrastructure is not an adequate measure of its environmental impact. Even though considerable efforts are now devoted to landscaping, a major road may completely change the character of a rural scene (as would a new railway). Plates IV and V show the effect of constructing a new trunk road through a rural landscape. Noise affects a wide area around a major road, as does lighting at night. A motorway in a quiet rural area can be heard up to 10 km away. A survey carried out for the Council for the Protection of Rural England and the Countryside Commission has drawn attention to the combined effect of road building, increased traffic on existing roads, the growth of air traffic and the construction of electric power lines in reducing the areas of unspoilt countryside that can be enjoyed in peace and tranquillity. The reduction has been especially dramatic in south-east England over the last 30 years, as plate III shows.

4.50 There are also wider repercussions for wildlife. Road traffic is responsible for the majority of wildlife casualties caused by transport, although in a few localities, bird strikes by aircraft are significant.[56] The importance of such casualties is difficult to assess. The former Nature Conservancy Council estimated in its evidence that 47,000 badgers, 3,000–5,000 barn owls and 20–40% of breeding amphibians may be killed on the roads each year. It is not possible to say with confidence that these deaths have had a significant effect on species numbers, except possibly in a few localities; the disappearance of suitable habitats, in which transport developments are but one contributory factor, is probably more important.

4.51 In lowland areas of Britain, linear features such as hedgerows and road verges tend to contain a very high proportion of the plant and animal species present in a locality.[57] The width of verges is

BOX 4B	**TWYFORD DOWN AND OXLEAS WOOD**

At *Twyford Down,* an extension to the M3 has been built on a route which affects parts of two SSSIs (5.45 hectares will be lost out of a total area of 140 hectares — about 4% from each SSSI), and three ancient monuments. It is also in an area of outstanding natural beauty. It replaces the former Winchester bypass, which is being torn up. Following public inquiries, the Secretaries of State for Transport and the Environment jointly rejected alternative proposals, one of which would have involved building a tunnel, on the grounds of cost (it was said that the tunnel would add £86 million to the overall cost of the project — DOT estimated that its preferred option would cost £30 million). This decision was later challenged by environmental groups which claimed that the route had not been subjected to an environmental assessment and that the possibility of adding a further lane to the existing road had not been publicly examined. A formal complaint was subsequently made to the European Commission. The Commission began formal proceedings against the UK government in 1991 but the government was able to demonstrate that it had followed procedures equivalent to those provided for in the EC Directive on environmental assessment (9.59). Plate VI shows excavation of the cutting through Twyford Down.

Oxleas Wood is an ancient woodland and SSSI, through which it was originally proposed that the East London River Crossing should run, designed to join the A13 to the A2 by means of a new road and bridge across the Thames. Following a public inquiry, a decision to allow the proposed route was taken jointly by the Secretaries of State for Transport and the Environment in November 1991. Opposition to the scheme, with action at local, national and EC level, continued from 1981, when it first went to public consultation, until summer 1993, when DOT agreed to set aside the original proposal and consider alternative routes for the river crossing.

correlated with the abundance of species such as butterflies, and the more diverse the habitat within the verges the greater the total number of species present.[58] Protecting existing verges is therefore an important factor in maintaining biological diversity in the wider countryside. Road improvements have led to the loss of many old hedgerows and species-rich verges. Generally speaking, newly planted verges are not an adequate replacement, although wide verges alongside new roads can provide cover for birds, mammals and insects and in the longer term a reservoir for native plant species. We welcome DOT's recognition of the environmental importance of roadside habitats and the need to ensure a diversity of plant species, and the care it now takes to reseed or plant verges with appropriate species.

4.52 Where a new road cuts across natural or semi-natural habitat, the effects will depend on the habitat's sensitivity to disturbance and the precise siting of the road, as well as the area of land taken. If the road separates populations of wildlife, individuals of some species may be unable to cross it.[59] The interchange of genes between the separated populations is then reduced or eliminated.[60] Rarer organisms tend to disappear from isolated areas because the population within them becomes too small to be viable.[61,62] Recolonisation is possible only if sufficient numbers of individuals can reach the areas from elsewhere.

4.53 Areas of land which have national or international conservation value are designated as sites of special scientific interest (SSSIs) under the Wildlife and Countryside Act 1981. SSSIs (including for this purpose the 48 areas of special scientific interest in Northern Ireland) cover about 8% of the UK. They include 304 national nature reserves, 86 areas designated as special protection areas (SPAs) under the EC Birds Directive and 76 wetlands designated as of international importance under the Ramsar Convention.[63] England has some 3,700 SSSIs which cover 6.6% of the land area (3.6% of the UK land area).[64] Designation as an SSSI was not intended to provide absolute protection against development but the government's stated policy is to keep roads away from such areas wherever possible. A number of road schemes which affect SSSIs have been approved and put into effect, although some other schemes which would have affected SSSIs have been suspended. The course of events in two recent cases is summarised in box 4B.[65]

4.54 A larger proportion of the UK (nearly a fifth) is designated because of its fine landscape as a national park, as an area of outstanding natural beauty (AONB) in England and Wales or as a national scenic area in Scotland.

4.55 In their evidence, statutory conservation bodies and the National Trust expressed considerable concern about the past effects of road building and the prospect of further damage. English Heritage said that road building has destroyed historic buildings and areas of historic interest and damaged archaeological features. The National Trust said in 1992 that there were 50 road schemes in England and Wales which could affect its properties to varying degrees through visual intrusion, noise, disruption of access, or loss of habitats or archaeological sites; it regarded 30 of these as significant threats. English Nature estimated that 150 SSSIs were potentially at risk from the trunk road programme as it existed in 1992 (2.6% of the UK total at the time). This is a much larger number than the DOT estimate, made at the same time, that 48 SSSIs would be affected by those trunk road schemes for which a preferred route was known.[66] A survey in 1990 by county wildlife trusts in south-east England identified 372 'important wildlife sites' in nine counties which were under threat from road schemes of all types; of these, 50 were SSSIs and 125 were protected by some other form of listing.[67] It was calculated that trunk road schemes in the South-East would each affect more than 8 important wildlife sites on average. In Wales, the proposed motorway to relieve the M4 between Newport and the Severn would cross the Gwent Levels SSSIs.

4.56 In announcing the outcome of the recent review of the trunk road programme DOT emphasised that only a small proportion of the programme now consists of roads to be constructed on new lines across open country. However, six important schemes still fall into that category. DOT estimated that, following the review, about 40 SSSIs are affected by schemes for which a preferred route is known.[68] The review has not therefore had a substantial effect on the position. Motorway widening schemes, which form a substantial part of the programme, can have a significant effect on areas of conservation importance. County wildlife trusts have calculated that the proposals for widening the M25 could affect 10 SSSIs, 47 ancient woodlands and 24 other important wildlife sites.

4.57 Under the EC Habitats and Species Directive[69] the UK is required to select special areas of conservation (SACs) to form part of a European network (Natura 2000) and submit its proposed areas to the European Commission by June 1995. The statutory conservation bodies, co-ordinated by the Joint Nature Conservation Committee, are providing advice to the government, which is consulting other member states with similar habitats. The deadline for agreement on areas is 1998 and the deadline for designation of areas is 2004: UK candidate sites will be protected in the interim period before designation. Areas already designated as SPAs under the EC Birds Directive will become SACs. Under the terms of the Directive a development which would damage an SAC is allowed to proceed only if there is no alternative and if considerations of human health and safety can be shown to override conservation considerations. The European Commission will be able to ask for the inclusion in the network of any important areas not put forward by the UK government; and the UK government will be answerable to the Commission for any damage to SACs.

4.58 The economic and cultural value of biological diversity is now widely recognised. The UN Biodiversity Convention signed at Rio de Janeiro in 1992 acknowledges the importance of biological diversity in the context of sustainable development and seeks to ensure that species and habitats are protected for the benefit of future generations. The maintenance of diversity depends on the continued availability of habitats and of genetic diversity within species. If major road building programmes continue, they are likely to come into conflict with the UK's national and international commitments to protect species and habitats. If they proceed in the form which can be expected at present they would be in conflict with the aim of sustainability because they would materially erode the stock of natural and semi-natural habitats and areas with amenity or cultural value (1.15).

4.59 We share the concern expressed to us by the Countryside Council for Wales about the implications of the government's statement in the 1990 White Paper on the Environment[70] that 'most damage [to SSSIs] has been more than compensated for by the continued expansion of the SSSI network.' New designations cannot compensate for sites that have been lost to roads or other forms of development. Nor can man-made habitats compensate fully for the loss of natural habitats. It is nevertheless desirable that, if development is permitted which destroys a natural or semi-natural habitat, the developer should be required to create a new area of habitat of comparable size and containing a diversity of species, as a contribution towards maintaining the overall level of biological diversity.

4.60 Losses of land for the construction of new infrastructure and the associated effects are important issues which must be dealt with in a sustainable transport policy. The effect of known plans on areas of environmental importance would already be considerable. If a further major programme of road building were undertaken after 2000, which is the logical implication of present policies, the effect could be very much greater. We have concluded that the policy objective must be:

TO HALT ANY LOSS OF LAND TO TRANSPORT INFRASTRUCTURE IN AREAS OF CONSERVATION, CULTURAL, SCENIC OR AMENITY VALUE UNLESS THE USE OF THE LAND FOR THAT PURPOSE HAS BEEN SHOWN TO BE THE BEST PRACTICABLE ENVIRONMENTAL OPTION.

4.61 Statutorily designated areas ought to be given more effective protection against the construction of transport infrastructure than they receive at present. That does not imply that there should be a weakening of protection in relation to other areas of land, because the aim is to give a higher overall priority to environmental protection. **We recommend that the following general principles should apply:**

- a. **strict protection for the special areas of conservation to be designated under the EC Habitats and Species Directive;**
- b. **any further loss or damage to natural and semi-natural habitats or archaeological features must be reduced to the absolute minimum;**
- c. **where loss of a natural or semi-natural habitat cannot be avoided, the developer must be required to provide some restitution by creating an appropriate new habitat in the vicinity;**
- d. **where other land which has significant amenity value is used for transport infrastructure, the developer must be required to provide an equivalent area of land of equivalent amenity with equivalent access for the public;**
- e. **where a proposed road or railway would cause serious environmental damage, careful consideration should be given to placing it in a tunnel.**

Tunnelling is an approach which may have considerable environmental advantages for a new road, both in sensitive rural areas and in urban areas.

4.62 It would be wrong to put an absolute bar on the building of new roads. There may be cases in which a new or widened road would represent the best practicable environmental option. The procedure for selecting the 'best practicable environmental option' (BPEO) was described in the Commission's Twelfth Report, in the context of controlling emissions from industrial sites; and we describe later in this report (box 9B) how the same principles can be applied to transport problems. An essential part of the BPEO concept is that options should be identified and evaluated at an early stage in the decision-making process. Before the BPEO concept can be applied, the procedures for considering and taking decisions on new transport infrastructure projects must be considerably improved. In particular they must include an examination of options which do not involve new construction or involve constructing infrastructure of a different type (for example, the upgrading of a railway line as an alternative to construction of a new road). We make recommendations for that purpose in chapter 9.

Demand for road building materials

4.63 An important dimension of the sustainability of the transport system is the extent to which it uses up finite resources of critical materials (1.15). The construction and repair of roads requires very substantial resources, including about 90 million tonnes of primary and secondary aggregates a year, a third of the total used in Britain.[71] It is estimated that 43% of the high quality aggregates produced from rock are used in road construction.[72] About 120,000 tonnes of aggregates are needed to build a kilometre of motorway.[73] The manufacture of vehicles gives rise to a large and growing demand for non-renewable resources. It is estimated that a fifth of world steel production and a tenth of world aluminium production are used for vehicle manufacture. The extraction, production and transport of the raw materials for vehicles uses significant amounts of energy, perhaps 5% of total energy use resulting from the UK transport system.[74] Further significant amounts of energy are used during vehicle manufacture (although probably less than a tenth of the average lifetime energy use for a car[75]). Scrapped vehicles represent a major waste management issue. We have concluded that, in order to move towards sustainable development and protect the environment from damage, there needs to be a general policy objective:

TO REDUCE SUBSTANTIALLY THE DEMANDS WHICH TRANSPORT INFRASTRUCTURE AND THE VEHICLE INDUSTRY PLACE ON NON-RENEWABLE MATERIALS.

We first discuss road building materials and then the use and disposal of materials in vehicles.

4.64 The greatest demand for aggregates is in south-east England. About half the region's requirements for all purposes come from local sources, mostly land-won sand and gravel; the remainder is marine-dredged sand and gravel, and rock material brought by rail from large quarries in the South-West and East Midlands. There is considerable concern amongst conservation bodies that new quarries could be opened on environmentally sensitive sites without the need for planning permission under the terms of about 600 interim development orders (IDOs) issued in 1947 to help post-war reconstruction. This issue has wider implications than we can explore in a study of transport, but the demand for aggregates for road construction may be a crucial factor in determining whether the rights conferred by IDOs will be used in practice.

4.65 Even if the rights conferred by IDOs were used, inland sources will not be sufficient to meet the forecast future demand for aggregates in south-east England. In 1976 the Verney Committee[76] proposed a new source, large coastal quarries in remote areas. More recent research has suggested that there are up to twenty potential sites in Scotland, Norway and Spain.[77] Rock is now being extracted from a superquarry on Loch Linnhe in the west of Scotland. A second superquarry has been proposed on South Harris in the Hebrides; Scottish Natural Heritage has lodged an objection on the ground that the site is within a national scenic area. Other sites are being investigated. The government's policy is that only a very limited number of superquarries will be permitted in Scotland.[78]

4.66 Rock from Loch Linnhe is being transported by sea to southern England, Germany and Holland. It will almost always be undesirable environmentally to use any other mode of transport. **If planning permission is granted for further coastal superquarries, we recommend there be a legal requirement that the quarried material must be transported by sea.**

4.67 Extraction of rock or gravel can severely damage natural habitats and scar the landscape. During extraction there is noise and disturbance from quarrying and transport of material. If sites are restored after extraction, they may in time come to have conservation value but they are more often used for purposes such as landfill or developed for watersports[79]; these developments in themselves generate traffic. We are concerned that extensive damage to the environment would be caused through extraction of the aggregates to carry out the present road building programme. We do not consider that the implied rate of consumption can be regarded as sustainable.

4.68 The UK has been slow to make use of waste industrial and demolition material or recycled asphalt in building roads and other transport infrastructure. There are huge volumes of waste materials available.[80] There has been a tendency for DOT and local authorities to overspecify; and because the price of primary material is low there has been little incentive to recycle. In contrast, the Dutch have made a considerable effort to replace imported roadstone and other primary materials. Over a ten-year period a government-led joint initiative has developed recycled substitutes. Three-quarters of demolition waste is now reused, together with a high proportion of incineration slag.[81] Waste asphalt is recycled to make up between 30% and 70% of the mix in newly laid asphalt, according to the conditions. Some highway authorities in Germany have also specified the use of high proportions of recycled asphalt.

4.69 The government's proposal in Minerals Planning Guidance Note 6 to increase the use of waste material in roads from 30 million tonnes a year to 55 million tonnes a year by 2006 is encouraging but will not be sufficient to slow down the rate of quarrying. A more challenging and longer-term target is required which takes account of the growing demand for aggregates. We propose that the government adopt the following target:

> **To double the proportion of recycled material used in road construction and reconstruction by 2005, and double it again by 2015.**

Manufacture and disposal of vehicles

4.70 There are several approaches to conserving the resources used in vehicles. One approach is to use fewer resources at the manufacturing stage. Vehicles can be designed to be smaller and lighter. This may involve using non-traditional materials, which could require less energy for their production and processing. We discuss the potential of this approach later in the report (8.41–8.42) in the context of the contribution that smaller and lighter vehicles can make to improved fuel efficiency. Another approach to the conservation of resources is for manufacturers to design and market vehicles with a longer life. The limitation on the life of cars has tended to be the body rather than the engine and mechanical parts, which can be reconditioned or replaced with newly manufactured spares or reused parts from dismantled cars. One significant factor limiting bodywork life has been the widespread use of salt to de-ice roads in winter. Some US states such as Oregon have now banned the use of salt for this purpose. Manufacturers can also contribute to prolonging vehicle life by ensuring that spares continue to be available. In practice, however, the most important contribution to conserving the resources used in vehicle manufacture has taken the form of reusing vehicle parts and recycling the materials in those parts which cannot be reused.[82] The Automotive Consortium on Recycling and Disposal (ACORD) estimates that 77% by weight of the materials in cars is now recycled in the UK.

4.71 A car typically contains thousands of separate parts and dozens of different materials. There is an assured market within the vehicle industry for lead from batteries and in future for precious metals from catalytic converters. In the traditional approach to recycling, reusable parts and non-ferrous metals are removed by firms of dismantlers and sold. The remainder of the vehicle is then sent to be shredded and turned into pellets, from which ferrous metals are separated. The residue is disposed of to landfill. At present about half a million tonnes a year of such residues are landfilled in the UK. More careful dismantling in order to maximise recycling would be labour-intensive and is not profitable on the basis of the prices that can be obtained for the materials.

4.72 The residues left after vehicle dismantling are expected to double in amount during the 1990s as the proportion of steel used in vehicles declines. This threatens to make the present level of recycling unprofitable. In addition to higher costs for disposal of residues, there will be less steel to sell and its value will be reduced because of the greater risk of contamination. As a result of tighter regulation, some substances which form part of the residues may no longer be acceptable for landfill, so raising costs further. Parts of the present recycling industry could disappear.

4.73 Of the new materials being used, plastics cause most concern. There is little recycling at present and they are expected to increase as a proportion of the materials used in cars from about 12% in 1990 to between 20% and 25% by 2000.[83] There are encouraging signs that new techniques will soon be available. An important principle is to design vehicles in a way that facilitates dismantling; this will require closer liaison between vehicle manufacturers and the waste management industry. There is a general trend, especially in Germany, for car manufacturers to take more responsibility for disposal. BMW intends to design cars for 90% recyclability by weight, has opened its own recycling plant in West Sussex to dismantle 2,500 cars a year, and plans up to 15 more such plants to handle all the BMWs scrapped in the UK. Rover has formed a joint venture with the Bird Group. The recycling of plastics will be facilitated if the number of types of plastic used in car manufacture can be substantially reduced: at present a car may typically contain 20 different plastics, each of them perhaps in a number of different grades.

4.74 Used tyres represent a major waste disposal problem, to which the NRA drew attention in its evidence. Of the 40 million scrap tyres requiring disposal each year, two-thirds have been landfilled or dumped illegally. Tyre dumps may catch fire and are then extremely difficult to extinguish: a dump in Wales has been burning for five years and could take ten years to burn out unless liquid nitrogen is injected or £4 million is spent on excavating it.[84] The leachate from such a fire is highly polluting. About 15% of car tyres and 40% of heavy goods vehicle tyres are reused after retreading. Rubber can be recycled in crumb form, and used for example in asphalt. The Commission's Seventeenth Report drew attention to the high calorific value of scrap tyres and identified incineration with energy recovery as the best option for used types after the scope for recycling has been exhausted. Specialised new power plants will have the capacity to burn a third of the tyres scrapped in the UK each year.[85] In the longer term more efficient chemical or physical processes may become available for recycling tyres.[86]

4.75 ACORD has suggested a system in which vehicle owners would be required by law to obtain a certificate of destruction from an authorised disposer. Although this suggestion deserves serious consideration, the priority is to ensure that the procedures and techniques used for disposal increase the amount of recycling, including energy recovery from the residues. ACORD have drawn up plans for this, but do not believe that this can be achieved solely through the natural development of the market. As part of the Priority Waste Streams Programme, the European Community is considering proposals by a working group to reduce the end-of-life vehicle waste going to landfill to 5% by weight by 2015.[87] The complexity of vehicle and tyre recycling demands government involvement to improve the market for recycled material by raising the price of virgin material and the cost of landfill.

4.76 We believe that a combination of approaches must be adopted to reduce the demands on non-renewable resources. This will involve smaller, lighter vehicles, built to last longer. Recycling of tyres and the non-metallic element in vehicles must be given much greater priority. Government must intervene to provide a stable framework that will encourage investment and use economic instruments to encourage the growth of markets for recycled materials. **We recommend that the Department of Trade and Industry and DOE work with vehicle manufacturers and dismantlers to develop a cradle-to-grave strategy for recycling. Where necessary they should use economic instruments and other forms of regulation to implement this strategy.** The Motor Vehicle Dismantlers Association should be involved at the formative stage to ensure that government initiatives take full account of industrial realities and to raise environmental awareness among its members.

4.77 In pursuit of our general objective of reducing the demands on non-renewable resources we propose that the government adopt the following target which has already been formulated by ACORD:

> **To increase the proportion by weight of scrapped vehicles which is recycled, or used for energy generation, from 77% at present to 85% by 2002 and 95% by 2015.**

There should be a separate, and more ambitious, target for tyres. We propose that this should be:

> **To increase the proportion of vehicle tyres recycled, or used for energy generation, from less than a third at present to 90% by 2015.**

Chapter 5

AIR TRANSPORT

5.1 Air transport accounts for almost a sixth of the energy used by the transport sector in the UK (3.60). It is the most rapidly growing mode of transport, both in the UK and globally. Some of the environmental effects, such as noise from aircraft and the requirements for land to extend airports and construct surface links, were touched on in the previous chapter. The full extent of emissions from aircraft is not apparent from UK environmental statistics, which normally include only emissions during ground movements, take-off and landing. Because the relevant regulations and most of the traffic flows are international, aircraft emissions are discussed separately here, rather than in chapter 3. We consider first the growth of air transport and then the key trends in noise and in emissions of pollutants and greenhouse gases. The possible effectiveness of technical and other measures to reduce emissions and noise is assessed in the light of the expected further growth in air traffic.

Growth of air transport

5.2 There are various measures of the growth in international passenger traffic as it has affected the UK. Between 1974 and 1992 passenger-kilometres travelled on international flights by UK airlines increased by 263%.[1] Between 1982 and 1992 the number of passengers on international scheduled services to and from UK airports more than doubled, while the number on non-scheduled flights increased by 62%.[2] International flights made by UK residents also doubled over the same period; about two-thirds of these were for tourism (figure 5-I[3]). Domestic traffic represents less than a quarter of aircraft movements at UK airports and the number of domestic passengers has increased less dramatically.

5.3 Air freight, which is carried predominantly on passenger aircraft, has increased less rapidly than passenger traffic.[4] In 1992 1.1 million tonnes of freight were carried by air to or from the UK. Domestic air freight movements remain insignificant (52,000 tonnes in 1993).[5]

5.4 At the global level, growth has also been very rapid: traffic on scheduled airlines doubled over the last decade. The latest available information about scheduled airline traffic is given in table 5.1.[6] Table 5.2[7] shows the present size of the civil aircraft fleet worldwide and in the UK. Europe (39%) and North America (35%) account for almost three-quarters of world aviation activity at present, but the potential for growth is greatest in the Asia and Pacific region.[8]

5.5 At present business travel accounts for 40% of passenger traffic globally, but tourist traffic is increasing more rapidly. The other principal growth area is intercontinental air freight. Air freight has been predicted to grow by 6–8% a year globally until the turn of the century and by 3–4% a year within Europe. While the number of dedicated cargo aircraft worldwide is expected to be about 1,420 by 2000, compared with 811 in 1990[9], the distribution of freight between passenger and cargo aircraft is not expected to change dramatically. The International Air Transport Association (IATA) is concerned about the implications of increasingly congested air space for the safe operation of the European air traffic control system, which will be at or near capacity by 1995. However, steps are being taken to improve the situation by harmonising procedures and upgrading the system. In the medium term, there are likely to be significant improvements in the technology of air traffic control through the use of satellites, which will considerably increase capacity.[10]

5.6 Because larger aircraft are being used, aircraft movements have increased less rapidly than the number of passengers. Even so aircraft movements at UK airports increased by more than half over the last decade (figure 5-II[11]). The most important airports in the UK are shown in table 5.3.[12] The past trend in passenger numbers and aircraft movements at Heathrow is shown in figure 5-III. The Department of Transport's (DOT) latest forecasts for the medium term are that the number of passengers passing through UK airports will increase by between 73% and 163% between 1992 and 2010; the proportion of passengers using regional airports, rather than London, is expected to increase from 35% in 1992 to about 40% in 2010.[13]

Figure 5-I
Growth in air transport: visits abroad by UK residents 1982-92

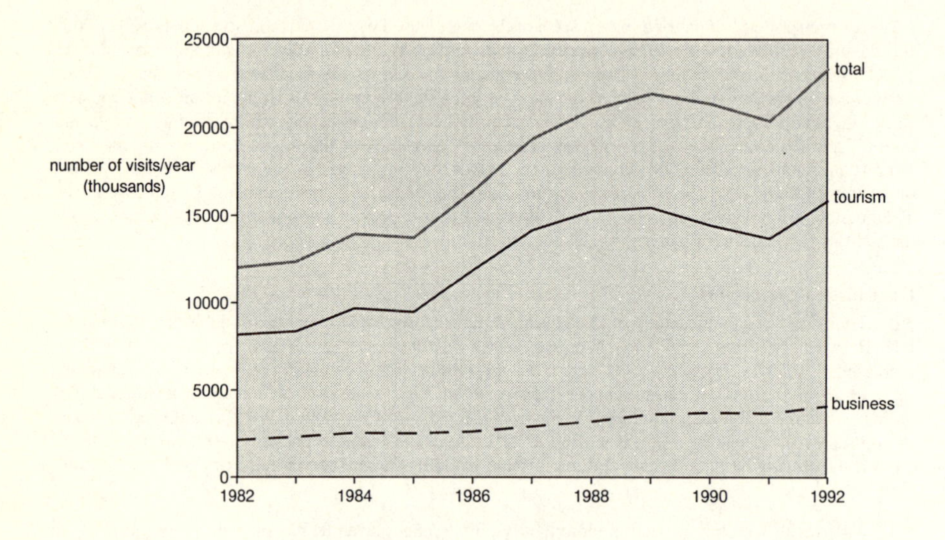

Tourism includes both package holidays and independent holidays. In addition to visits for business and tourism, the total includes visits to friends and relations and a miscellaneous category of visits not defined further.

5.7 There are disagreements about the extent and location of surface development required to handle the growth in air traffic. A public inquiry has considered a proposal for a second runway at Manchester airport. In south-east England a government study[14] concluded there would be a case for a new runway at Heathrow or Gatwick by 2010 and a second runway at Stansted by 2015. The Council for the Protection of Rural England has questioned the need for a new runway in the South-East, arguing that airport capacity is increasing as a result of improvements in air traffic control technology.[15] BAA plc, which operates the main London airports, believes that terminal capacity, rather than runway capacity, is the principal constraint on further growth at Heathrow and Gatwick; and has applied for planning permission for a fifth passenger terminal at Heathrow, which would increase capacity by more than half. The London Planning Advisory Committee opposes this proposal; it is particularly concerned about the implications in terms of road links and the effect on noise levels in the area.[16]

5.8 At the international level, British Airways said in oral evidence that they expect passenger travel worldwide to grow at 5–6% a year between now and 2010, which means it would more than double. This reflects a general view in the industry that growth will be about 5% a year for the next 20 years, declining to about 2–2.5% a year over the following 20–25 years.[17] A report prepared for IATA suggested that passenger numbers in Europe will treble by 2010 (which implies that they will more than treble in the rest of the world) and that the number of passengers using UK airports will more than treble by 2025.[18] An alternative view is that the rapid worldwide growth in air traffic during the 1970s and 1980s will prove to be part of a non-linear process and will reach saturation after 2010 at 290–335 billion tonne-kilometres for passenger, freight and mail services, an increase of 30–50% on the 1992 level shown in table 5.1.[19]

Table 5.1
Global air transport: operations of scheduled commercial carriers in 1992

total aircraft kilometres		15,310 million
total passengers carried		1.07 billion
total passenger-kilometres		1,773 billion
average distance travelled per passenger		1,650 km
average passenger load factor		65%
average number of passengers per flight		120
total freight carried	(all planes)	15,710,000 tonnes
	(cargo planes)	4,950,000 tonnes
total tonne-kilometres	(all services)	225,650 million tonne-km
	(freight services only)	23,170 million tonne-km

These figures do not include the Commonwealth of Independent States.

Table 5.2
Civil aircraft on ICAO register worldwide and in the UK (1992)

	World	UK
fixed-wing turbo-jets	10,750	489
other heavier aircraft	7,850	195
light aircraft	360,780	6,725

Light aircraft are those with a maximum take-off weight of less than 9,000 kg. These figures do not include the Commonwealth of Independent States or China.

5.9 Because of its geographical position and historical links, the UK has a disproportionately large share of international air traffic. Other major European airports are expanding their capacity, including:

Amsterdam Schiphol, where a new terminal will shortly be completed and there is approval for a further (fifth) terminal; passenger numbers are expected to exceed 40 million a year by the end of the century;

Frankfurt, where a new terminal is planned which will enable it to handle more passengers than Heathrow does now;

Paris Charles de Gaulle, where there are plans for up to five runways, allowing up to 130 movements an hour, supported by increased terminal capacity, for which space is available.[20]

Figure 5-II
Growth in air transport: movements at UK airports 1982-93

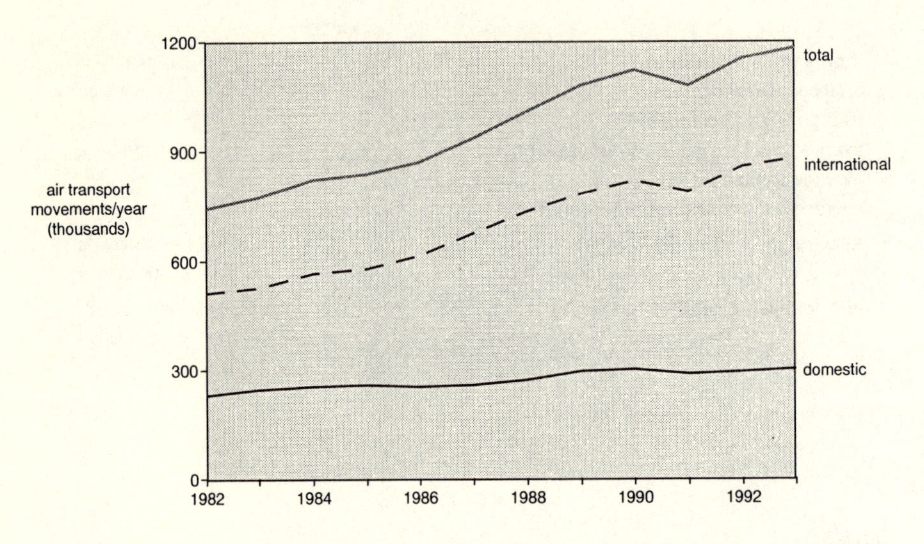

Both scheduled and non-scheduled flights are included. Flights to and from oil rigs are counted as international movements. Domestic flights are counted as a single movement.

The effect of continued growth in international air traffic without increases in capacity at the main London airports might be that some traffic would divert to competing airports on the continent. This would transfer to other countries some part of the environmental consequences of continued growth, such as noise and traffic on surface links, and might possibly (if holding delays were eliminated as a result) bring some net benefit in lower fuel use and emissions, but at the cost of a reduction in economic activity and employment in the UK.

5.10 A significant proportion of the world's aircraft are military aircraft. There is published information about the numbers of such aircraft[21], but it is difficult to assess the environmental implications of their use; neither fuel efficiency nor the reduction of emissions is likely to be a primary consideration. One measure of their environmental impact is the amount of fuel they use. Estimates of the proportion of aviation fuel used by military aircraft worldwide have varied from as little as 5% to as much as 25%.[22] The analysis in this chapter limits itself to civil aircraft.

Aircraft noise

5.11 The main environmental concern of the aviation industry and the public has been with noise. Exposure to noise from transport was discussed in general terms in the previous chapter (4.5–4.7). Aircraft are noisier than other forms of transport, and noise is the most intrusive aspect of air transport. Box 5A describes how aircraft noise is controlled at the three main London airports. In future DOT intends that greater day-to-day responsibility and accountability for noise control measures at these airports should pass to BAA plc as the operator (the position which already exists at other UK airports).

5.12 At Heathrow, Gatwick and Stansted, mean noise levels of 57–63 dBL_{Aeq} over the daytime period

Table 5.3
Busiest UK airports (1993)

	Air transport movements (aircraft landing or take-offs)	Passengers (million)
Heathrow	396,000	47.6
Gatwick	175,000	20.1
Manchester	136,000	12.8
Aberdeen	93,000	2.3
Glasgow	77,000	5.0
Birmingham	69,000	4.0
Edinburgh	59,000	2.7
Stansted	48,000	2.7
Newcastle	36,000	2.1
Belfast	35,000	2.2
East Midlands	27,000	1.4
Luton	20,000	1.8

Figure 5-III
Growth in air traffic at Heathrow and population affected by aircraft noise 1974-90

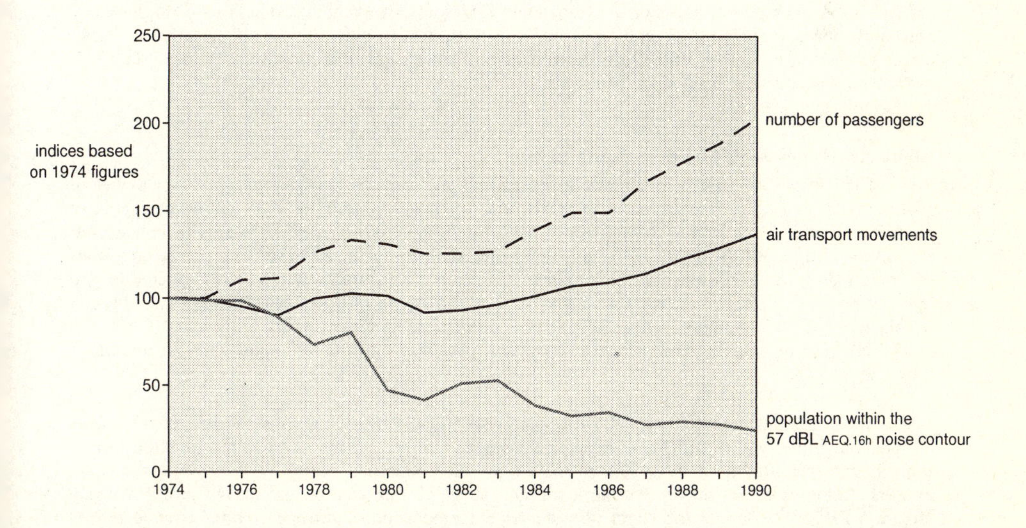

from 6 am to 10 pm are defined as 'low', levels of 63–69 dBL_{Aeq} as 'medium' and levels of more than 69 dBL_{Aeq} as 'high'. Table 5.4[23] shows the areas of land and numbers of people within each noise contour at each airport. Despite the increase in aircraft movements at Heathrow, the number of people within the 57 dBL_{Aeq} contour (or the 35 Noise Number Index contour under the system used previously) declined from 2 million in 1974 to less than half a million in 1991 (figure 5-III[24]).

5.13 The reduction in the area and number of people affected by high levels of noise at Heathrow is the result of a major reduction over the last 20 years in the noise produced by subsonic commercial aircraft. The noisiness of aircraft engines is a direct result of the enormous thrust required to lift aircraft off the ground. In large modern jet engines, fans direct low velocity air from the low pressure compressor round the outside of the turbines to mix with the high velocity air from the main jet; the resulting reduction in the overall velocity of the air flow makes the aircraft quieter. While further noise reductions are technically possible, they may carry economic and environmental penalties in terms of fuel consumption and efficiency.

5.14 The first international regulations on aircraft noise were adopted in 1971 by the International Civil Aviation Organization (ICAO), the body responsible for regulating aviation worldwide, and took the form of a technical annex to the Chicago Convention on International Civil Aviation.[25] Noise certification standards were subsequently introduced for subsonic jets and propeller-driven aircraft, other than short take-off and landing (STOL) aircraft; and guidelines were developed for noise certification of future supersonic and STOL aircraft and for aircraft systems operating on the ground. With the exception of a few older aircraft designed before October 1977, which were built to the standards in Chapter 2 of Annex 16 to the Chicago Convention, commercial jets comply with the more stringent Chapter 3 standards.

5.15 ICAO's policy is that, as well as environmental concerns, operating restrictions should take account of the balance of interests between developing and developed countries. The ICAO resolution that Chapter 2 aircraft should be phased out and withdrawn from service by 2002 is not therefore binding. However, it is binding on member states of the European Community[26], and action is already being taken by some countries including the USA. According to studies carried out by the Civil Aviation Authority the phasing out of Chapter 2 aircraft should substantially reduce noise levels around airports despite the expected growth in air traffic.

5.16 There is a trend towards using larger and heavier aircraft for international flights. Boeing's forecast is that average size will increase from about 190 seats now to about 230 in 2013.[27] Because they need more power for take-off larger aircraft are likely to be noisier, other things being equal. The design target for an airliner to carry 600–800 passengers is for a noise level close to those of the newest aircraft now in service, such as the Boeing 747–400.[28]

Emissions of pollutants and greenhouse gases

5.17 The substances emitted by aircraft are essentially the same as those emitted by other modes of transport powered by petroleum (see figure 3-I). The amounts of carbon dioxide and water vapour in aircraft emissions are related to the amount of fuel burnt. Fuel consumption is lowest when an aircraft is cruising at high altitude; short flights are energy-intensive and give rise to relatively large emissions of carbon dioxide. Emissions of the products of incomplete combustion (carbon monoxide and hydrocarbons) are greatest during taxi-ing or idling. Emissions of particulates are at their highest during take-off and climbing, when the engines are producing most power. Emissions of nitrogen oxides are also at their highest during take-off and climbing but are still high when an aircraft is cruising.

5.18 The emissions during a particular flight are also affected by the air speed of the aircraft and its altitude and route, which will be selected in the light of forecast wind speeds and directions. On a typical short-haul flight, whereas 80% of the carbon monoxide and nearly all the hydrocarbons are emitted during taxi-ing, about two-thirds of emissions of nitrogen oxides occur during cruising (figure 5-IV[29]). The longer the flight, the greater the proportion of nitrogen oxides emitted during cruising.

BOX 5A CONTROL OF AIRCRAFT NOISE AT MAIN LONDON AIRPORTS

DOT sets noise limits for departing aircraft at Heathrow, Gatwick and Stansted under section 78 of the Civil Aviation Act 1982. The limits for all three airports are 89 $dB(A)_{Lmax}$ at night (from 11 pm to 7 am) and 97 $dB(A)_{Lmax}$ during the day ($dB(A)_{Lmax}$ being the highest instantaneous sound level recorded during an aircraft's take-off or landing). These noise limits are being reviewed.

At Heathrow there were 226 infringements in 1991/92 and 71 in 1992/93. In 1993 the airport operator, BAA plc, introduced supplementary charges at all three airports for breaches of the night-time limits: £500 for every departure over 89–92 $dB(A)_{Lmax}$ and £1,000 for every departure over 92 $dB(A)_{Lmax}$. Since April 1994, fines have also been imposed in the first two and last two hours of the daytime period.

DOT sets a limit on the number of movements at each airport at night (from 11.30 pm to 6 am). As a supplementary control there is a quota system based on the noise certification of aircraft. The noisiest types of aircraft are prohibited from landing or taking-off during the night-time period and are subject to restrictions in the periods 11 pm–11.30 pm and 6 am–7 am. In the 1994 summer season the movements limits and quotas were:

Heathrow	3,250 movements	7,000 quota
Gatwick	7,000 movements	9,550 quota
Stansted	6,500 movements	4,200 quota

In the winter of 1994/95 the limits and quotas will be:

Heathrow	2,550 movements	5,000 quota
Gatwick	3,000 movements	6,820 quota
Stansted	4,640 movements	3,000 quota

Over the next four years the quotas will be unchanged. However, the movements limits for Gatwick and Stansted will increase gradually to 11,200 at Gatwick and 7,000 at Stansted in the summer of 1998, and 5,250 at Gatwick and 5,000 at Stansted in the winter of 1997/98. There will be no increase in the movements limits for Heathrow.

There have been several schemes to provide noise insulation at Heathrow and Gatwick. A scheme was introduced in 1991 at Stansted to provide insulation for people living within an area defined by reference to the forecast noise levels when traffic reaches 8 million passengers a year. Compensation for a reduction in the value of property is payable under the Land Compensation Act 1973.

Table 5.4
Daytime noise at main London airports: areas and numbers of people affected (1991)

dBL_{Aeq16h} contour	area within contour (km^2)			number of people within contour		
	Heathrow	Gatwick	Stansted	Heathrow	Gatwick	Stansted
57-63	165	89	15	319,000	21,000	900
63-69	40	23	6	88,000	2,000	300
>69	30	14	3	22,000	1,000	100

5.19 The effects of some of the substances emitted by aircraft are strongly dependent on the altitude at which they are released. Above the boundary layer (3.13), two regions of the atmosphere need to be distinguished:

> the *troposphere*, which is subject to relatively rapid vertical movements, associated with major weather systems, and extends up to a transitional layer, the tropopause, at the base of the stratosphere;
>
> the *stratosphere*, a stable, dry region where vertical motion and transfer is slow, which extends up from the tropopause to a boundary with the mesosphere at a height of about 50 km.

The height of the tropopause varies with latitude, from about 8 km in the polar regions to about 12 km in middle latitudes and about 16 km in the tropics. It also tends to be higher in summer than in winter and, to a lesser extent, higher during the day than at night. Although most commercial jets fly within the troposphere, those on routes through high latitudes cruise in the stratosphere for part of the time. About 20% of the fuel used by commercial airlines is burnt in the stratosphere[30]; it has been estimated that the proportion on North Atlantic routes may be as much as 75%.[31] The cruising height for supersonic aircraft is well within the stratosphere.

5.20 Greenhouse gases have been discussed earlier (3.55–3.65), together with the possible extent of climate change (3.66–3.67). On the basis of an estimate by the International Energy Agency that 176 million tonnes of aviation fuel were burnt in 1990[32], aircraft emit about 450 million tonnes of carbon dioxide a year in total (about 122 million tonnes of carbon).[33] This is less than 3% of the total amount released as a result of human activities and less than a fifth of the amount emitted from road vehicles. Aircraft also release about 220 million tonnes of water vapour a year and the effects in the upper troposphere are often visible from the ground: the local concentration of water is raised to saturation point and a persistent condensation trail of ice particles forms. Persistent high thin ice clouds have a global warming effect. Such clouds can also form in the stratosphere but only at very low temperatures; because the lower stratosphere is very dry, aircraft condensation trails formed there are much less persistent than those in the upper troposphere.

5.21 Emissions of nitrogen oxides from aircraft are estimated to be about 3.5 million tonnes a year or 2–3% of the emissions which result from human activities. Together with the carbon monoxide and hydrocarbons emitted, nitrogen oxides become involved in a complex series of photochemical reactions (some of which were described in box 3A) in which ozone and its main precursor, nitrogen dioxide, are both formed and destroyed. The net effect depends on the concentrations of the primary pollutants and the secondary pollutants formed from them. Conditions in the troposphere are such that release of nitrogen oxides there almost always leads to increased production of ozone and, in this way, contributes to the greenhouse effect. However, further research is needed to improve understanding of the processes and assess the likely extent of the effect.

5.22 In contrast, release of nitrogen oxides into the stratosphere at middle and low latitudes is thought to lead to a loss of ozone. This is because at these altitudes nitrogen oxides are involved in rapid catalytic cycles which destroy ozone. In the polar stratosphere there will likewise be a loss of ozone in summer; in winter the outcome will depend on the extent to which activated chlorine compounds are present (see below).

5.23 Although the sulphur dioxide and particulates emitted by aircraft are negligible proportions of global emissions from human activities, emissions in the stratosphere may have particular significance. Sulphate aerosols, which are predominantly droplets of sulphuric acid, and for which the other main source in the case of stratosphere is volcanic eruptions, are known to play an important role in stratospheric chemistry, especially at lower temperatures. At about −55°C the droplets support reactions leading to the formation of nitric acid, as well as reactions which enhance the destruction of ozone by bromine compounds; at about −70°C reactions enhancing ozone destruction by chlorine compounds such as chlorofluorocarbons (CFCs) become more important. At lower temperatures the droplets act as condensation nuclei for cloud formation. In polar stratospheric clouds there is very rapid activation of chlorine compounds and, in the presence of sunlight, rapid destruction of ozone (at the rate of up to 3% a day over Antarctica).[34]

5.24 The effects of aircraft emissions on stratospheric chemistry are complicated and difficult to

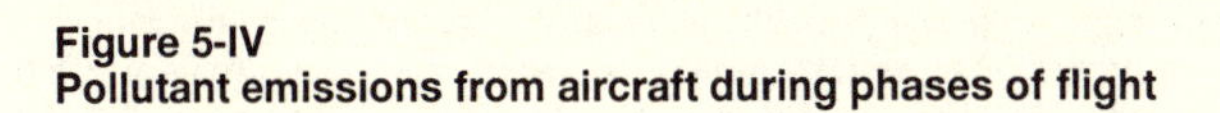

Figure 5-IV
Pollutant emissions from aircraft during phases of flight

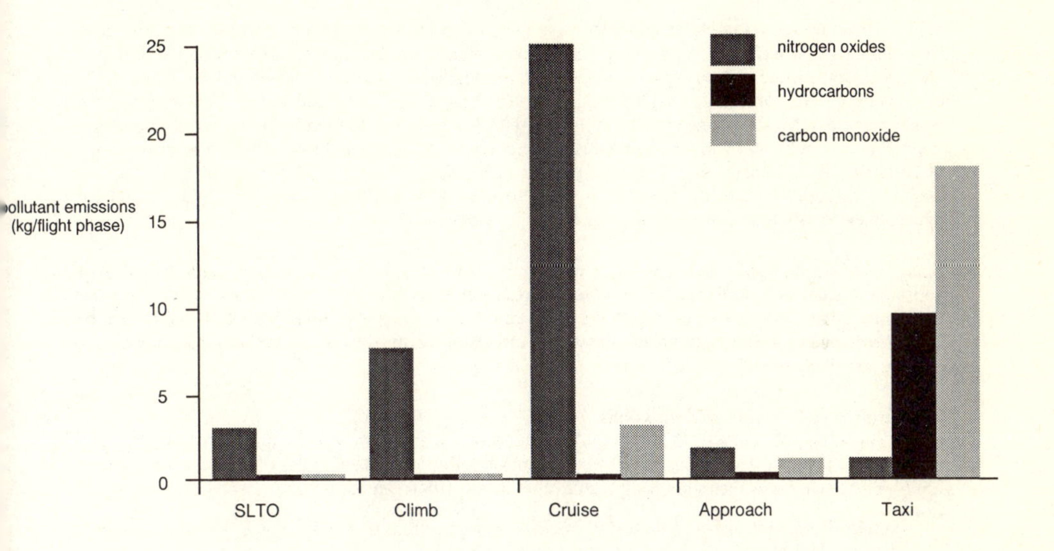

SLTO = ICAO Standard Landing and Take-Off cycle

quantify. For example, sulphuric acid is involved in the conversion of nitrogen dioxide to nitric acid and this partially offsets the direct effect of nitrogen dioxide on ozone destruction. On the other hand at very low temperatures nitric acid droplets, like sulphuric acid, contribute to the formation of stratospheric clouds, which promote the destruction of ozone by other substances such as CFCs. Very large losses of stratospheric ozone have recently been observed in the northern hemisphere at mid-latitudes and over Antarctica. The net effect of aircraft on stratospheric ozone however is far from clear.[35]

5.25 The Stratospheric Ozone Review Group[36] has considered the most recent information on the possible role of aircraft emissions and concluded that:

a. there are areas of uncertainty which make it difficult to quantify the effects of emissions of nitrogen oxides from aircraft on ozone formation and destruction in the troposphere and stratosphere;
b. more information is needed about the exchange of air between stratosphere and troposphere, about the many different chemical processes involved and about the role of other substances emitted by aircraft (particularly water vapour and the small amounts of sulphur dioxide and soot);
c. it is extremely difficult to draw up detailed emissions inventories for aircraft.

The difficulty of drawing up inventories reflects the diversity of aircraft in terms of engine size and efficiency and the grade of fuel used. The diversity of the routes flown also makes it difficult to assess in any detail where emissions take place.

5.26 An assessment of the overall significance of aircraft emissions for global warming must take into account the effects they have in the upper troposphere on the production of ozone and the formation of ice clouds. There is considerable uncertainty about the size of these two effects, but it has been suggested that the contribution of each of them to global warming is as important as or more important than the emissions of carbon dioxide from aircraft.[37,38] Their impact would be offset to some extent if the net effect of aircraft emissions in the stratosphere is to destroy ozone.

5.27 Collaborative research programmes have been established to draw up inventories of emissions from aircraft.[39] AERONOX, a two-year international programme to which the European Commission and UK government and industry are contributing, is intended to develop a clearer understanding of the effects of emissions of nitrogen oxides taking place at altitudes between 8 and 15 km. Another European research project, MOZAIC, has the use of five Airbus A340 aircraft to monitor atmospheric concentrations of ozone and water vapour continuously over a two-year period and make in-flight measurements of emissions. It is hoped this will enable models to be built of emissions from global air traffic and chemical reactions in the atmosphere, in order to provide a scientific basis for regulation and for management of air traffic to minimise the effects of emissions.[40]

5.28 As there are still many areas of uncertainty, better quantitative data about aircraft and their emissions are needed, combined with further environmental monitoring and experimental work, in order to produce better assessments of the effects of air transport on the global environment. We welcome the UK's involvement in international studies of the effects of aircraft emissions in the upper troposphere and the stratosphere.

Measures to reduce aircraft emissions

5.29 For some pollutants ICAO sets limits on emissions, which form part of annex 16 to the Chicago Convention. The primary purpose has been to control pollution in the vicinity of airports. The present limits reflect what was technologically achievable in the 1980s, and apply to:

> **smoke** from turbo-jet and turbo-fan engines manufactured after February 1982 (for supersonic aircraft) and after January 1983 (for subsonic aircraft);
>
> **carbon monoxide, unburnt hydrocarbons and nitrogen oxides** from turbine engines manufactured after January 1986 for subsonic aircraft.

In November 1993 ICAO ratified a recommendation to reduce the limits for nitrogen oxides by 20%. Annex 16 also prohibits the intentional venting of raw fuel to the atmosphere from turbine-engined aircraft manufactured after February 1982.[41] Although testing to establish compliance with the limits is at present confined to a landing and take-off (LTO) cycle, with no account taken of emissions at cruising speeds or high altitudes, ICAO is studying the relationship between the LTO cycle and the cruise phase of flight, and the technical feasibility of certification standards for the cruise and climb phases. No attempt has been made to regulate the fuel efficiency of aircraft; the assumption is that airlines already have the financial incentive to improve fuel efficiency.

5.30 Although there is uncertainty about the rate at which air transport will grow, it is clear that its contribution to adverse changes in the Earth's atmosphere will increase as the levels of air traffic increase unless significant improvements are possible in both the amounts of pollutants emitted and fuel efficiency. Energy use per seat-kilometre has fallen significantly over the last 20 years and, because load factors have improved, energy use per passenger-kilometre has fallen even faster.[42] There is scope for obtaining further improvements in fuel efficiency through better engine performance, weight reduction in engine and airframe, and the use of new materials (developed for military aircraft) to reduce drag. An ICAO study has suggested that the reduction in fuel consumption of up to 2.5% a year from all causes seen in recent years could continue to 2000, but is likely to decline to about 2% a year in the following decade.[43] This rate of improvement would not be sufficient to offset the growth of traffic forecast by the industry; emissions of carbon dioxide and water vapour can therefore be expected to increase by 2–3% a year between now and 2010.[44,45]

5.31 Whereas more efficient combustion improves fuel efficiency, and also reduces emissions of carbon monoxide and hydrocarbons, it generally requires higher engine temperatures, which favour the

Figure 5-V
Carbon dioxide emissions from aircraft to 2041: effects of different measures

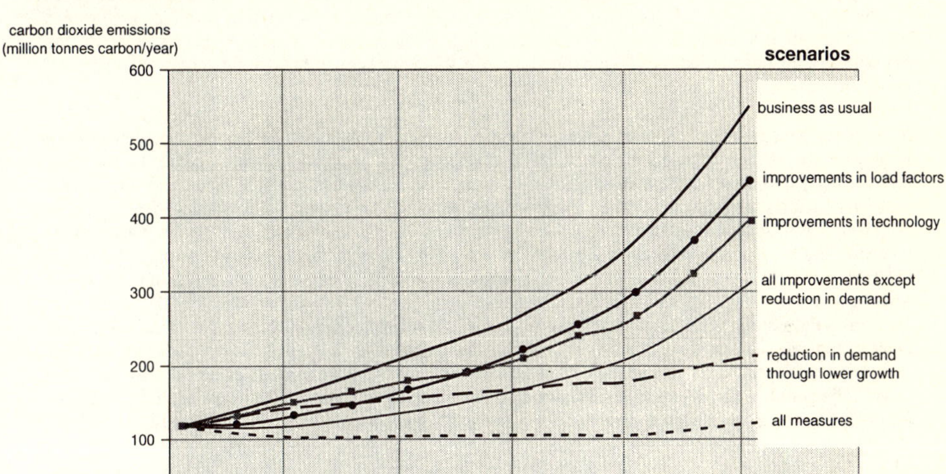

formation of nitrogen oxides. Post-1982 jet engines in the USA, for example, were found to produce 70–85% less carbon monoxide and unburnt hydrocarbons than earlier engines but at least 12% more nitrogen oxides.[46] The potential for reducing emissions of nitrogen oxides using present technology appears to be limited. The development of lean-burn engines (which operate with a high ratio of air to fuel) is being explored but they are likely to emit more smoke and hydrocarbons. Staged combustion, with alternating lean-burn and rich-burn, can cut nitrogen oxides by 60–70%[47] and may be a solution in the longer term, but there are inherent problems of reliability and safety to overcome.

5.32 Some studies have been optimistic about the prospects of reducing emissions of carbon dioxide, water vapour and nitrogen oxides in the period from 2010 to 2040, on the basis that successive generations of advanced aircraft engines will be introduced, with lower fuel consumption and improved combustion chambers.[48,49] Another study[50] assumes that technical and operational changes could achieve a 50% reduction in fuel use per passenger-kilometre by 2041 and that engines producing smaller amounts of nitrogen oxides would be introduced by then. It suggests, however, that this gain in efficiency would be overwhelmed by the growth of demand, and that carbon dioxide emissions from air transport worldwide would triple by about 2040 (figure 5-V). The conclusion reached is that management of demand will be the most critical factor in the long term in limiting carbon dioxide emissions from air transport: it is suggested that this might be achieved through a progressive reduction in business travel and air freight, and slower growth of tourist travel, so that the growth rate of air transport globally would be limited to 2.3% a year on average. Figure 5-V shows that, even with growth limited to that extent and a very large efficiency gain assumed, an upward trend in carbon dioxide emissions would be re-established after 2030.

5.33 The objective of the Climate Change Convention is the stabilisation of greenhouse gas concentrations in the atmosphere (3.71). There are disturbing indications that carbon dioxide emissions from air transport may increase substantially, and that there would be a parallel increase in the contribution air transport makes to global warming in other ways. The impact of aircraft on the chemistry of the stratosphere is also likely to increase, but it is not clear in the present state of knowledge what the net outcome will be in terms of destruction of ozone. There is a powerful case on environmental grounds for regulatory action to avert what could be irreversible damage to the Earth's atmosphere from the growth of air transport, or at least serious damage of a long-term nature (1.15). However, in view of the uncertainties that still exist about the size, and in some cases the direction, of the effects that have been identified, it is very difficult to reach a judgement at present about the relative priorities as between improving fuel efficiency (in order to reduce emissions of carbon dioxide and water vapour), reducing emissions of nitrogen oxides, and reducing noise levels. As has been indicated, the latter two objectives may carry a penalty in terms of fuel efficiency.

5.34 Because of the international nature of aviation, regulatory action to protect the atmosphere must clearly be global in scope, and the objectives and targets must be internationally agreed. ICAO is the obvious body to undertake the task of regulation. Its environmental role needs to be broadened from the original concern with pollution in the vicinity of airports to encompass a specific commitment to protect the Earth's atmosphere against damage caused by aircraft. In the case of nitrogen oxides, for example, measures are being taken to reduce substantially emissions from surface transport because of concern about their effects on health, including nitrogen dioxide's role as an ozone precursor. ICAO sets a limit on emissions from aircraft, but in effect this applies only to landing and take-off. Figure 5-IV shows that, even in quite a short flight, the great bulk of nitrogen oxides is emitted during the climb and cruise phases; and the content of regulation required for these phases is likely to be different from what has been required for landing and take-off. **We recommend that the government press for the extension of ICAO's regulatory role to cover emissions from aircraft engines at all phases of flight, with the aim of protecting the Earth's atmosphere against irreversible or long-term changes.**

5.35 The priorities for action must be established in the light of findings from further research. The introduction of more advanced engine technology may well carry a cost penalty, but the additional cost is likely to be justified in view of the seriousness of the risks of environmental damage involved. If the scientific case is established for reducing emissions of sulphur dioxide from aircraft, this can probably be achieved most cost-effectively by placing limits on the sulphur content of aviation fuel, which would bring additional benefits in the form of more efficient combustion and reductions in other types of emissions. The trend in aircraft noise levels appears to give less ground for concern. Nevertheless **we recommend that the government support more stringent noise certification standards for new aircraft, if these can be met without a significant fuel penalty.** For both noise and pollutants ICAO should not confine itself to technology-based standards, but should adopt a technology-forcing approach.

Supersonic aircraft

5.36 The 13 civil supersonic aircraft in use at present were designed in the 1960s. They use far more fuel than modern subsonic aircraft and produce correspondingly greater amounts of pollutants. They fly almost exclusively in the stratosphere. There are also larger numbers of military supersonic aircraft in use. A new generation of civil supersonic aircraft is being planned and may well find a wider market, for example for routes across the Pacific. The US National Aeronautics and Space Administration (NASA) is investigating ways of designing supersonic aircraft that will minimise their effects on the stratosphere.[51]

5.37 There is reason for concern at the prospect that a new, and much more numerous, generation of civil supersonic aircraft will be flying some time in the first half of the next century. **We recommend that the UK collaborate in research into the possible effects of supersonic aircraft on the stratosphere and ways of minimising those effects (including the possible imposition of route restrictions in relation to latitude and altitude), so that a comprehensive environmental assessment can be produced and considered on an international basis before decisions are taken to build and operate a new generation of commercial supersonic aircraft.**

Influencing the growth of demand

5.38 A reduction in the rate of growth of air travel would help considerably towards reducing, or at least stabilising, emissions from aircraft. It would also reduce the scale of some of the other environmentally damaging effects of air transport, such as noise and the loss of land for airports and surface links.

The rapid growth in air transport has been the result of the world's increasing wealth and interdependence. Industry in general is increasingly multinational. Most important in this context is the rapid growth of the tourist industry, now second only to the oil industry as a contributor to world trade. Within Europe, high-speed rail services provide an alternative to air travel and their potential has increased with the opening of the Channel Tunnel; the scope for transferring UK traffic from air to rail is discussed in chapter 12. Even if all possible passenger traffic were to be transferred to rail, however, and all intercontinental freight were to be carried by sea, the international growth in air traffic would still be substantial.

5.39 An unquestioning attitude towards future growth in air travel, and an acceptance that the projected demand for additional facilities and services must be met, are incompatible with the aim of sustainable development, just as acceptance that there will be a continuing growth in demand for energy would be incompatible. Another parallel is with the forecast growth in road traffic. We have already referred to the widespread doubts about whether it would be practicable to accommodate such growth, and we discuss that issue in detail in the next chapter. A comparable change in attitude towards the growth of air transport is needed, only in this case on an international scale. The demand for air transport might not be growing at the present rate if airlines and their customers had to face the costs of the damage they are causing to the environment. We discuss in chapter 7 how economic instruments can be used to ensure that decisions take account of such costs.

5.40 Deregulation of the airline industry in the USA in the 1980s had damaging effects on the environment. It led to the use of smaller aircraft and to lower load factors, and thus to an increase in emissions per passenger-kilometre.[52] The price reductions it produced, at least initially, also brought about a further increase in air travel in the USA. The introduction of the single European air transport market at the beginning of 1993 has had only minor effects so far. **We recommend that proposals for further measures to promote competition in air services in Europe be accompanied by a full assessment of the environmental implications.**

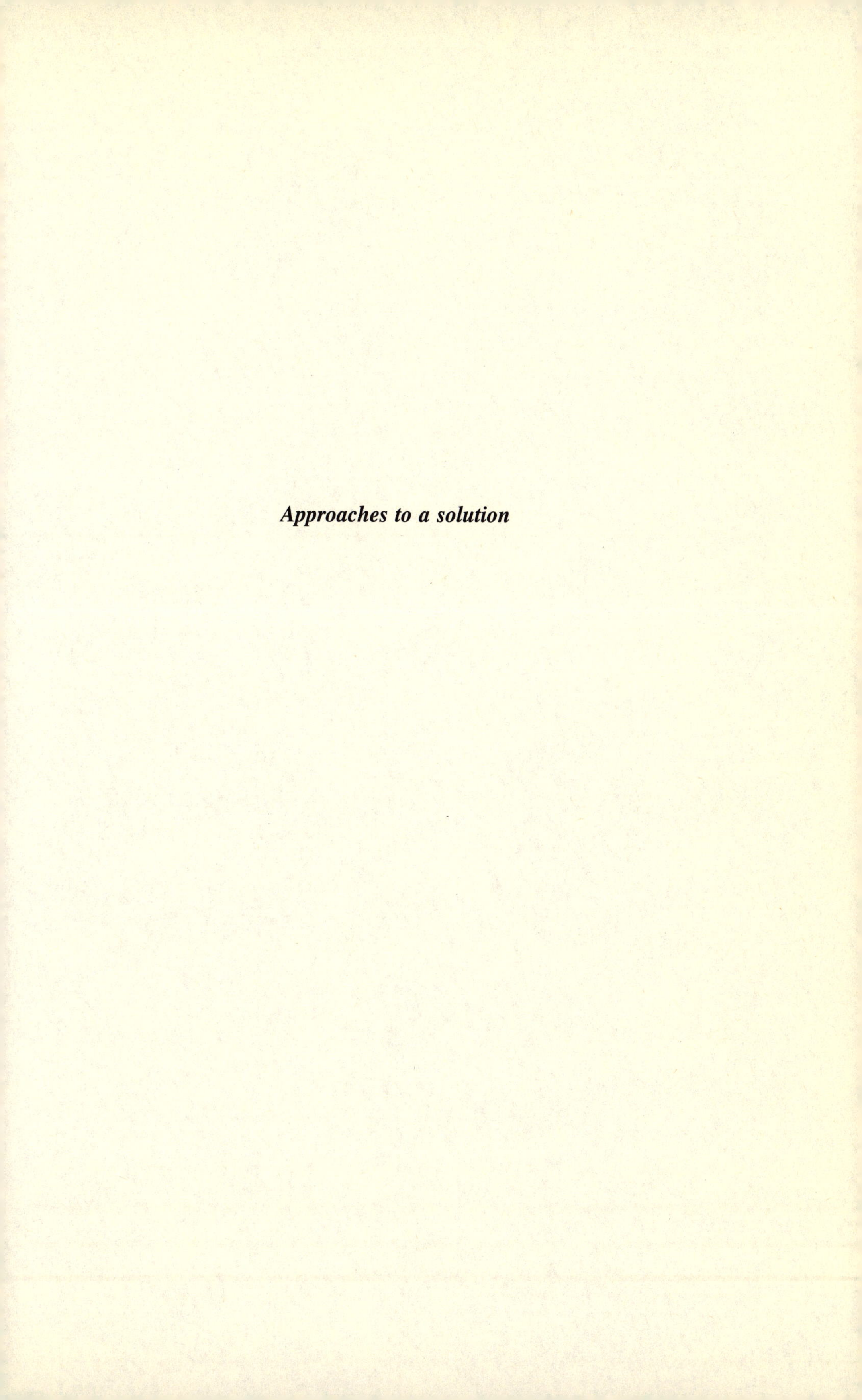

Approaches to a solution

Chapter 6

PERSPECTIVES ON TRANSPORT POLICY

Introduction to chapters 6–9

6.1 The recent expansion of transport described in chapter 2, and in the case of air transport in chapter 5, is the latest stage in a process of growth which has been shown to be exponential over the last two hundred years.[1] As a result transport has acquired a greater relative importance within developed countries, not only as an aspect of society, but also as an element in the economy. This contrasts with overall energy use, which is becoming steadily less important in relation to the economy as a whole. To be more precise, the additional use of primary energy associated with a given increase in gross domestic product (GDP) has shown a long-term decline[2], whereas the extent of additional movement associated with a given increase in GDP has tended to increase. It was shown earlier that the rapid growth has been in movements of people (2.8). At least in the UK, freight transport has grown at a similar rate to the economy over the last 40 years (2.7) with a broadly constant weight of goods being carried over progressively longer distances (2.5).

6.2 It has been suggested that the *transport intensity* of an economy can be measured by relating changes in mass movement (which, as explained in box 2A, cover movements of both people and goods) to changes in economic activity.[3] Between 1952 and 1992 the *net transport intensity* of the UK economy remained fairly constant at about 0.7 tonne-kilometre equivalent for each £ GDP. In contrast the UK's *gross transport intensity*, which takes into account movements of carriers as well as the people and goods moved, increased by a fifth, from 2.6 tonne-kilometres equivalent for each £ GDP to 3.1 tonne-kilometres equivalent for each £ GDP. In relation to the size of the economy, in other words, the scale of the task performed by the transport system was roughly the same in 1952 as in 1992; but performing that task in 1992 required 2.4 tonne-kilometres equivalent in movements of carriers for each £ GDP, an increase of more than a quarter (0.5 tonne-kilometre equivalent) since 1952. This contrasts sharply with the 40% decrease in the *primary energy intensity* of the UK economy over the same period.

6.3 Freight accounted for the greater part of net mass movement in the UK (about 85%) in both 1952 and 1992. Gross mass movement, on the other hand, was split roughly equally between movement of people and movement of goods.[4] Movements of people also require more energy in relation to the total mass moved. Thus they now account for about 70% of the total energy used in transport in developed countries.[5]

6.4 This increase in the gross transport intensity of the UK economy over the last 40 years has presumably been matched by benefits obtained by users of the transport system, whether individuals or firms. Whether the trend has been a favourable one from the overall viewpoint of the UK economy depends on the scale and nature of the external costs which users of the transport system have been imposing but have not themselves had to meet. It has been shown in chapters 3, 4 and 5 that there are major grounds for concern about the damage being caused to human health and the environment. There is even more cause for concern about the effects in terms of vehicle emissions, losses of land and reduced quality of life if past trends were to continue to the extent shown in the 1989 forecasts of road traffic (2.29). There may well be serious implications for the Earth's atmosphere if air transport continues to grow at the past rate (5.8).

6.5 In this second part of our report (chapters 6–9) we examine some of the key factors which underlie recent trends in transport and look more closely at the prospects for the future. Chapter 7 considers the external costs of the present transport system and ways in which those costs might in future be reflected in decisions by users of the system. Chapter 8 assesses the implications for emissions if road traffic continues to grow in line with the 1989 forecasts and reviews the scope for reducing environmental damage through changes in the design and performance of road vehicles. Chapter 9 discusses the way in which land use and travel patterns have reinforced each other in the emergence of a dispersed and mobile society, and the possibilities for modifying land use policies and strengthening the land use planning system.

6.6 In this chapter we consider six different perspectives on the future of surface transport which were represented in the evidence we received and the views expressed to us, and which we have called:

— letting congestion find its own level
— predict and provide
— greening the way we live
— collective action
— selling road space
— relying on technology.

We assess the environmental implications of each perspective, and draw where appropriate on experience in other countries. The main focus is on road transport because it is by a long way the dominant mode in the UK, both in present share and in projected growth. From the analysis in this chapter we identify a number of critical issues which are examined in detail in the following chapters.

Letting congestion find its own level

6.7 The first of the six perspectives we have identified is based on two assumptions: that congestion is an inescapable feature of road traffic, and is ultimately self-regulating. The approach to transport policy which follows logically from these assumptions is that congestion would be left to find its own level, government action to influence the demand for travel would be inappropriate and unnecessary, and governments would take only limited action to provide facilities for additional travel.

6.8 It is true there will always be some congestion in road networks. It would never be environmentally acceptable, or indeed practicable in a densely populated country like the UK, to construct a system of roads which would allow traffic to flow freely at all points and at all times. Quite apart from the disruptions caused by adverse weather, accidents or special events, there will always be daily, weekly and seasonal fluctuations in traffic flow. Attempts have been made to estimate the 'cost of congestion' by taking the legal speed limits on roads or the standards to which they were designed as baselines from which to calculate and cost delays. A study by the British Road Federation in 1988 produced an estimate of £3 billion a year for UK conurbations.[6] A survey of member companies of the Confederation of British Industry in 1989 produced a figure of £15 billion for the UK as a whole. To the extent that some congestion is inevitable, however, there is a fallacy in exercises of this kind.

6.9 Congestion certainly imposes substantial costs on road users. Or rather, as congestion is the result of too many vehicles trying to use the same stretch of road at the same time, road users impose substantial congestion costs on each other. As only parts of the road network are congested, and for only part of the time, drivers seek to avoid the costs of congestion by varying their route, sometimes making considerable detours, or varying their time of travel. The use of electronics to enable drivers to avoid congestion more successfully is a rapidly growing activity. Bulletins heard on car radios are now being supplemented by experiments with signs beside motorways warning of delays ahead (already in use on a large scale in Japan) and with displays on car dashboards of information about traffic conditions on motorways and other trunk roads in south-east England and the Midlands. (illustrated in plate XII). Research into more advanced driver information systems is being carried out on an extensive scale in Europe, the USA and Japan.

6.10 Individual self-interest and ingenuity often serve to maintain congestion at a level which stops short of the intolerable. That is not invariably the case. There are some cities in the world, for example Bangkok[7], which have completely outgrown their present road networks and in which fearful traffic jams are a permanent feature. In most cities, however, congestion is in a sense self-regulating: despite enormous economic changes, and with only limited new road construction, the average speed of traffic shows relatively little change over many decades. The average speed of motor vehicles in Australian cities did not increase significantly between the 1920s and the 1980s.[8] In central London, traffic management schemes have allowed the volume of traffic to double, but there was no appreciable change in traffic speeds between the 1930s and the 1980s, and probably for long before that.[9]

6.11 However, the self-regulating nature of congestion eventually has other effects. At one time it was argued that maintaining high levels of congestion in urban areas might be the best way of reducing the dominance of the car.[10] It can now be seen that the opposite effect occurs. If people find that varying their route or time of travel does not provide an acceptable solution, they are likely to change their destination and/or starting-point. If someone's journey to work is regularly and seriously affected by congestion, for example, he or she has an incentive to find a job in a less congested area, or pressure the present employer to move to a less congested area. 'Edge cities' in the USA provide a vivid example of the drive to escape congestion. Business centres have grown up at road junctions in open country in many parts of the USA. When an edge city becomes large enough to generate its own congestion at the morning and evening peaks, it loses its attraction, further development tends to slow to a halt, and a new edge city may be developed elsewhere.[11]

6.12 Even the increase in demand for mobility implied by the 1989 forecasts could be met, it has been argued, if homes and jobs were redistributed away from older urban areas. Whether or not that would be the case, it is clear that the outcome of such an approach to transport policy would be very damaging to the environment. In addition to the increases in emissions if traffic grew to this extent, the amounts of land required in order to carry out such a redistribution could be orders of magnitude greater than the stretches of land threatened by present road building proposals: the amount of undeveloped land in southern Britain would be seriously reduced, and almost the whole of south-east England and the Midlands would be covered with low-density urban development.

6.13 Apart from the long-term consequences of letting congestion find its own level, the more immediate results of a transport policy based on this approach must also be regarded as unacceptable, for a number of reasons:

a. by creating stop-start driving conditions, congestion has the effect of reducing fuel efficiency and increasing emissions.
b. There is a substantial proportion of journeys (particularly those to or from or in the course of work, or for education) over which people have relatively little discretion in the short term.
c. Where (as is often the case) the main, or only, form of public transport is buses sharing road space with other traffic, these are not only affected by the congestion, but more affected than cars. By increasing the perceived advantage of car over bus, and discouraging people from using buses, this can make congestion worse.
d. Congestion severely affects other types of vehicle which have little discretion about the trips they make. This is most obviously true of emergency vehicles, but also applies to the delivery and collection of goods.
e. As a higher and higher proportion of the maximum theoretical capacity of the road network is taken up, the effects if anything goes wrong are likely to be more severe and more widespread. At worst, several minor incidents occurring within a short time at key points in a complex road network can bring traffic to a standstill in a 'gridlock' covering a large area.
f. More generally there is an important limitation on the extent to which individual self-interest and ingenuity can be relied upon to make congestion self-regulating. In congested traffic the time an additional road user takes for a journey is less than the costs he or she imposes on other road users in the form of additional delays. The self-regulating nature of congestion will not lead to an efficient outcome in economic terms unless the costs of such delays are met by those who cause them.[12]

This last consideration is one of the main factors which have led to proposals for pricing road space. These are discussed later in this chapter, after other perspectives have been considered.

Predict and provide

6.14 The traditional approach to transport in the UK has had the aim of avoiding or limiting road congestion, based on the perspective of *predict and provide*. It has been assumed that there will be a continuing growth in road traffic, and that a continuous programme of building new roads and improving existing roads is required in order to accommodate that growth. The methodology used to produce forecasts of road traffic has been described earlier (box 2B); we indicated at that stage that we do not regard those forecasts as representing the only possible future. We focus here on the government's road building programme and its relationship to the forecast growth in traffic.

The trunk road programme

6.15 Road programmes are drawn up by central government and published separately for each part of the UK. They cover the construction and improvement of a network of roads for through traffic. For simplicity the discussion here relates to England. There are about 1,500 kilometres of *trunk roads* in England (including 2,700 km of motorways) for which the Secretary of State for Transport is the highway authority. These represent about 4% of the total length of surfaced roads and carry a third of road traffic and more than half of heavy goods traffic. As well as trunk roads, the *primary route network* defined by the Secretary of State includes about 9,300 km of major roads managed and maintained by local authorities. Proposals by local authorities to construct or improve these and other roads for which they are responsible are considered by the Department of Transport (DOT) in the context of applications for Transport Supplementary Grant (TSG). The Highways Agency established on 1 April 1994 is now responsible to the Secretary of State for the trunk road network including delivery of his programme of capital schemes; its role is discussed in chapter 13.

6.16 Road transport is the only mode of transport for which there is a long-term programme of investment set out in government policy statements. It is also a field in which public expenditure has risen sharply: planned expenditure in 1994/95 by central government on capital schemes on the English trunk road network is over £2 billion, an increase of more than 50% in real terms on the level in the 1980s. Total investment in road infrastructure in Britain in 1992/93 was over £4 billion (figure 2-V). Table 6.1 compares this with investment in other transport modes.[13] In 1985/86 investment in road infrastructure was almost two-thirds of transport investment (excluding purchase of road vehicles) and in 1992/93 it was rather more than half.

6.17 The size of the trunk road programme was doubled by the 1989 White Paper 'Roads for prosperity'[14], which was presented as a response to the sharp increases in road traffic during the 1980s and published simultaneously with the National Road Traffic Forecasts. The stated objectives of the programme were to:

assist economic growth by reducing transport costs;

improve the environment by removing through traffic from unsuitable roads in towns and villages;

enhance road safety.

The White Paper also indicated the government's willingness to consider private sector proposals for roads not included in the programme.[15] No period of time was specified for completion of the programme.

6.18 It was accepted that 'in urban areas there are severe limits to the amount of road space that can be provided.'[16] The overall programme nevertheless contained a substantial amount of improvement and construction in urban areas. Indeed the White Paper claimed that over 20% of the programme would directly benefit inner cities. In addition, it was estimated in 1990 that over a hundred major inner city highway improvements, costing in all £1 billion, would be carried out by local authorities with the help of TSG.[17] Other major road schemes in urban areas have been undertaken by development corporations, the largest being the £250 million Limehouse Link in London Docklands.

6.19 A major theme of the 1989 White Paper was that congested roads outside urban areas would be widened where possible, in preference to building relief roads on new alignments. The emphasis on relieving congestion on interurban roads was regarded as consistent with the existing objectives for the programme quoted above, rather than as a new objective. Even so it substantially affected the content of the programme. Motorway schemes and improvements to the A1 made up about three-quarters of the additions to the programme, and about three-quarters of these in turn were improvements to existing motorway routes. DOT has said it expects no more than 140 km of new routes to be added to the motorway network in Britain before 2000, an increase of less than 5%.[18] In addition to the revised programme the White Paper announced 25 additional studies, of which 11 were categorised as 'major routes/corridors'. Three of the latter, in conjunction with the Aylesbury–A12 scheme added in 1990, amounted in effect to a new London outer orbital road, beyond the M25.

Table 6.1
Investment in transport (excluding road vehicles) 1985/86 - 1992/93

percentages

	1985/86		1989/90		1990/91		1991/92		1992/93	
road infrastructure	65.7		59.3		56.1		55.3		54.5	
rail infrastructure	19.4		25.5		27.0		28.5		29.2	
British Rail		14.5		11.4		9.9		11.7		12.5
other public		4.9		6.2		8.2		6.9		8.9
Channel Tunnel		-		7.9		8.9		9.9		7.8
rail rolling stock	4.5		6.2		6.4		7.8		9.3	
British Rail		2.5		3.8		3.9		4.7		5.0
other rail		2.0		2.4		2.5		3.1		4.3
ports infrastructure	3.6		2.1		1.9		1.7		1.2	
airports infrastructure	6.8		6.9		8.6		6.7		5.8	
	100		100		100		100		100	

6.20 A review of the trunk road programme was announced in August 1993 in order to produce a clear and manageable programme of work for the new Highways Agency, as part of a package of measures to speed up delivery of trunk road improvements, and was published in March 1994.[19] Out of 371 schemes considered in the review, 22% have been given priority 1 and 47% priority 2 status; 19% are judged not to be needed until later in the programme and preparatory work on them will be suspended at the next suitable point. The remaining 13% of schemes have been withdrawn, either on environmental grounds or because they were unlikely to be progressed for the foreseeable future. Some of the additional studies for 'major routes/corridors' announced in 1989 have also been withdrawn. It remains the case that no period of time is specified for completion of the programme, but press comment indicated that DOT expects the priority 1 and priority 2 schemes to be completed within 10 years.

6.21 The review produced the following restatement of the emphases in the 1989 White Paper:

the majority of resources should be devoted to improving sections of existing key routes likely to experience congestion in the near future, primarily by adding lanes to existing motorways;

priority should be given to providing urgently needed bypasses of towns and villages;

proposals for building new trunk routes, particularly those which go through open countryside, should be reduced still further;

the programme of major urban road improvements should be a very limited one, in the light of continuing substantial investment in urban transport initiatives.[20]

Many of the schemes dropped as a result of the review were small: their estimated cost at 1989 prices was £550 million, about 2% of the total. Against that, motorway widening schemes have turned out to be considerably more expensive than was expected in 1989.[21] The overall scale and nature of the trunk road programme is therefore little changed as a result of the review.

6.22 The outcome of a review by the Welsh Office of the trunk road programme in Wales, covering strategic priorities as well as its detailed content, was also announced in March 1994. Resources are to be concentrated on improving key east-west routes serving South and North Wales. A number of planned improvements to roads elsewhere in Wales, which had aroused concern on environmental grounds (mainly because of their effects on rural landscapes), have been deferred (and may not be resurrected) or deleted. Of the 51 schemes considered in the review 25% were placed in the latter category, although these represented only 5% of total estimated costs.[22]

6.23 The conclusions of a review of the strategic road network in Scotland were published by the Scottish Office in 1992.[23] Following traffic growth in the 1980s, expenditure on new construction on national roads had increased by 30% in real terms between 1988 and 1992, and further small increases were planned. A long-term programme of major schemes was announced with the emphasis on upgrading existing dual carriageway roads to near-motorway standards and dualling some single carriageway roads, and complemented by improvement programmes on long-distance single carriageway routes.

6.24 In Northern Ireland the Roads Service of the Department of the Environment is responsible for all roads. As a result of a large programme of road building in the 1960s and 1970s, with further additions in the 1980s, there are 111 km of motorways and 2,200 km of class I roads. Expenditure on major new construction schemes is now about £11 million a year, sufficient for 4–5 km of new carriageway on interurban roads and some improvements in urban areas.[24]

Inadequacy of provision for forecast growth

6.25 Although the 1989 White Paper on roads in England was published at the same time as the National Road Traffic Forecasts, it did not explain the road programme's relationship to those forecasts. DOT subsequently said that the schemes listed in the White Paper were 'not intended to cater for all forecast demand to 2025. There will be cases where on economic or environmental grounds it is neither practicable nor desirable to meet the demand by road building, for example in city centres.'[25] In oral evidence to the Commission, DOT officials were not able to express any view about the proportion of forecast demand which might be restrained or transferred to public transport nor to explain convincingly what the effects would be if capacity is not provided to meet the forecast demand. The 1989 White Paper gave a more specific justification for the schemes which it added to the programme: the need for improvement had been considered wherever it was already the case, or forecast to be the case, that in weekday peak hours traffic on an interurban road 'is regularly forced into stop-start operation and minor incidents cause considerable delay'.[26] We have examined the extent to which the schemes in the trunk road programme would provide for growth in traffic on the scale shown in the 1989 forecasts. In the absence of analyses by DOT we have drawn on two studies using different approaches, one by Friends of the Earth and the other carried out for the British Road Federation.

6.26 The study by Friends of the Earth[27] examined the situation at the automated national traffic census points on the motorway network, which are selected to represent free-flowing traffic conditions and do not therefore include the most congested sections of motorway. Figure 6-I shows that there was peak hour congestion at 14 of the 29 national traffic census points in 1990, and at another 5 points congestion extended beyond peak hours. Even with the additions to capacity which form part of the trunk road programme, it was calculated that congestion at the national traffic census points will have worsened significantly by 2000. There would be a further major deterioration by 2010; by 2025, as figure 6-I shows, there would be chronic congestion at all but one of the 29 points.

6.27 This study also calculated the number of lanes that would notionally be necessary at each point in order to prevent chronic congestion in 2025, if traffic increases in line with the high variant of the 1989 forecasts. On the M1 between the M25 and Luton, ten lanes would be needed in each direction, and in another eight cases six or more lanes would be needed in each direction. Finally table 6.2 shows the national traffic census points directly affected by existing motorway widening schemes (which are not, as explained above, the most congested points in the motorway network) and the date by which it was calculated that traffic on the widened motorway at each of these points will return to the category

Figure 6-I
Congestion on motorways to 2025 if traffic grows as forecast in1989

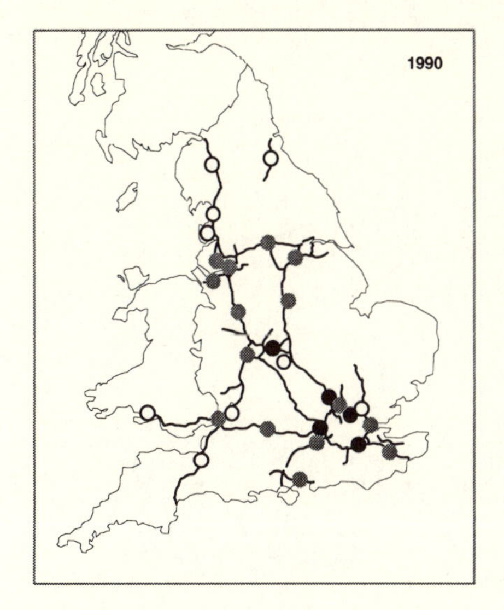

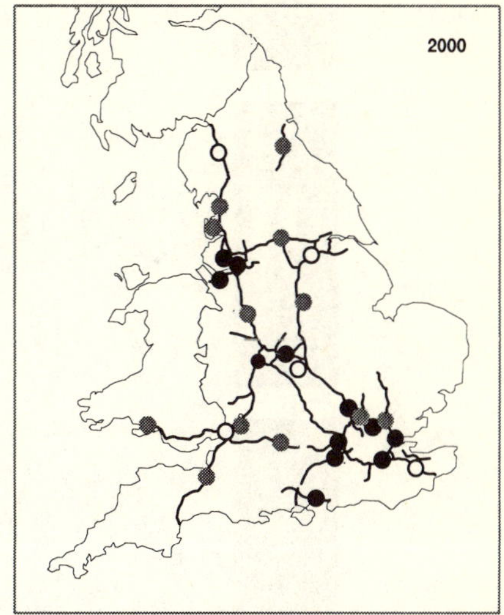

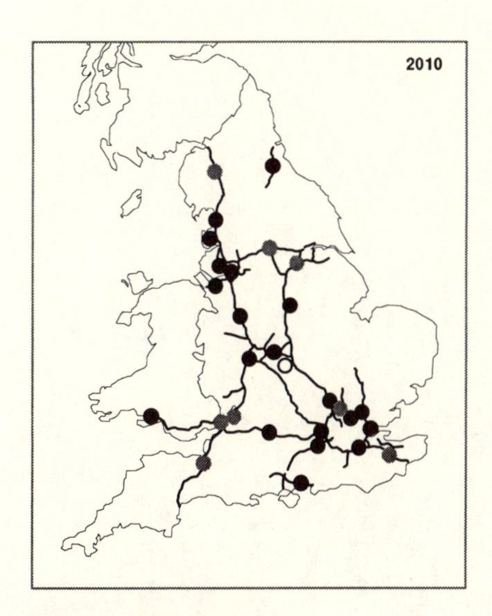

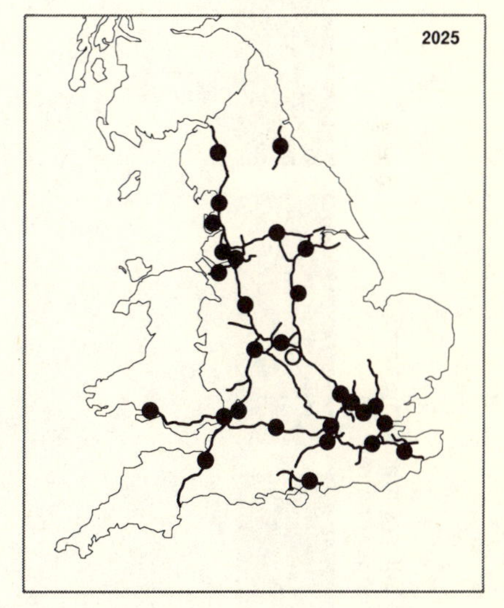

National traffic census points
- ● chronic congestion
- ● peak hour congestion
- ○ points not experiencing congestion

Table 6.2
Effects of the current road programme: time before congestion returns

section of road		1990 congestion	present width	after widening	return to 1990 congestion	reach chronic congestion
			dual			
M4	(J7-9)	Chronic	3	4	1993	1993
M25	(J26-27)	Chronic	3	4	1995	1995
M6	(J2-3)	Chronic	3	4	1995	1995
M25	(J5-6)	Chronic	3	4	1996	1996
M3	(J3-4)	Peak	3	4	1990	2000
M6	(J14-15)	Peak	3	4	1990	2001
M4	(J14-15)	Peak	3	4	1993	2004
M1	(J25-26)	Peak	3	4	1995	2006
A1(M)	(M25-A1001)	Peak	3	4	2000	2011
M20	(J6-7)	Peak	2	4	2003	2014
M4	(J21-22)	Peak	2	5	2011	2023

of congestion experienced in 1990: in 8 of the 11 cases this will happen well before 2000. For the purposes of this study Friends of the Earth made the cautious assumption that, although motorway traffic has shown above average growth in the past, it will grow at the national average rate in the future.

6.28 The study carried out for the British Road Federation examined the effect of various levels of expenditure on the trunk road programme in the period to 2010.[28] It did so by calculating an index of congestion: broadly speaking, this represents the ratio between the level of road traffic and the capacity of the road network, with some allowances made for differences between 13 regions of England and between peak and off-peak conditions. There was no attempt to identify the locations at which congestion would appear. The conclusion reached was that, if expenditure on the trunk road programme continues at the present level, congestion will increase by 14% between 1993 and 2010 and the average speed of traffic will fall by 5%. Even with a 50% increase in expenditure on the trunk road programme, it was calculated that congestion would increase by 7% and speeds fall by 3%. (The explanation for this relatively small effect from a large increase in expenditure seems to be that, after the best available schemes already in the trunk road programme have been carried out, further road schemes quickly become more expensive.) The effect of a 50% reduction in expenditure on the trunk road programme was also examined: it was calculated that congestion would increase by 23% and speeds fall by 9%.

6.29 These two studies show clearly that, if the past trend of road traffic growth continues, the present programme of road building and road widening would not prevent a serious worsening of congestion on the trunk road network as a whole after 2000. The Friends of the Earth study indicates that the present programme would not even come near meeting the 1989 White Paper's primary aim, which was to eliminate peak hour congestion on motorways. To provide enough capacity to prevent a serious worsening of congestion would require a very large increase in expenditure on the trunk road programme (substantially more than a 50% increase, according to the British Road Federation study), over and above the large increase which has already occurred. Many major new road schemes would have to be designed and carried out which, as well as not being in the present trunk road programme, have not even been identified. One estimate is that, following the addition of 12,500 miles of new lanes to the trunk and local road network by 2000, another 1,700 miles of new lanes would have to be added each year between 2000 and 2025, a total of 42,000 miles of additional lanes.[29] This would be a more rapid rate of road construction than ever achieved up to now in the UK. It is an obviously unrealistic scenario, even if it were acceptable in environmental terms.

6.30 We have also seen a study by Oxfordshire County Council which covered the Oxfordshire section of the M40, non-motorway trunk roads in the county and the council's own principal roads.[30] It was assumed that until 2012 traffic will show uniform growth at a rate midway between the high and low forecasts made by DOT in 1989. By 2002 the increased traffic would exceed the capacity of a substantial proportion of the trunk roads (only some stretches of which are due to be improved under the present programme) as well as some principal and B roads. By 2012 a much larger proportion of the trunk roads would be over capacity (especially existing dual carriageways), as would a stretch of the M40. By 2022 most of the M40 in Oxfordshire would have a severe capacity problem. The conclusion must again be that, if DOT's 1989 forecasts are valid, a large number of major schemes which are not included in the present trunk road programme would have to be carried out to prevent a serious worsening of traffic conditions, especially after 2000.

6.31 In a survey carried out 18 months after the National Road Traffic Forecasts were published, two-thirds of metropolitan and London borough councils said that, regardless of the money available, it would be undesirable to provide road space to meet the projected traffic levels. Among county councils only a fifth shared that view at the time (another fifth shared it in relation to urban areas), but local authority views have hardened subsequently against a large programme of road building and in favour of what has been called the 'new realism' in transport policy.[31] In a letter to the Secretary of State for Transport in August 1993 the Chairman of the Environment Committee of the Association of County Councils said: 'The continuing attempt to build our way out of trouble has become self-defeating. We need to decide how large a road system and what volume of traffic we can afford to accommodate without unacceptable damage to the environment and people's lives.'[32]

6.32 Over and above the inevitable difficulties of forecasting the future, several aspects of the methodology used to produce the 1989 forecasts can be questioned. One is the assumption made about the saturation level of car ownership. DOT's assumption (box 2B) was that this would be 90% of the population aged 17–74, broadly equivalent to the ratio of 600 cars per 1,000 people which existed in the

USA as a whole in 1990. The 1988 ratio in Britain was 331 cars per 1,000 people. The low forecast of traffic growth made in 1989 implied that this ratio will increase to 529 cars per 1,000 people in 2025, the high forecast that it will increase to 608 cars per 1,000 people by that date. It is doubtful whether the saturation level will necessarily be the same in Britain as in a wealthier and less densely populated country such as the USA. On the basis that the car is now a mature technology, it has been argued both that saturation level has nearly been reached in the USA and that saturation level in other countries will be much lower.[33]

6.33 Another questionable aspect is the fundamental assumption that the level of road traffic is not affected by the capacity of the road network. There is a strong case for believing that the extension and improvement of the road network leads to an increase in the total amount of road traffic, as distinct from redistributing a predetermined amount of traffic onto the new and improved roads. This is because it becomes attractive to make more trips and longer trips. DOT's view has been that any effect the road programme has in generating additional traffic is of minor importance, and that comparisons between forecast traffic levels and actual traffic levels a year after road schemes have been completed support this.[34] DOT sought advice on the issue from the Standing Advisory Committee on Trunk Road Assessment (SACTRA), whose report has been submitted but not published.

6.34 Last year DOT commissioned a review of the methodology used for forecasting road traffic[35], which will cover in particular the effect on future traffic levels of limitations in the capacity of the road network. Pending the outcome of the review, no revised forecasts have been produced.

6.35 The inability of any foreseeable trunk road programme to cope with the forecast growth in traffic destroys the rationale of the 'predict and provide' perspective. In its Sustainable Development Strategy for the UK the government has distanced itself from this approach and accepted that, if present trends were to continue, the resulting traffic growth would have unacceptable consequences for both the environment and the economy of certain parts of the country. It has also accepted the need for government measures to influence the overall level of traffic growth. At the same time it has emphasised individual choice by saying: 'It is not the Government's job to tell people where and how to travel.'[36]

Greening the way we live

6.36 The third perspective discussed here highlights the importance of the way people choose to live. It diagnoses the environmental damage caused by transport as an inherent consequence of present lifestyles in developed countries, and makes the assumption that the situation can be radically changed as a result of initiatives by local communities. In part this would involve reversing the changes in lifestyles which were identified earlier in this report (2.17–2.27) as being closely associated with the recent growth of transport, but many people envisage a process going much further than that.

6.37 The central aim would be to make the inhabitants of each town or neighbourhood as far as possible self-sufficient by 'redesigning the urban fabric and also the urban metabolism'.[37] A major reduction in energy use for transport would be one contribution to that; and greater self-sufficiency in other respects would greatly reduce demand for the transport of goods and people. Although urban areas are where most people live, the same aim could also be pursued by similar methods in rural areas.

6.38 Advocates of self-sufficiency nevertheless differ in the way they view transport. One writer has envisaged a society emerging towards the end of the next century based on local self-sufficiency, but in which people will travel a great deal and almost the only remaining requirement for government on a large scale will be 'a world agency . . . to provide traffic control'.[38] More usually, a new approach to transport has been seen as one of the crucial steps towards a self-sufficient community. As another writer puts it, 'turning too much living space into road space destroys the inner form and balance of the city'. The city is viewed as 'an organism, an eco-system, with its own internal life, creative energy and interdependence', and the criterion used in planning and transport decisions should be 'how well the city encourages or discourages creative relationships between people'.[39] Reassessments of transport policies from this perspective have often been associated with opposition to proposals for large-scale road construction or, in one case (the German city of Freiburg-im-Breisgau), opposition to construction of an underground light-rail line.

6.39 A central feature of the self-sufficient lifestyle is the possibility for any able-bodied person to make all everyday journeys within the neighbourhood on foot or by cycle. These modes of travel use no fossil fuel, and very little energy: someone weighing 70 kg has been estimated to use 0.14 megajoule/kilometre when walking and 0.035 megajoule/kilometre when riding a 20 kg bicycle[40], whereas cars on commuting journeys in built-up areas have been estimated to use 2.8 megajoules/passenger-kilometre on average and cars on other urban journeys 1.8 megajoules/passenger-kilometre.[41] Ideally there would also be a communal same-day delivery service (as in Freiburg and in Zurich[42]) so that, even when shopping for bulky goods, people can travel on foot, by cycle or by public transport rather than by car.

6.40 A number of cities claim to be pioneers of greener transport policies: as well as Freiburg (see plate XI), they include York (which the Commission visited), and Delft (which the Commission also visited), Groningen and Münster on the continent. In all of them, between a half and a third of journeys not made on foot are made by cycle, compared with less than 3% in the UK as a whole. Large areas of these cities exclude motor traffic, and the rest of the city has safe and attractive cycle routes. Safe storage for cycles has been established at key points. Other policies include better provision for pedestrians and linkages between other transport modes, for example through park and ride schemes.

6.41 The cities which have been most successful in encouraging journeys on foot or by cycle tend to be of a particular type: medium-sized, free-standing, with a population in the range 150,000 to 450,000, and usually flat. In economic terms they are relatively self-contained, and not in serious competition with other urban centres nearby. They are compactly built, with their central area on a medieval street-plan. Cities like this provide a pleasant lifestyle for many people, with a rich social and cultural life. They are often university cities, and the non-student part of their populations is to a degree self-selected: people who like such a lifestyle gravitate to them, and those for whom it would not be attractive choose to live elsewhere.

6.42 One view of experience so far would be that radical policies aimed at reducing car use are inherently more suitable for cities of the kind just described. In that case it would be unwise to assume that the same policies could be equally successful elsewhere. Another view would be that, while (for historical and cultural reasons connected with civic identity) cities of this kind have been able to reach a consensus about such policies more quickly and more easily, they represent the leading edge of a general trend. Certainly there seems to be a much wider movement in local communities towards systematic policies for reducing reliance on cars, even if the eventual limits of such policies remain unclear.

6.43 Changes in lifestyles have an important role to play in creating a sustainable transport system for the UK. They need to embrace, not only greater resort to walking and cycling, but far-reaching changes in the way people perceive and use other modes of transport. New lifestyles cannot be imposed by governments. It is not likely they could be brought about solely by education or persuasion, or by other promotional measures, on a sufficient scale to resolve the basic dilemmas of present transport policies. Greener lifestyles will have to spread over a period of time. They will have more appeal to people at certain periods of their lives, and may be taken up more readily in some areas of the country than in others. Their eventual success will depend on the action taken by central and local government to provide frameworks within which individual choices can be exercised in an environmentally responsible way. We discuss in later chapters the measures needed for this purpose, particularly in chapter 11 on local journeys.

Collective action

6.44 For those journeys which it is impracticable to make on foot or by cycle, use of public transport rather than cars has been advocated for environmental reasons. The primary reason advanced is that public transport is more energy efficient. Where public transport is electrically powered, it also has obvious advantages over petrol or diesel cars in terms of local air quality. Even where it is not electrically powered, it may be perceived as causing less exposure to airborne pollutants. Use of public transport is also advocated on the ground that it minimises the requirements for land for transport infrastructure. It is argued that greater use of existing railway lines (or possibly in some cases new underground railway lines) may avoid the need for road building or road widening schemes; and that, even where public transport uses the roads, it takes up less space than cars carrying the same number of people and will therefore help to avoid the need for losses of land for new infrastructure.

6.45 In this perspective on transport policy the emphasis is on the role of the community in ensuring through collective action that an attractive and comprehensive system of public transport is available. A precondition for individual choice is that there should be an alternative to the car for those journeys which it is impracticable to make on foot or by cycle. For those people who are not sufficiently active to walk or cycle and do not have a car, some form of public transport is the only way of making journeys; the need for a system of public transport is therefore advocated on social grounds. However, the

Table 6.3
Modes of personal travel in developed countries (1981-91)

percentage (to nearest whole number)

	car		bus		rail		public transport (bus and rail)	
	1991 share	change in passenger-km since 1981	1991 share	change in passenger-km since 1981	1991 share	change in passenger-km since 1981	1991 share	change in passenger-km since 1981
Great Britain	88	+51	7	-11	6	+12	12	-2
Belgium	82	+20	11	+10	7	-8	18	+2
Denmark	79	+43	14	+26	7	+9	21	+20
France	85	+28	6	+12	9	+11	15	+12
Germany	84	+52	8	-27	7	-12	16	-21
Italy	79	+47	14	-0.5	7	+13	21	+4
Netherlands	84	+21	8	+18	8	+30	16	+24
Portugal	81	+58	13	+22	7	-3	19	+12
Spain	73	+10	19	+36	8	-3	27	+22
Sweden	85	+39	9	+27	6	-14	15	+7
USA	98	+30	2	+52	1	-24	2	+44
Japan	53	+64	10	+23	36	+23	47	+23

Shares are of total personal travel by car, bus and rail, measured as passenger-kilometres and disregarding travel by other modes.

environmental argument goes beyond that to advocate government action to increase considerably public transport's share of travel, so reversing its long-term decline.

6.46 On the ground that carrying freight by rail or sea is more energy efficient and environmentally less damaging than carrying it by road, the aim would also be that a high proportion of the long-distance freight at present carried by road should transfer to rail or coastal shipping for as large a part of its journey as possible.

6.47 A crude assessment of the relative energy efficiency of different transport modes can be made by comparing their shares of energy consumption for transport with their shares of movement of people (figure 2-I) and goods (figure 2-III). Road, rail and water use respectively 94%, 2.4% and 3.5% of the energy consumption by surface transport in the UK.[43] In contrast their shares of net mass movement by UK transport are 69% for road, 7.5% for rail (7.8% for goods and 5.7% for people) and 24% for water.[44] Even allowing for distorting factors (such as the statistical convention that trips are assigned to the main mode used), it is clear that present use of rail transport in the UK is significantly more energy efficient than present use of road transport, and that use of water to move freight is considerably more energy efficient than other modes. Moreover, although road transport cannot readily be sub-divided to make a separate comparison for bus and coach travel, other calculations (see for example table 12.2) confirm that this in turn is significantly more energy efficient than travel by private car.

6.48 Table 6.3[45] shows the shares of personal travel held by car, bus and rail in a number of countries in 1991, and the changes since 1981 in the passenger-kilometres travelled by each mode annually. The most striking features are the importance of public transport in Japan and its insignificance nationally in the USA. However there are other significant differences. Whereas travel by bus declined between 1981 and 1991 in Britain (and also in Germany and marginally in Italy), it grew in all the other countries listed, and in Spain and the USA increased its share of personal travel. Public transport's total share of personal travel is lower in Britain than in any of the other countries listed except the USA. Although it is not markedly lower than in other northern European countries, we have been particularly impressed by the success of several cities in continental Europe over the last 20 years in increasing the proportion of journeys made by public transport. The case of Zurich is described in box 6A. Another striking example of success in making public transport attractive, described in box 6B, is Curitiba in Brazil.

BOX 6A **PUBLIC TRANSPORT IN A SWISS CITY**

Zurich, a medium-sized city of 500,000 people without an underground railway, has developed a surface public transport system which is one of the most efficient in Europe. The policy is to make public transport faster and more attractive than travel by car for many journeys in the city. To ensure there is adequate space on the roads for buses and trams and for pedestrians and cyclists, the number of cars in the city at any time is controlled by a computerised system of traffic lights.

The existing tram services have been made faster and more reliable. Where possible segregated track has been provided. Automatic traffic lights give trams priority over road vehicles at all intersections. In areas not served by trams, bus services have been improved. A timetable is displayed at each stop, and is closely adhered to. Tickets obtained from machines at all tram stops and railway stations are valid for all forms of public transport within a specified area for a specified period. Because of high usage, over 70% of costs (including interest on loans) are covered by fare revenue.

Outside the city a regional railway network, integrated with the national railways, serves the canton of Zurich. The Zurich Transport Authority controls the timetables and fares of rail and bus operators in the canton; fares cover travel on all modes within a specified zone.

Use of public transport has always been relatively high and showed little change between 1965 and 1985. Between 1985 and 1990 the number of trips by public transport increased by a third. In 1990 public transport trips averaged 470 per person in the canton of Zurich, and 690 in the city. This can be compared with 290 per person in London, 340 in Tyne and Wear, and 130 in Manchester.

Although public transport has expanded considerably, car traffic in the city has remained roughly constant. In 1992, of trips within the city which were not made on foot, 51% were made by public transport, 39% by car and 10% by cycle.

Table 6.4
Modes of freight transport in developed countries

percentages (to nearest whole number)

	road		rail		water	
	1991 share	change in tonne-km since 1981	1991 share	change in tonne-km since 1981	1991 share	change in tonne-km since 1981
Great Britain	65	+40	8	-12	28	+5
Belgium	75	+70	13	-29	12	-4
Denmark	70	+41	13	+19	17	+39
France	64	+20	31	-17	5	-24
Germany	48	+45	37	+14	16	+7
Italy	82	+30	1	-88	17	+24
Netherlands	38	+31	5	-9	58	+12
Spain	79	+31	6	-5	16	+5
USA	29	+50	36	+2	35	-3
Japan	50	+51	5	-20	45	+16

Shares are of total freight transport by the modes shown, measured as tonne-kilometres and disregarding transport by other modes.

6.49 The shares of freight transport held by road, rail and water in other developed countries show wide variation (table 6.4[46]). Road transport has been the growth sector. Nevertheless, in contrast to Britain, large amounts of freight are still carried by rail in Germany, the USA and France. Water transport dominates in the Netherlands, and is very important in Japan. Although in both cases this reflects geographical factors, it is nevertheless striking that Japan has been able to achieve a high level of economic growth without a modern road system and with a maximum lorry weight of 20 tonnes on nearly all roads until recently (the limit has now been raised to 25 tonnes).[47] If the calculation is extended to include pipelines, their share ranges up to 17% in the USA and 11% in France.

6.50 The relative energy efficiency of public and private transport is heavily dependent on occupancy. Because one mode of transport is more energy efficient than another on average, it does not follow that it will necessarily have the same advantage for a marginal journey at lower occupancy. There is a trade-off between the quality of service provided by public transport and its energy efficiency. Providing services at regular intervals from early morning to late evening will almost certainly produce low occupancies for some of the time. Moreover, there may be more scope for increasing the energy

BOX 6B **BUS SERVICES IN A BRAZILIAN CITY**

Curitiba, a city of 1.6 million people in Brazil, has radically redesigned its bus system in order to cut overall journey times. The key features of the present system are reserved lanes for buses; priority for buses over other traffic at junctions; a hierarchy of frequent, interconnecting routes each with a clearly defined function (express, inter-district, feeder); issuing of tickets from a machine at each stop before passengers join the bus, in order to minimise stopping times without requiring an additional member of staff; and a shelter providing adequate weather protection at each stop. As a result of these measures Curitiba combines car ownership of about 300 per 1,000 people (comparable to the UK and the highest of any Brazilian city) with relatively low car usage.

efficiency of some modes than of others, and the relative attractiveness of different modes in environmental terms may therefore change in the longer term.

6.51 The density of population and the pattern of development are important in determining whether traffic flows are sufficiently large to make any particular form of public transport cost-effective. To take the extreme case, Hong Kong is so densely populated that very expensive forms of public transport, such as the Mass Transit Railway, can be built and operated profitably without government meeting any of the costs.[48] Population density is a factor in the viability of public transport systems in some cities in continental Europe. This aspect is further discussed in chapter 9.

6.52 In Britain, public transport is more important in London and the surrounding region than in the country as a whole. More than half of passenger rail travel in Britain in 1992 was in south-east England, 15% on the London Underground and 36% on what was then the Network SouthEast sector of British Rail.[49] Because of the share taken by the London Underground, bus travel in Greater London on the other hand is somewhat less important than in the rest of Britain, representing 9% of passenger-kilometres travelled by bus nationally and 13% of vehicle-kilometres. There is widespread agreement that London depends heavily on its public transport system for its efficient functioning, and that very substantial capital expenditure is required in order to raise the system to an acceptable standard of efficiency. There is concern that, without such expenditure, London's economy may be seriously damaged.[50] From the transport point of view, a decline in the economic importance of London would exert a downward pressure on public transport use in Britain.

6.53 There are problems in making public transport attractive to potential passengers. The car has an enormous initial advantage in terms of convenience, privacy and perceived security. The countries of eastern Europe were heavily dependent on public transport systems under Communism, but many of those systems are now showing serious deterioration as ownership of a car becomes a crucial symbol of the new freedom.[51] In western Europe investment in cleaner, faster and more comfortable forms of transport, such as metro lines, has often had the effect of generating additional travel, and attracting passengers who previously cycled or used buses, without persuading substantial numbers of people to give up using their cars. The outcome often seems to have been to prevent further decline in public transport use rather than reverse the decline.

6.54 Attractive and efficient public transport systems are essential for the functioning of very large modern cities. Many (even Los Angeles, the prime example of a car-based city) are constructing new rail systems. Considered as a substitute for journeys hitherto made by car, public transport's fundamental limitation is that, however ingeniously it is managed and adapted, collective provision can never be completely reconciled with the individual's choice about what journeys to make and when to make them. A major direction for innovations in technology and operating practices is to make public transport considerably more flexible and responsive. The pricing of public transport is also a crucial issue; we discuss in the next chapter whether government contributions to the cost of public transport are justifiable on environmental grounds.

Selling road space

6.55 Another important perspective on transport policy places the main emphasis on more efficient use of the existing road network. One of the reasons why the approach of letting congestion find its own level discussed at the beginning of this chapter would be unsatisfactory is that the time an additional road user takes for a journey in congested conditions is less than the costs he or she imposes on other road users in the form of additional delays. It can also be assumed that the values of individual road journeys differ when expressed in money terms. The possibility therefore exists of increasing the efficiency of the road network by levying charges for road use. These will deter people from making journeys which are of low value in relation to the price payable, or for which a less costly alternative is available, and so allow journeys of higher value to be made under less congested conditions.

6.56 Although originally proposed as a solution to congestion, road pricing could also be used as a method of reducing road traffic for environmental reasons or, in principle, as a way of ensuring that road users make payments which reflect the costs of the damage they cause to the environment or other social costs they impose. It could be used in support of a decision not to undertake further road building, in order to ensure that use of the existing road network is optimised. It could also be used to raise money for the building of new roads, to make public transport a more attractive alternative to car use, or as a contribution to the government's general revenue. The objectives of a road pricing scheme determine what structure and levels of charges are appropriate.

6.57 The most successful existing road pricing scheme is in Singapore. One important factor contributing to success is that Singapore is geographically isolated, as are the three Norwegian cities which have introduced charges for entry (see box 6C[52]). A second important factor in Singapore is that other economic instruments are used to discourage car ownership (7.33). In other countries the political unattractiveness of road pricing has prevented progress towards it[53]; a scheme in Hong Kong was never implemented, even though the pilot stage had been reasonably successful.[54] Demonstration trials of road pricing equipment were carried out in Cambridge in 1993 as part of the ongoing Adept project under the auspices of the EC Drive programme.[55]

6.58 Despite the intellectual attractions of road pricing, it presents a number of difficulties. There is now quite extensive experience of using electronic systems in the more straightforward situations of collecting tolls for the use of tunnels, bridges or motorways. A system on the Dartford Bridge identifies vehicles by means of transponders fitted to windscreens and debits the tolls from pre-paid accounts. In other cases a pre-paid smart card has been used. However, the technology for large-scale electronic road pricing has yet to be demonstrated. Development of a system that would operate nationally appears to be technically feasible but would be expensive and take a number of years. International compatibility of systems will become an important issue, especially in Europe.

BOX 6C — ROAD PRICING IN SINGAPORE AND NORWAY

In **Singapore** an area licensing scheme operates in the central area during the morning and evening peaks. It requires display of a supplementary licence which can be bought at booths on the approaches to the area at a cost of $3 (about £1.25) a day. The government intends to replace this by an electronic system by 1997 and extend the coverage to other congested areas. In the meantime, the existing scheme is to be converted to use smart cards and extended throughout the day. To act as a sufficient deterrent fees were originally set at a high level ($3 a day, subsequently raised to $5, reflecting inflation). After traffic had initially been reduced by 50% the daily fee was lowered to $3. The reduction in traffic has now stabilised at 30–35%. Inconvenience, rather than cost, is the main deterrent to car use: daily licences in particular have to be bought at special off-road sites. When the electronic system is introduced, the inconvenience and thus the deterrent effect will be less.

The main objective of road pricing in **Oslo** is to part-finance public transport and road construction (including the further development of tunnels under central Oslo to create traffic-free areas). Tolls are collected from vehicles passing through a cordon. Similar schemes exist in **Bergen** and **Trondheim.** Charging is planned to cease at a date between 2003 and 2007, when the projects will be fully paid for. The toll company's liabilities and share capital will be repaid; the toll booths removed; and the company dissolved. On average, over 200,000 vehicles a day pass through the Oslo toll ring. Mainly as a result of the introduction of 'electronic clip-cards', the share of traffic accounted for by subscribers increased in 1992 from 52% to 60%.

6.59 Concern has been expressed about the possible environmental effects of introducing road pricing in local areas or on particular roads. Drivers who do not want to pay for road use may make longer journeys to avoid the area or road for which a charge is being made, thus using more fuel and producing more emissions. Use of alternative routes may cause environmental damage in areas which were not previously affected by heavy traffic. There has also been concern about the social consequences of road pricing. In order to deter a significant number of journeys it will probably be necessary to charge a high price. This is particularly likely where discretionary car journeys have already been deterred by congestion and the availability of public transport. People with high incomes will find it easier to pay such prices; but some people on low incomes will nevertheless have to undertake certain car journeys for work or family reasons. There might well be pressure for a wide range of exemptions for 'essential users', which if granted would undermine the scheme's effectiveness.

6.60 If road use has been underpriced, road pricing is not the only possible instrument for remedying that situation. Other methods of ensuring that road users are faced with the costs of their journeys may be preferable, or at least easier to implement. A view about the desirability of road pricing must be based on an examination of the various types of cost associated with transport and the options for meeting those costs in ways that are both efficient and equitable. We undertake this task in the next chapter.

Relying on technology

6.61 The last of the six perspectives on the future of transport considered in this chapter emphasises the role of technological progress. There are three broad directions such progress might take. It could lead to the emergence of new modes of transport which would displace existing modes. It could take telecommunications to a level of sophistication which would substantially reduce the demand for transport. Or it could make existing modes of transport less damaging to the environment. The assumption which forms the basis for this perspective is that technological innovations in one or more of these three directions can resolve the conflict between the rapid growth in transport and the aim of sustainable development.

6.62 Two modes of transport which are being investigated for possible widespread use in the next century are a new generation of supersonic aircraft (perhaps to be followed eventually by hypersonic aircraft) and magnetically levitated (maglev) surface vehicles using superconducting magnets and linear motors.[56] In both cases the purpose would be to move people more quickly over long distances, and in all probability larger numbers of people over such distances. In order to achieve high speeds these modes of transport will have to use substantially more energy than present-day trains or aircraft, and in that respect will be more damaging to the environment. The environmental implications of supersonic aircraft have already been briefly discussed (5.36–5.37). The relative desirability in environmental terms of conventional jet aircraft and very high speed trains is considered in chapter 12.

6.63 In the past, transport and other forms of communication have tended to grow at much the same rate.[57] In other words they have tended to complement each other rather than substitute for each other. More recently telecommunications has grown much more rapidly than transport, and there has been considerable interest in the possibility that it might provide a way of reducing future demand for mobility. DOT is studying the implications but has not yet reached any conclusions.[58] Office work, shopping and meetings are the three main areas in which it has been suggested telecommunications might eliminate the need for journeys.

6.64 Although working from home over telephone lines ('teleworking') is well established in the computer services industry, past growth has failed to fulfil optimistic predictions.[59] For most people teleworking reduces, rather than eliminates, the need for travel to and from the office. In pilot schemes for government employees in California and the Netherlands about 40% of the working week was spent at home and commuting journeys fell by rather less. There is also the possibility that a reduction in commuting journeys may be offset by an increase in the number or length of journeys made for other purposes. Some people, perhaps a majority, dislike teleworking because it takes away the personal satisfactions and career advantages of face-to-face exchanges with colleagues (although the introduction of videophones may help to overcome such objections). A US survey estimated that about 7% of the working population was teleworking for all or part of the time in 1992 and predicted growth of 20% a year. There seems to be wide agreement, however, that, even if teleworking becomes established on this scale, the net benefit in terms of reduced carbon dioxide emissions is likely to be small.[60] The national transport plan for the Netherlands produced in 1990 saw the potential benefit of teleworking as less congestion with the possibility of a 5% reduction in peak hour car traffic, equivalent to a reduction of a little over 1% in total road traffic.[61]

6.65 Another application of advanced communications is 'teleshopping'. Although this has not so far had any success in the UK, there are local schemes in the USA, and a facility called Caditel on the Minitel videotext system in France.[62] The availability of videophones and virtual reality could make teleshopping more attractive and extend its scope from routine items to discretionary purchases such as clothes. However the attraction of teleshopping for customers and its advantage in reducing energy use are both dependent on provision of an efficient delivery service. As in the case of public transport for passengers, there is a trade-off between the quality of a delivery service from the customer's point of view and its energy efficiency. Customers may find teleshopping unattractive because they value shopping as a social activity, or because they like the entertainments and attractions the largest shopping centres now offer. Teleshopping is a desirable development in principle, especially as a way of reducing the congestion problems associated with the largest shopping centres, but will probably spread quite slowly. More specialised applications, such as telebanking and telephone booking for arts and sporting events and travel, will achieve a much higher penetration, but lead to only a small reduction in the number of journeys made.

6.66 It has been suggested that video links and virtual reality could substantially reduce the need for journeys to business and academic meetings, and even the need for travel to school or college. So far videoconferencing has spread only slowly.[63] If it gathers momentum in the next couple of decades its main effect is likely to be to moderate the rapid growth in air travel. The growth in leisure journeys is unlikely to be reduced by any foreseeable developments in telecommunications. Virtual reality systems will replace travel brochures rather than travel.

6.67 As to the third direction for technological development, considerable effort is being devoted to developing alternative methods of propelling road vehicles of broadly the present types, but there has been little tangible progress so far. We have concluded that some form of combustion engine, similar to those used at present, is likely to remain the dominant form of motive power until at least 2020, and probably much longer. We consider in chapter 8 the prospects for making combustion engines more energy efficient than at present and reducing their emissions of pollutants, and assess the progress so far made in developing electric propulsion and the use of fuel cells. Even the universal use of electric cars would not eliminate environmental problems. The environmental impact of the power stations producing the electricity still has to be considered. Fuel cells will be a more efficient way of using fossil fuels, and will emit minimal amounts of pollutants, but they will produce as much as 60% of the carbon dioxide that would be emitted by a conventional car.[64]

6.68 Another major element in worldwide research programmes is the development of advanced electronic systems for controlling highways and vehicle movement. The aims are to increase the capacity of existing roads, and eventually to increase safety by fitting vehicles with automatic guidance and collision-avoidance systems.[65] This could bring environmental benefits if it removes the need for road building or widening; but could be environmentally damaging if it stimulates a greater growth in vehicle use than would otherwise have occurred.

6.69 Technologies which would make transport more energy-intensive than at present are clearly undesirable in environmental terms. The desirable approach from an environmental point of view is 'technology forcing'. This requires explicit objectives and targets for preventing environmental damage and overcoming present pollution problems, and the translation of these into clear and challenging standards. Given adequate time to meet such standards, industry will have a framework within which to develop sustainable technologies, and the incentive to develop efficient and innovative solutions.

Conclusion

6.70 In this chapter we have described and discussed six different perspectives on the future of transport. Letting congestion find its own level would cause irretrievable long-term damage to both urban and rural areas, and would ultimately have the effect of reinforcing car-dependent lifestyles. The 'predict and provide' approach is now discredited. The other four approaches however all have an important contribution to make in overcoming the problem of reconciling people's desires for mobility with the need to achieve sustainable development. In the next three chapters we look more closely at the nature of these contributions.

6.71 Chapter 9 discusses the dynamic and reciprocal relationship between transport and land use, and the extent to which land use planning policies can influence the demand for personal travel, including changing its nature in ways that will be favourable to public transport.

6.72 Some forms of advanced technology would be more energy-intensive than existing methods of transport and could be environmentally damaging in other ways. However, further developments in technology are of crucial importance in overcoming the environmental problems of transport, and chapter 8 investigates and assesses the technological potential for improving the environmental performance of road vehicles.

6.73 We all have the power to make modifications in our own behaviour, and the cumulative impact of such modifications could be very large. Chapter 8 describes the major benefits that could be achieved for the environment, without any reduction in the amount of personal travel, from changes in driving style, in the preferences of car purchasers, and in the importance which vehicle owners attach to proper maintenance. Although such modifications in behaviour are ultimately the individual's own responsibility, there is action which the government can and should take to encourage them. The extent to which people make journeys on foot or by cycle in future will also depend on the action taken by central government and local authorities to encourage them to do so: that aspect is dealt with in chapter 12.

6.74 The government can influence individual behaviour by means of economic instruments. Road pricing is one option, and is pursued further in chapters 11 (urban road pricing) and 12 (charges for motorway use). There are other possible instruments however. To clarify what the objectives of using economic instruments should be, and which instruments are preferable, chapter 7 looks at all the types of costs associated with transport and at the most equitable and efficient ways of meeting them.

Chapter 7

ECONOMIC ASPECTS OF TRANSPORT

Introduction

7.1 The enormous growth in mobility described in chapter 2 has brought many economic and social benefits to individuals, firms and the community as a whole. The present network of roads, railways, airports and ports, and the consequent ability to move people and goods swiftly, are important national assets. They have been essential for the development of the UK economy and its manufacturing and commercial base, and remain essential for maintaining the UK's competitive position. They have also provided access for very large numbers of people to wider social contacts and to many forms of recreation and leisure in the countryside and elsewhere.

7.2 The costs of the transport system include the costs of the diverse forms of environmental damage described in chapters 3, 4 and 5. It is difficult, if not impossible, to produce a comprehensive measure of the money value of either the costs or the benefits of the present transport system. It is clear, however, that the costs involved are substantial, and that they will become much greater if further growth in road and air traffic occurs on the scale that has been forecast.

7.3 Our aim is to propose a package of measures that will retain the advantages mobility has brought while reducing environmental damage. Any attempt to design such a package by identifying and balancing against each other all the benefits and costs involved would inevitably be incomplete and unsatisfactory. We explain below (7.30–7.31) why the approach we prefer is to establish targets for environmental improvement and the reduction of environmental damage. Although such targets are not defined in terms of costs and benefits, action to achieve them will bring about a closer correspondence between the benefits and the costs of the transport system than would exist if past trends were to continue unchecked. This action must include adjusting the prices paid by users of the transport system, by means of 'economic instruments', so as to provide a better reflection of all the costs imposed by such use.

7.4 The various types of cost involved in use of the transport system are distinguished in box 7A. This chapter focuses initially on road transport as the dominant mode and considers the relationship between the revenue from road taxation and the infrastructure costs incurred by the public sector in providing roads (7.5–7.9). Next, it reviews attempts made to place money values on environmental costs (7.10–7.18); and examines the relationship between the revenue from road taxation and the combination of infrastructure costs and those environmental costs of road transport to which we have attached a money value (7.19). We discuss the factors that affect decisions by transport users (7.20–7.25), and what is likely to be the most effective way of ensuring that environmental costs are taken into account (7.26–7.31). We examine various possibilities for using economic instruments to reflect the environmental costs of road transport (7.32–7.49) and the implications of raising fuel prices (7.50–7.59). We then consider the justification on environmental grounds for using public expenditure to increase the role of public transport (7.60–7.71). We discuss more briefly the scope for using economic instruments to obtain reductions in emissions from aircraft (7.72–7.76). The final section discusses the likely effects on the wider economy of the measures recommended earlier in the chapter (7.77–7.83).

Tax revenue and road infrastructure costs

7.5 Government policy, as stated by the Treasury in evidence to the Commission, is 'to ensure that the taxes paid by road users at least cover the full economic costs of road provision and road use.' The Department of Transport (DOT) publishes an annual table showing the relationship between the revenue from road taxation and the public sector costs of road infrastructure in Britain.[1] The two taxes in question are vehicle excise duty (the annual tax which has to be paid before a vehicle is used on the road) and fuel duty (which can be regarded as a tax on the amount of use made of a vehicle). The practice of publishing this table reflects the historical origins of vehicle excise duty in payments to a statutory fund to finance road improvements.[2]

BOX 7A	ECONOMIC COSTS OF TRANSPORT

In discussing the economic costs of transport, the following terms are employed:

infrastructure costs:
the costs of investment in roads, railways or other transport infrastructure, together with the costs of maintaining and operating such infrastructure;

vehicle costs:
the costs of acquiring and operating the vehicles (cars, lorries, buses, trains, aircraft, ships) using such infrastructure;

public sector costs:
those costs paid by central and local government or other public sector bodies;

external costs (or externalities):
costs which are side-effects of an activity (for example, the costs of damage caused by vehicle emissions), which are not borne by the people responsible for the activity and are not therefore taken into account in their decisions. In the case of transport the main external costs are:

> the costs of damage to the environment caused either by the construction or existence of transport infrastructure or by the use of vehicles;
>
> certain costs of accidents (see definition of 'environmental costs');
>
> the costs of the additional time taken for journeys when congestion occurs.

Although an individual user of the transport system may be said to impose external costs by contributing to congestion or causing an accident, the costs of congestion and a large part of the costs of accidents can be regarded as both imposed and borne by the relevant group of transport users;

environmental costs:
the external costs of transport borne by the wider community as a result of environmental damage or in the form of those costs of accidents (such as injuries to pedestrians or the grief of relatives) which are not borne by transport users either directly or through payment of insurance premiums.

7.6 Table 7.1 shows corresponding information for 1994/95 on a modified basis. The allocation of infrastructure costs between different classes of road user involves uncertainties. DOT allocated 8% of infrastructure costs to pedestrians (for footways, sweeping and street lighting), but none to cyclists. These costs have been allocated to vehicles in table 7.1. Heavy goods vehicles (HGVs), particularly the heaviest goods vehicles, require roads and bridges built to a much higher specification than would be necessary for cars. Some of the structural damage vehicles cause to roads increases as the fourth power of the axle weight.[3] It is questionable whether DOT's allocation of costs makes sufficient allowance for these factors: a study in 1990 suggested that, to cover their appropriate share of infrastructure costs, HGV operators would have to pay at least 30% more on average in road taxation, and almost 60% in the case of the heaviest rigid lorries.[4] The HGV share of costs has therefore been increased by 30%. The revenue from value added tax (VAT) on transport-related expenditure was not included in DOT's calculations on the ground that the same tax is payable on most forms of expenditure, but VAT on fuel duty may be considered a tax on transport and is included in table 7.1. For buses and coaches, fuel duty rebates are deducted from gross fuel duty revenue.

7.7 With these modifications, total revenue from road taxation (£20.4 billion) is projected to be 2.9 times road infrastructure costs (£6.9 billion), the bulk of the revenue (£16.7 billion) coming from fuel duty. The ratios between revenue and costs range from 4.5:1 for cars (including taxis) and goods vehicles under 3.5 tonnes to 1.1:1 for HGVs and 1.0:1 for buses and coaches (reflecting the rebate given on fuel duty).

7.8 There are also various ways in which infrastructure costs can be calculated and expressed. DOT has used for this purpose the public expenditure on roads in the relevant year. There is an implicit assumption that annual expenditure on maintenance and renovation is sufficient to prevent any long-term depreciation in the quality of the road network, and it is not clear whether this is the case. No allowance is included for the value of the land occupied by roads. DOT is now considering the possibility of an

Table 7.1
Tax revenue from road users in relation to infrastructure costs (1994/95 projected), by vehicle class

£ million

		revenue from road taxation[a]					
	number of vehicles (thousand)	fuel duty	vehicle excise duty	total	infrastructure costs	tax revenue less costs	ratio of tax revenue to costs
cars and light goods vehicles[b]	24,560	13,240	3,360	16,600	3,697	12,900	4.5: 1
motorcycles	690	60	20	80	20	60	4.4: 1
buses and coaches	80	210	30	240	240	0	1.0: 1
heavy goods vehicles	430	2,610	530	3,140	2,830	310	1.1: 1
other vehicles	1,310	360	20	380	140	240	2.6: 1
all vehicles	27,060	16,660	3,950	20,420	6,930	13,590	2.9: 1

This table differs from the annual tables published by DOT in the following respects:

a. revenue from fuel duty includes £2.5 billion VAT payable on the amount paid in fuel duty. VAT payable on other transport-related expenditure is not included.
b. for buses and coaches net revenue from fuel duty appears in the table, after deducting the £190 million rebate paid to bus operators.
c. no deduction has been made from infrastructure costs for the costs (£480 million in 1994/95) which DOT allocates to pedestrians.
d. the infrastructure costs allocated to heavy goods vehicles have been increased by 30% to reflect higher costs of structural damage to roads by HGVs than estimated by DOT, with a corresponding reduction in the infrastructure costs allocated to other classes of vehicle.

alternative approach favoured by the Treasury and recommended by some commentators, in which the road system would be valued as a capital asset and the revenue from road taxation would be required both to cover depreciation and to yield a rate of return on that asset. This is similar to the approach that has been required of Railtrack in relation to railway infrastructure costs. It is likely to produce a considerably higher figure for the capital costs of road infrastructure than has been used hitherto: DOT considers that the contribution from revenue needed to cover capital costs under this approach might be twice as large as average capital expenditure in recent years (£6.3 billion instead of £3.2 billion a year at 1994/95 prices).[5] The effect would be to increase the total infrastructure costs shown in table 7.1 from £6.9 billion to £10 billion. Another estimate, based on an assumed asset value of £100 billion, is that annual interest charges would be £7.2 billion, implying total infrastructure costs of £10.9 billion.[6]

7.9 A comparison between infrastructure costs, irrespective of how these are calculated, and revenue from road taxation does not provide a basis for formulating a sustainable transport policy because it does not take into account all the relevant factors. The next two sections consider the money values of the environmental damage caused by transport and the ways in which people take decisions about their use of the transport system.

Putting a money value on environmental costs

7.10 We have reviewed various attempts to place a money value on the environmental costs of transport. There seems to be agreement that in industrialised countries generally such costs are equivalent to several per cent of gross domestic product (GDP)[7] but estimates are subject to very large uncertainty. There are also some forms of environmental damage to which it is not at present possible, and may never be possible, to assign money values.

7.11 The more common techniques for placing a money value on environmental damage are described in appendix C. There are a number of difficulties in using such techniques, which result in widely differing valuations of similar effects:

a. people differ in their perception of, and relative valuation of, particular environmental gains and losses;

b. there is limited scientific information about the effects the transport system has on the environment and on human health, especially in the case of some pollutants;

c. knowledge changes over time. For example, estimates of the external costs of road transport have risen substantially as increased understanding of global warming caused by the greenhouse effect has led to its inclusion in estimates of environmental damage;

d. the complexities of dispersal pathways and chemical reactions make it difficult even to apportion known effects of pollutants between known sources. One example is the proportion of crop damage from tropospheric ozone that is attributable to transport. Another relates to acid deposition: although emissions of nitrogen oxides from transport make a significant contribution, the amount of damage caused to fisheries and forests depends on local factors such as geology and rainfall.

7.12 Prominent among the effects to which it is hard to assign any money value are those which affect sustainability or the interests of future generations. Loss of land and access to the countryside, visual intrusion, community severance, and loss or disruption of habitats are significant effects of both existing and new transport infrastructure, and some of these effects extend well beyond the area of land which the infrastructure occupies. Land taken for transport infrastructure rarely reverts to any other use, although some of the structures built have themselves eventually justified conservation (for example, some railway viaducts and canals). We set out in chapter 4 the general principles we believe should be applied in considering any proposal to use land for transport infrastructure. Compliance with those principles will add to the cost of transport projects (for example, through expenditure on adopting less damaging alternatives or on habitat creation or landscape restoration) and will therefore feature directly in the infrastructure costs of new projects. Even in cases where our recommended principles are applied, however, some environmental costs will be imposed by the taking of land for transport infrastructure. The value of the best alternative use to which the land could be put, generally indicated by its market price, reflects the value of the land to its owner, but does not reflect the important amenity, cultural and nature conservation dimensions of the value it has for the whole community.

7.13 One view is that the techniques summarised in appendix C should be employed, in spite of the practical difficulties, to provide a best estimate in money terms of the environmental costs of taking land for transport infrastructure. That may be appropriate in certain circumstances. But putting money values on irreversible loss of habitat, degradation of landscape or destruction of cultural assets may be objectionable on grounds of principle as well as practicality. The protection of some environmental assets for future generations is seen by many as a moral obligation and it is often not appropriate to trade such assets off against other things like the benefits of mobility.

7.14 Despite the resulting incomplete coverage, there is nonetheless a good argument in many circumstances for monetary valuation where possible: measures to reduce environmental damage use resources that could be used for other purposes, and decisions on the use of resources will generally be helped by whatever quantification of information is possible. It is important, however, that those effects on which it is not possible to put a money value are also fully recognised.

7.15 The concept of 'environmental costs' used here does not include congestion costs, which were discussed in chapter 6. Although they are sometimes included in estimates of the social or external costs of road transport, and given a very high money value[8], congestion costs (as explained in box 7A) are not external to road users as a group.[9] Estimates of the money value of environmental damage already cover in principle the additional damage brought about when congested conditions increase vehicle emissions and nuisance from noise and vibration.

7.16 Table 7.2 contains estimates of money values at 1994/95 prices for certain environmental costs of the transport system in Britain. These estimates are based on a number of studies, described in appendix C. They are expressed as ranges so as to attempt to take into account the large uncertainties associated with them. For the forms of environmental damage shown in table 7.2 they come out at between £5 billion and £15 billion a year in total. We regard the environmental costs of the transport system as also including those costs of accidents which are not borne by users of the system (see box 7A); these are estimated to be £5.5 billion a year. In all, therefore, estimates of those environmental costs of the transport system on which we have placed money values come out at between £11 billion and £21 billion a year, the equivalent of between 2% and 3½% of GDP.

7.17 The money values given in table 7.2 all relate to costs imposed by use of the transport system. It is essential that any overall assessment also recognises the forms of environmental damage caused by the construction of transport infrastructure, even though we have not attempted to place money values on them. Any allowance made for them in money terms must inevitably be notional, but would certainly raise the environmental costs of the transport system to levels well above the estimates in the table.

7.18 Table 7.2 also contains (in parentheses) estimates of the environmental costs of road transport. It has been assumed that road transport gives rise to almost all the accident costs included in these calculations and that its share of the costs of environmental damage is proportional to its share of transport-related carbon dioxide emissions (3.58). On that basis, the quantified environmental costs of road transport are estimated to be between £10 billion and £18 billion or between nearly 2% and over 3% of GDP. This is broadly compatible with another recent estimate, which was £9 billion to £12 billion (for 1991)[10], although the make-up of our estimate is rather different.

7.19 Table 7.3 shows the relationship between tax revenue from road users and the combined total of infrastructure costs and those environmental costs to which we have attached a money value. For road vehicles as a whole, tax revenue exceeds this combined total of costs at the lower end of the estimated range of environmental costs, but falls short of it at the upper end of the range. For cars and light goods vehicles, tax revenue is equal to total costs if environmental costs are near the upper end of the estimated range. For HGVs on the other hand, tax revenue falls short of the combined total of costs even if environmental costs are at the lower end of the estimated range. In considering the implications of these relationships, two points must be emphasised. First, the tables do not include some important environmental costs for which it has not been possible to estimate a money value, particularly those imposed by the construction of transport infrastructure. Second, in any revised treatment of infrastructure costs by DOT (7.8) they are likely to emerge as larger than shown in table 7.1.

Table 7.2
Estimates of environmental costs of the transport system (1994/95)

£ billion a year

costs attributable to road transport are shown in parentheses	lower end of range		upper end of range	
air pollution	2.4		6.0	
climate change	1.8	(4.6)	3.6	(12.9)
noise and vibration	1.2		5.4	
accidents	5.5	(5.4)	5.5	(5.4)
total quantified environmental costs	**10.9**	**(10.0)**	**20.5**	**(18.3)**

among environmental costs for which it has not been possible to estimate a money value and which are not therefore included above are:

- losses of land
- loss of access to land
- visual intrusion
- severance of communities
- loss or disruption of habitats

Table 7.3
Tax revenue from road users in relation to quantified environmental and public costs, by vehicle class (1994/95)

£ billion

	cars and light goods vehicles	heavy goods vehicles	all vehicles
infrastructure costs[a]	3.7	2.8	6.9
external costs of accidents[b]	3.7	0.9	5.4
other quantified environmental costs[c]	3.5 - 9.6	0.9 - 2.7	4.6 - 12.9
total quantified costs	10.9 - 17.0	4.6 - 6.4	16.9 - 25.2
revenue[a]	16.6	3.1	20.4
revenue as percentage of quantified costs	152% - 98%	68% - 49%	121% - 81%

a - from table 7.1.
b - from table 7.2. The external costs of accidents have been allocated between vehicle classes in proportion to the involvement of each type of vehicle.
c - from table 7.2. HGVs have been allocated 30% of noise and vibration costs; in other respects environmental costs have been allocated between vehicle classes in proportion to fuel used.

Figure 7-I
What it costs to run a car

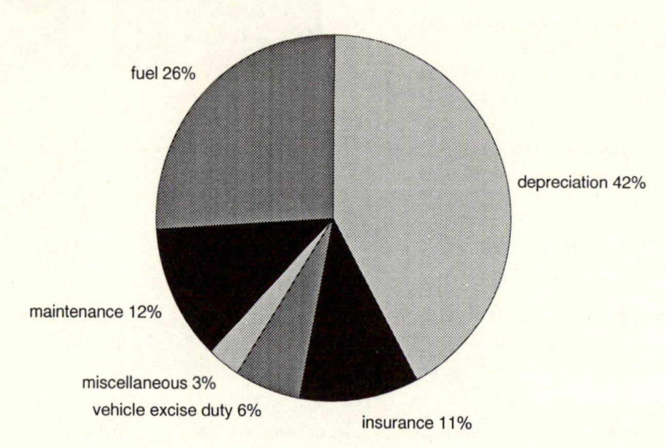

Factors affecting decisions by transport users

7.20 About half the population, and two-thirds of households, now has access to a car. Of the costs faced by car owners almost three-quarters are incurred because of their initial decision to acquire and operate a vehicle, as figure 7-I[11] shows. Almost all of those costs are vehicle costs, although one small element, vehicle excise duty, could be regarded as a payment towards infrastructure costs. When car owners consider whether to make a particular journey, and whether to make it by road, the costs they have already incurred are not relevant. If they are taken into account at all, they may even be perceived as a reason to make maximum use of the car. Moreover, owners often do not in practice take account of all their other costs; the cost of additional wear and tear on a vehicle as the result of an additional journey is likely to be small and may generally be overlooked, as may any impact of extra mileage on the date of the next servicing or the resale value. The amount of fuel needed for a long journey will usually be taken into account, but in the case of shorter journeys even this cost may be disregarded. Car owners complete many journeys without making any actual payments, unless they have to pay a toll or parking charge. A substantial proportion of drivers uses company cars. Because of the present tax rules, they are even less likely to take either fixed or variable costs into account in their decisions, and in some circumstances may have a financial incentive to make additional journeys, as explained in box 7B.

7.21 Freight hauliers and other commercial vehicle operators have to take into account both fuel costs and other vehicle costs associated with additional mileage. Drivers' wages are another significant element in their calculations. However, commercial operators likewise incur substantial costs as a result of their initial decision to acquire and operate a vehicle. In their case, these costs already incurred certainly constitute an incentive to use the vehicle for as many journeys as possible, in order to maximise their return on capital.

7.22 In contrast to private road users, users of rail, bus and air transport typically make a single payment for the service provided, which has to cover both vehicle costs and infrastructure costs. Standard fares and freight charges are set to cover average total costs, including both fixed and variable costs.

7.23 A fare fixed to recover a share of total costs will be considerably higher than the marginal cost to the operator of carrying an additional passenger (until the point is reached at which an additional bus or train has to be run). Public transport operators offer discounted fares alongside standard fares in order to achieve a higher occupancy on their services and maximise their revenue. They also offer season tickets and travelcards, which in certain circumstances reduce the cost of an additional journey to zero. Total fare revenue must still cover all their costs, including any profit margin sought (but less any contribution from public funds, an aspect discussed later in this chapter).

7.24 Public and private transport differ in attractiveness and convenience, and that aspect is discussed elsewhere in the report; but the differing bases on which users pay towards the costs of public and private transport have an important distorting effect on decisions about which mode to use. The relative costs of additional journeys therefore typically favour car use against public transport, even if the total costs of using a car are at least as large as the costs of using public transport. A smaller proportion of journeys will therefore be made by public transport than if decisions were based on the actual additional resource costs of alternative modes of travel.

7.25 To the extent that users of the transport system are unaware of the environmental and other external costs imposed by their journeys or movements of goods, their private decisions will be even further removed from what would have been desirable on environmental grounds. More journeys and movements will be undertaken than would be the case if individuals and firms had to take full account of the additional environmental costs involved; and, because public transport is less environmentally damaging than private transport in many circumstances, those journeys and movements will harm the environment more than would have been the case if those making them had been faced only with the full additional costs of using different modes of transport.

How environmental costs can best be taken into account in decisions

7.26 A journey is likely to be undertaken (or goods moved) if the benefit to the person making the decision is not less than the perceived additional cost. A method of ensuring that decisions of this nature represent an appropriate balancing of costs and benefits from the viewpoint of the community would be to require individuals or firms to pay an additional charge for each journey (or movement of goods), at a rate which reflects the environmental costs that journey or movement imposes on the community; and then leave those individuals or firms to decide for themselves on their preferred level of travel (or movement of goods).

7.27 In principle, such decisions should also take into account external benefits which journeys (or movements of goods) have for the rest of the community. In practice, however, nearly all the benefits of private road use are gained directly by users. Some of those benefits are redistributed to other firms or individuals through market processes; and other benefits may spill over to other firms or individuals in a way which is common with all kinds of market transactions and non-market activities[12], and could not appropriately be represented in any form of payment. There are few, if any, true external benefits from private road use to set against the external costs.[13]

7.28 The more directly and conspicuously an additional charge of the kind envisaged above bears upon decisions by individuals or firms to make a journey (or move goods), and if so by which mode, the more effective it will be in influencing behaviour. Moreover, the extent of environmental damage from individual journeys or movements varies with location, time of day, type of road, and other factors. In the long run, electronic systems may well make it possible to levy a charge on an individual vehicle as it makes a particular journey, at a rate which reflects the relevant circumstances, as well as the nature of the vehicle itself. Preferably it should also be possible to ascertain in advance what charge will be made. We believe the possibility of this form of electronic pricing is worth pursuing; but, as large-scale systems with the necessary degree of sophistication will not be practicable for some time, we discuss in the next section what other forms an additional charge might take.

7.29 There are considerable difficulties, as was shown above, in assigning money values to all the environmental costs of transport. The more that disaggregation is attempted, the greater the difficulties become. It will not usually be possible to identify the additional environmental effects of a particular journey. Even if that were possible, it would be impracticable to determine uniquely what level of additional charge would be appropriate for that journey. Charging must therefore inevitably involve a degree of averaging.

7.30 In view of these difficulties it seems preferable to adopt a pragmatic approach to transport policy. Rather than attempting to balance costs and benefits at the margin, this approach is based on setting targets for environmental improvement and the reduction of environmental damage, on the lines of those already proposed in chapters 3 and 4. Such targets stem from a recognition that the present and prospective use of transport is creating environmental damage which is in the long term unacceptable. The targets proposed have been formulated in the light of the information at present available about the environmental costs of the transport system, and will need revision in the light of experience. We also consider it appropriate to set targets for a shift from private to public transport, on the ground that use of public transport is likely to be less damaging to the environment; and we discuss that aspect in chapters 10–12.

7.31 Although the targets proposed elsewhere in this report are not defined in terms of costs and benefits, action taken to achieve them is likely to bring about a situation in which the benefits from the use of the transport system correspond more closely with the costs of such use. It is also important to find efficient and cost-effective methods for achieving such targets. A number of aspects of the environmental effects of transport are subject to direct regulation; and we consider later (in particular, in chapter 8 for vehicle emissions and noise, in chapter 10 for road haulage, in chapter 11 for traffic in urban areas, and in chapter 12 for speed limits outside urban areas) whether such regulation needs to be made more stringent. It is desirable, however, to use the price mechanism as the method of achieving targets wherever possible. Although economic instruments utilising the price mechanism are not a complete alternative to direct regulation, they tend to be more efficient for reasons which have been analysed in the Commission's Sixteenth Report (paragraphs 8.1–8.17). They should make it possible to minimise the cost to the community of achieving a particular target because they provide incentives to users or manufacturers to modify their behaviour in a particular direction but leave them with the flexibility to decide which modifications will be most cost-effective. If the structure of prices is related to environmental objectives and targets, it will provide continuous incentives for behaviour which is compatible with them, and thus reinforce the effectiveness of any other instruments used.

Using economic instruments

7.32 In looking at the scope for using economic instruments to achieve environmental objectives in relation to transport we look first at how they might influence decisions about acquisition of a car and then at how they might influence decisions about whether, and how, journeys (and movements of goods) should be made. The two aspects are discussed separately here in order to simplify the exposition; in practice the decision whether to acquire a car will be influenced by expectations about the cost of using it.

Instruments affecting car ownership

7.33 Some countries with very restricted land areas, such as Hong Kong and Singapore, have for some years used economic instruments to deter people from buying and operating cars. In Singapore a predetermined number of permits for car ownership is auctioned each year, and there is also a high annual tax on cars. Hong Kong also levies a high annual tax, but that has not been sufficient to prevent a recent growth of 10% a year in the number of cars.[14] A car sales tax has been applied in the UK in the past, but not with the avowed intention of limiting car ownership.

7.34 An alternative approach, used in Tokyo, is to require everyone applying for registration of a car to show that there is an off-street parking space available for it. While this is not strictly speaking an economic instrument, one effect is to drive up the cost of off-street parking spaces and increase the cost of operating a car very considerably (by £2,500-£5,000 a year at present).[15]

7.35 Increasing the level of vehicle excise duty in the UK, or reintroducing a car sales tax, would slow the spread of car ownership, which has been one of the main causes of the growth in road traffic and is the main factor in DOT's forecasts of future traffic growth (see box 2B). However, it would be inequitable to impose high barriers against the acquisition of cars by families who have not been able

BOX 7B **COMPANY CARS**

About an eighth of all cars in the UK (almost 3 million) are company cars. In 1988 the then Chancellor of the Exchequer described company cars as 'far and away the most widespread benefit in kind' and estimated that 'an employee with a typical car may be taxed on only about a quarter of its true value'. In 1992 it was estimated that the tax yield on company cars was about £2 billion a year less than if the value of the benefits involved had been paid to employees in cash. Following increases in scale charges made in subsequent budgets it is likely that the corresponding figure now would be of the order of £1.5 billion.

Before April 1994 the tax levied on an employee who had a company car was a fixed charge based on bands of price and engine size. The benefit is now assessed for tax purposes as 35% of the manufacturer's list price. A stated objective of the changes was to eliminate detrimental effects on fuel economy. The change was presented by the Inland Revenue as having a neutral overall effect on revenue. There is as yet little evidence that employees faced with an increased tax charge are trading down.

The benefit on which tax is assessed is reduced by one-third if business mileage exceeds 2,500 miles a year, and by two-thirds if it exceeds 18,000 miles a year. This feature of the tax rules provides an incentive for some drivers to make journeys unnecessarily by car in order to exceed a threshold.

Over two-fifths of employees with company cars also receive free fuel for some or all of their commuting and private use. The tax charge for this benefit is a fixed sum determined by engine size and whether the car uses petrol or diesel. It is not affected by the use made of the car.

In the 1989/91 National Travel Survey two-thirds of the much higher annual mileage of company cars (19,900 miles against 8,500 miles for privately owned cars) was for private use. About 6% of company cars did no business mileage.

to afford them until now. Increasing taxes on car ownership would not affect the decisions those with access to cars make about additional journeys (7.20).

7.36 It might be argued that increasing vehicle excise duty would be an appropriate way of requiring road users to make a payment to cover external costs associated with the presence of road infrastructure, and associated public expenditure that is not dependent upon the level of use. However, it seems preferable to recover at least the greater part of those costs through payments that bear on use rather than ownership, because wear and tear to infrastructure depends on the extent to which a given vehicle is used and it is congestion which stimulates proposals for road building and widening.

7.37 There is nevertheless a case for retaining an annual tax corresponding to vehicle excise duty. Some characteristics of vehicles are important in determining the amount of damage they cause. For example the extent of damage to road surfaces is very dependent on the weight of a vehicle. This factor would not be sufficiently reflected in differences in fuel use, but is reflected in principle in the steep graduation in vehicle excise duty for goods vehicles. There are other options for graduating or varying vehicle excise duty in order to influence behaviour in environmentally beneficial ways. We discuss the potential for that in chapters 8 and 10.

7.38 The income tax rules for 'company cars' provided by an employer for an employee or director make it attractive to take a car in preference to other forms of remuneration. They also have a particularly strong influence on the characteristics of new cars sold in the UK: more than half of new cars sold are registered in the name of an employer, compared with only 37% in 1975, and a further proportion are provided by an employer but registered in the employee's name. The tax rules also give the drivers of company cars incentives to make more use of their cars and (in conjunction with subsidised parking) to commute by car rather than by public transport. The effects of the tax rules are described in box 7B.[16]

7.39 Although the Treasury claimed in evidence that the new rules introduced in April 1994 will contribute significantly to reducing distortions as between company cars and cash remuneration, we have concluded that they continue to distort decisions about both the acquisition of cars and their use. They do not take the form of assessing and taxing the private benefit an employee obtains from a company car. At present only a small proportion of employers offer a cash alternative and, where one is offered, its cost to the employer is usually less than the cost of a car. **We recommend further modifications to the tax rules, including abolition of the mileage thresholds and taxation of the actual value of benefits in the form of fuel, to remove the incentives for environmentally damaging behaviour.**

Instruments affecting vehicle use

7.40 Although it was concluded above that attempting to balance costs and benefits for road users at the margin was not a practicable or appropriate approach to transport policy, charges related to use of a vehicle have the general advantage that the amounts paid will bear a relationship to the environmental costs imposed by journeys or movements. The main forms such a charge might take are:

a. charges related to use of road space ('road pricing')
b. charges related to occupation of parking space at the destination of a journey
c. charges related to distance travelled
d. charges related to pollutants emitted
e. charges related to fuel used.

These options are not mutually exclusive, and in some cases overlap; for example, both the quantities of pollutants emitted and the distance travelled are related to the amount of fuel used. Some of them would be technically very complex to implement. However, recent and foreseeable advances in electronics may tip the balance of advantage away from the administratively simpler schemes and towards approaches in which the charges made would be more directly related to the environmental damage caused by a vehicle.

7.41 **Road pricing** has been discussed already (6.55–6.60). Its great attraction is that it can be applied on a local basis. The rates of charge can be readily adjusted to reflect differences in the scale of external costs in different areas, and as between different days of the week or times of day. With a local system it is also easier to assess whether the benefits justify the costs of installing equipment. It would nevertheless be desirable to establish technical standards for road pricing equipment which are at least EC-wide, in order to minimise the problems created by out-of-area cars and lorries. Road pricing has so far been contemplated for areas or stretches of road where there is heavy congestion. In principle, charges could be set not only by reference to congestion costs but also by reference to the environmental costs of vehicle movements made in a particular area at a particular time.[17] Road pricing could therefore be used to supplement a general charge such as fuel duty by giving a broad indication of particularly high environmental costs in certain areas (mainly, but not exclusively, urban areas). The drawback of road pricing is that environmental damage may be caused if the effect is to divert traffic onto nearby roads or into adjoining areas. The case for introducing road pricing in urban areas is discussed in chapter 11 and the case for tolling of motorways in chapter 12.

7.42 Another, and much simpler, form of charging which can vary according to the location and time of day in charges for **parking**. Charges for public parking on and off-street have been used since the 1960s to influence both the total level of traffic in urban areas and its composition (in particular, by favouring either shoppers or commuters). In principle, charges for parking could also be used to reflect external costs or to achieve environmental targets. As with road pricing, the use of increased parking charges in this way should be a supplementary measure in areas where external costs are particularly high. We discuss parking policies in chapter 11.

7.43 Charges related to **distance travelled** have been used hitherto only for HGVs, for which they can be varied to reflect the weight or the capacity of an individual vehicle. Sweden charges a tax on tonne-kilometres which currently recovers around 50% of HGVs' marginal social costs; the tax is easy to levy because HGV speed, distance and driving time are already recorded by tachograph. The European Commission is working on proposals for additional electronic data-processing equipment to be mandatory on new HGVs from 1997. Using smart cards, this has the potential to calculate tonne-kilometre or capacity tonne-kilometre charges. Charging by capacity would give operators the incentive to reduce empty and part-load running. Consolidation of loads and matching of return loads would be encouraged and the commercial balance would tilt back towards smaller vehicles. The competitive position of waterborne and rail freight would obviously improve. Total demand should fall, or at the least be restrained, because it would become less attractive to service the whole country from a handful of depots. The effect would be felt most keenly on long trunk hauls. For cars on the other hand a charge related to distance travelled would be more complicated administratively and technically than a charge related to fuel used, and would also be less directly related to environmental damage caused.

7.44 The German Council of Environmental Advisors has put forward proposals for a system of charges that would be directly related to quantities of **pollutants** emitted.[18] Data on the use of a car during the year would be stored in an electronic engine management system, read out as part of the annual test on emissions and passed on to the tax authorities. We see merit in this approach for the longer term. The Council recognises that it will take a considerable time to develop the necessary equipment, pass the EC legislation which will be necessary to make it compulsory, and fit it to a sufficient number of vehicles. For the time being therefore it will be necessary to use other approaches.

7.45 A duty is already charged on **fuel** for road vehicles. It will be apparent from the analysis above (table 7.3) that the revenue from this duty at its present level is not large enough to cover both that part of infrastructure costs not covered by revenue from vehicle excise duty (table 7.1) and the environmental costs of road transport (the quantified costs shown in table 7.2 and the other costs referred to in 7.19). For HGVs the revenue falls well short of covering even the quantified environmental costs; and that would also be the case for other categories of vehicles if quantified environmental costs are towards the upper end of the ranges we have given.

7.46 Fuel duty has a number of advantages as an economic instrument for influencing decisions about additional journeys:

the amount of tax paid varies with the environmental costs:
the amount of fuel used and duty paid is in the main proportional to the amount of carbon dioxide emitted, and (for any given vehicle) is closely reflected in the quantities of other substances emitted. Fuel consumption is substantially higher in congested urban traffic, and is therefore correlated to some degree with situations in which a vehicle is contributing to higher concentrations of pollutants, there is a higher exposure to the noise and vibration it is producing, and it is more likely to be involved in an accident. (Nevertheless, using the rate of fuel duty to reflect environmental costs implies a greater degree of averaging between localities and times of day. On the assumption that additional environmental costs were on average covered, road users in urban areas would probably be paying less than the amount of the environmental costs they impose and road users in rural areas would be paying more.)

It is simple to administer:
it costs little to collect, is difficult to avoid or evade, and can easily be modified.

Road users have discretion about how to respond:
road users may respond either by reducing the number or length of their journeys or by reducing their use of fuel in other ways, such as switching to a smaller or more fuel-efficient vehicle or driving in a more fuel-efficient way.

7.47 There has been much discussion of the case for a 'carbon tax' applied at a uniform rate to all sectors of the economy and based on quantities of **carbon dioxide** emitted from the combustion of fuel. Because fuel for road vehicles is much more heavily taxed at present than other forms of fuel, a general carbon tax would be at a lower rate than the present rate of fuel duty.[19] As it would be related solely to carbon dioxide emissions, it would not reflect the other major environmental costs imposed by the transport system. Moreover we have previously concluded (3.76–3.79) that it would be right to set a separate target for reductions in carbon dioxide emissions from the transport sector, and this entails that measures specific to the transport sector are likely to be necessary in order to achieve such a target. If a carbon tax is introduced across the whole economy, therefore, it will not remove the environmental case for levying a duty separately on fuel for road vehicles.

7.48 It is possible to vary the rate of fuel duty to provide an incentive to use environmentally less damaging forms of fuel, as in the existing small differentials in favour of diesel and unleaded petrol.[20] We discuss in chapter 8 whether petrol or diesel engines should be regarded as preferable on environmental grounds. The price differential in favour of unleaded petrol has been an important instrument for reducing the exposure of the population to lead in the atmosphere (which was considered in the

Commission's Ninth Report), and this has been especially beneficial in the case of children. About half the petrol now sold is unleaded; further increases in market share will be largely determined by the rate at which old cars are scrapped. Nevertheless, it is important that there should continue to be some difference in price in order to prevent any attrition of the present market share. **We recommend that the differential duty which favours unleaded petrol should be retained.**

7.49 The potential convenience of collecting an additional payment when fuel is purchased has led to proposals for switching charges at present levied in other ways to a 'pay-at-the-pump' basis. This has been advocated in the USA as a way of overcoming widespread evasion of legal requirements about insurance by collecting premiums as a levy on fuel prices. The resulting increase in the price paid for fuel would be particularly important as an incentive to fuel efficiency in a country in which there has been successful political opposition to anything more than a very modest tax-induced increase. The equivalent of the revenue raised in the UK from vehicle excise duty could also be recovered as an additional tax on fuel. One objection in the past has been that annual payment of vehicle excise duty is an essential component in enforcing legal requirements for third party insurance and MOT tests. However, technology is now advancing to the point where a pay-at-the-pump scheme could achieve a higher level of enforcement as well as ensuring that a higher proportion of the costs of using a vehicle would be variable costs. Box 7C provides an example of such a scheme. **We recommend that the government study the possibility of a pay-at-the-pump scheme.**

The effects of raising fuel prices

7.50 As a charge related to fuel used appears to be the most advantageous form of economic instrument for general application in present circumstances, we have considered what increase in fuel duty is justifiable on environmental grounds, and what the effects of such an increase will be. The price road users pay for fuel reflects both the rate of fuel duty and the market price of oil. The retail price of fuel is now lower than it has been for much of the last 40 years (figure 7-II). Although oil is a finite resource, there is no sign that, even with an economic recovery, market forces will lead to substantially higher prices in the medium term. Consumption has greatly increased over the last 20 years, but the ratios of known reserves of oil and gas to present production rates have also increased. In 1991 the world's known reserves of oil (990 billion barrels) were equivalent to 46 years' production; and it was estimated that there were in addition undiscovered reserves equivalent to between 13 and 44 years' production.[21] It is therefore unlikely that there will be a substantial and lasting increase in the market price of oil in real terms before 2020, at the earliest. This is also the view expressed to us by the oil industry[22], although some people have predicted temporary shortages before 2000 because of lack of investment.[23] A DOT forecast (also shown in figure 7-II) is that market factors will cause the retail price of fuel to increase by 20% in real terms between 1990 and 2000 and by 35% in real terms between 1990 and 2025.[24]

7.51 European Community (EC) legislation[25] sets minimum levels for fuel duty, but these are too low to have a practical effect on the duty charged by member states on fuel for road vehicles. A study carried out for the European Federation for Transport and the Environment estimated that most European countries would have to more than double taxes on petrol and diesel in order to reflect the external costs of road transport.[26] Proposals have been put forward in Germany for large increases in fuel prices: the Environment Minister has suggested a staged doubling in real terms; a Parliamentary Commission has proposed annual increases of 10% of 1994 pump prices; and the Council of Environmental Advisors has said that a fuel price of DM2.28 per litre (equivalent to about £1 per litre) would be desirable in 2005 (but only on the basis that parallel moves are made in other EC countries).[27] Table 7.4[28] shows fuel prices in December 1993 in a number of EC countries.

7.52 The UK government has adopted increases in fuel duty as the principal measure for limiting increases in carbon dioxide emissions from road transport. It increased fuel duty by 10% in March 1993 and by a further 3p a litre (about 10%) in November 1993, and is committed to further year-by-year increases in duty of 5% in real terms.

7.53 DOT has modelled the effect of tax-induced price rises on carbon dioxide emissions from road transport.[29] Its estimates imply that a 10% increase in the price of fuel in real terms would lead to a fall in fuel use of up to 3%, of which half would be the result of reduced vehicle use. The conclusion reached

BOX 7C **A POSSIBLE PAY-AT-THE-PUMP SCHEME**

It is desirable to separate the costs of owning and of operating a road vehicle: though both are related to the distance travelled, it is operating the vehicle that wears roads and produces pollution.

A vehicle can be purchased outright, perhaps by taking out a loan, or by instalments over a period, perhaps with an option to purchase. In any case owners are in general well aware of purchase costs and of resale values in relation to mileage.

The costs of operating a vehicle (see figure 7-I) are made up of fixed annual costs (insurance, vehicle excise duty, MOT test fee, etc), sporadic costs (repair and maintenance) and fuel costs. The result is that the annual average cost per mile decreases as the annual mileage increases and is frequently perceived as merely the cost of fuel (7.20). If vehicle operators are to make informed choices as to the relative merits of private and public transport, and if environmental costs are to be related to vehicle use (the polluter pays) it seems desirable that what are now fixed operating costs be related to road use and to the pollutants emitted. A method for doing this can probably be developed using existing technology, by adding an appropriate levy to fuel prices.

Vehicles would continue to be registered centrally, the fee covering only the administration involved. As well as a registration document and a conventional number plate each vehicle would be issued with a smart card or, preferably, fitted with a transponder, that could be interrogated at a fuelling station to identify the vehicle and its characteristics.

Inserting the card, or inserting the fuel nozzle if a transponder were used, would result in the display of the total cost associated with a litre of fuel. The levy on the fuel price would vary from vehicle to vehicle, depending on such things as axle weight and emission levels, supplemented by the cost of fuel consumption-related obligatory, and optional comprehensive, insurance. Such a pay-at-the-pump system would transfer most of what are at present fixed annual costs to an immediately apprehensible cost per litre which the vehicle owner could convert to an approximate cost per mile.

An additional advantage of such a scheme would be as a control of legislative requirements. A stolen vehicle, an unregistered vehicle, or a vehicle that had not passed the annual MOT/emissions test, would be identified at the fuelling station and appropriate action taken. If a transponder system were used it would be possible to identify rapidly vehicles exceeding speed limits and committing similar traffic offences as well as signalling entry to and exit from areas designated in any future road pricing scheme.

As with all computer-based financial systems much care would be needed to prevent fraud: in this connection it would be of great value if a tamper-proof odometer were obligatory on all vehicles. The co-operation of insurance companies would be essential: they may find it convenient to link such a pay-at-the-pump scheme to a credit card system. That the probability of a particular driver being involved in an accident increases with the use of their vehicle is likely to be accepted but as the risk of theft is less closely linked with use an annual premium may be a more appropriate way of covering it.

However, a pay-at-the-pump scheme would probably be worth developing only if comprehensive insurance were included and billed to individual policyholders. This would necessitate the unique identification of the vehicle concerned. There could be objectors to the scheme on the grounds that it invaded their privacy. In their case the levy on the fuel price would not include any component for insurance cover, which would be arranged directly with an insurance company. At a fuelling station the vehicle involved would be identified but its presence would not be recorded if all legal requirements had been met.

was that, in order to limit carbon dioxide emissions from road transport to 1990 levels by this means alone, the price of fuel would have to increase by 80% in real terms by 2000, and by 250% by 2025.

7.54 The long-term response to what are perceived as permanent changes in fuel prices may be significantly greater than DOT assumed. Over a longer period users have more scope to change their patterns of travel and choose more fuel-efficient vehicles and driving styles, and manufacturers have more scope to develop and market more fuel-efficient vehicles. On the basis of a review of 120 studies in the UK and elsewhere, Goodwin[30] suggests that in the long run a 10% increase in fuel prices might result in a 7% fall in the total amount of petrol used; and that reduced car ownership and reduced use of cars might account for about half the fall. This is not very different from the conclusions of another

Figure 7-II
Retail price of fuel in real terms 1950-2025

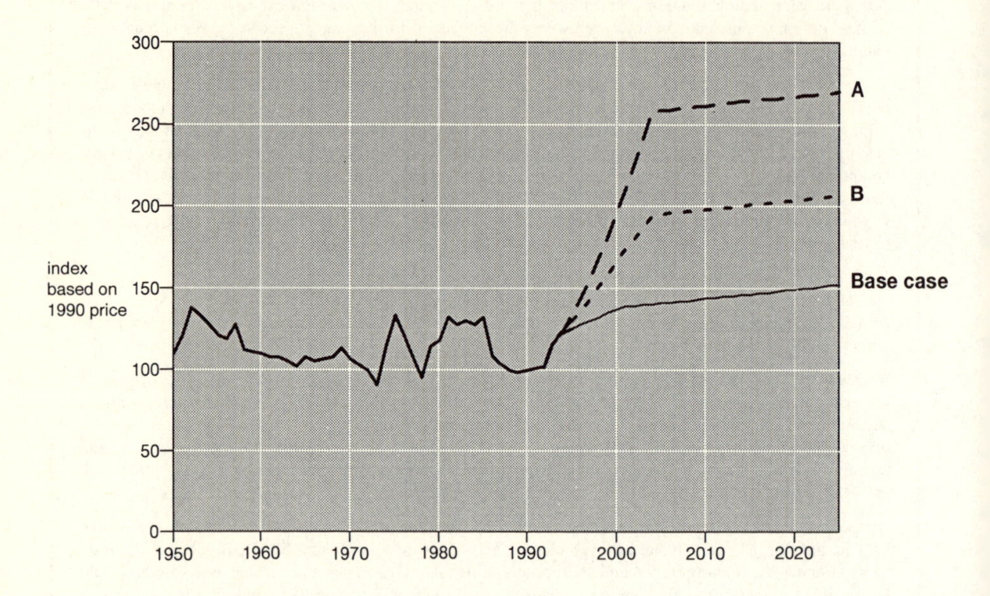

Base case is DOT forecast of market-induced price increases
A: additional effect of increasing duty by 9% a year 1995-2004
B: additional effect of increasing duty by 5% a year 1995-2004

review[31], which suggested that in the long term a 10% increase in fuel prices might result in a fall of 2% to 3% in car use. For road freight, studies show greater variations but suggest that long-term effects are similar to those for car use. On the basis of these reviews and DOT's modelling work, we have assumed that each 1% increase in the price of fuel in real terms will reduce fuel use by 0.3% in the short term and 0.7% in the medium term (that is, in the period to 2020).

7.55 Another way of analysing the long-term effects of price on fuel consumption is to compare different countries. A study of 14 developed countries[32] which had maintained broadly constant differentials between their prices for fuel for road vehicles showed a strong negative correlation between real fuel price and fuel use per head (figure 7-III). Other factors, particularly disparities in income, explained some of the differences between countries but fuel prices were considered to have had a significant influence on consumption.

7.56 There will remain considerable uncertainty about the size of the effect that an increase in the price of fuel will have on fuel use. The evidence from the past relates to relatively small changes against a background of fluctuating market prices. Sustained rises in the real price of fuel over a number of years on the scale now under consideration are a new phenomenon.

7.57 Figure 7-II shows the effect on the retail price of fuel of duty increases of 5% and 9% a year up to 2004, when added to forecast market-induced price increases. Table 7.5 shows the effect the resulting increases in the price of fuel would have on carbon dioxide emissions from road transport.

Table 7.4
Petrol prices in EC countries, December 1993

	pence per litre of 4 star petrol
Netherlands	67.3
France	65.5
Italy	64.6
Belgium	62.1
Germany	58.6
UK	56.9
Denmark	53.9
Spain	52.4
Luxembourg	49.1

Table 7.5
Effect of fuel price increases on carbon dioxide emissions from road transport

million tonnes carbon/year

	1990	2000	2020
base case	30.5	34.7	42.9
estimated reductions if fuel duty increases by:			
A: 9% a year 1995-2004		3.0	15.5
B: 5% a year 1995-2004		1.4	8.8
reductions required to achieve Commission's proposed targets		4.2	18.5

Figure 7-III
Relationship between fuel price and fuel use by road vehicles in developed countries (1988)

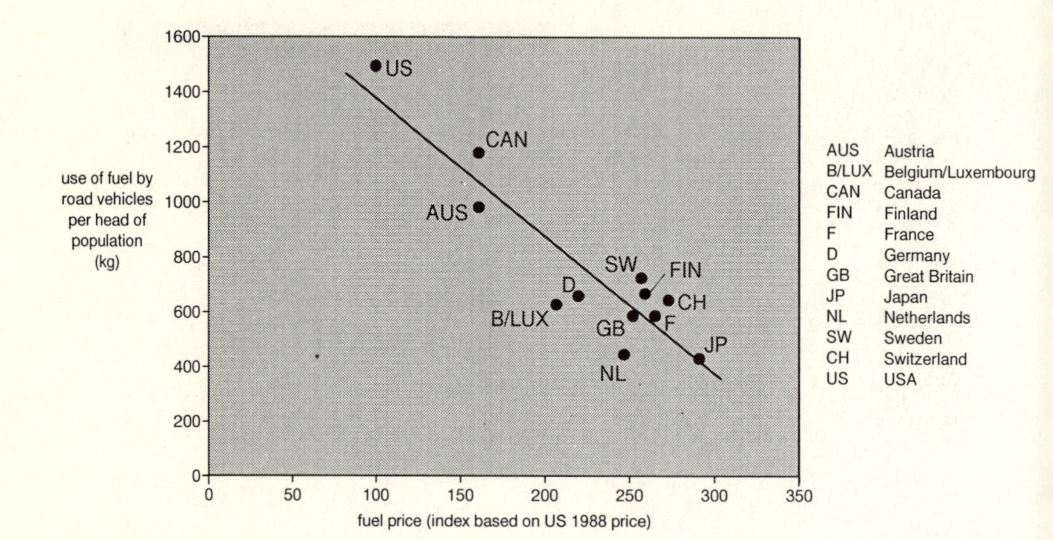

7.58 We welcome the government's commitment to a year-by-year increase in the real price of fuel for road vehicles. But table 7.5 suggests that a 5% increase in fuel duty each year is unlikely to be sufficient to achieve environmental objectives. In view of the limited scope for existing users to modify their car purchase and travel choices in the short term, too rapid a rate of increase would be an unreasonable burden. However, we consider a doubling of the price of fuel in ten years is necessary in order to bring about substantial reductions in fuel use and make a major contribution to achieving the targets we have proposed for carbon dioxide emissions. We also consider that, in order to alter users' expectations, it is important to make clear that the aim is a permanent and substantial increase in prices. Accordingly, **we recommend that fuel duty be increased year by year so as to double the price of fuel, relative to the prices of other goods, by 2005.** If market-induced price rises follow the base case shown in figure 7-II, the increase in fuel duty required to achieve this doubling in price would be 9% a year for ten years.

7.59 By 2000 it will be possible to make an assessment of the effectiveness of fuel price increases, in combination with other measures we recommend, in bringing about reductions in carbon dioxide emissions. That will provide the basis for judging whether further price increases may be needed after 2005 in order to achieve our proposed target of a 20% reduction in carbon dioxide emissions from surface transport between 1990 and 2020. **We recommend that a view be taken after 2000 about the case for further increases in fuel duty after 2005 in the light of a review of progress made towards our targets.**

Increasing the use of public transport

7.60 As well as having direct effects on private road use, increases in fuel prices and, where appropriate, increases in parking charges and the introduction of road pricing, will also affect the relative position of private and public transport. They will therefore go some way to mitigate the distortions in choices between private and public transport brought about by the different ways in which people make payments towards their costs (7.24). However, they will not in themselves bring about a situation in which transport users are faced only with the full additional costs of each mode of transport (7.25). Nor are they likely to bring about a switch to public transport on the scale that would be desirable on environmental grounds (7.30).

7.61 The inbuilt advantage of private transport would not be entirely removed. Fares for public transport are based on average costs. To achieve pricing for public transport comparable to the perceived cost of private transport, fares would have to be based on marginal costs (including marginal environmental costs). However, public transport fares based on marginal costs would show large short-term fluctuations (even larger than the differences at present between fares at peak times and special off-peak fares), and would not necessarily enable public transport operators to cover all their capital and operating costs.

7.62 A second important factor restraining any switch is that the quality of the service provided by public transport might not be sufficiently attractive. Relative price is important, but so are frequency, reliability, speed, comfort and convenience. Although public transport operators have made great efforts to improve reliability in recent years, and the speed of some railway services has significantly improved, frequency has often declined, especially at off-peak times and for bus services. This creates a vicious circle in which a less frequent service is used by fewer passengers, so justifying cuts in service. Moreover, if fewer people are travelling, there may be an increased incidence of crime and vandalism. With increasing usage, frequencies will rise, public transport operators will find it easier to cover their full costs, and hopefully a virtuous circle will emerge. And, at higher levels of occupancy, public transport will have a clearer advantage over private transport in terms of energy efficiency.

7.63 Most countries provide support to public transport. International comparisons can most readily be made for railways. Table 7.6[33] compares performance indicators for Great Britain with averages for 13 other European countries. As a percentage of GDP, the support British Rail (BR) received from public funds in 1992/93 was only 36% of the average for other countries. It has also shown a declining trend. To some extent the lower figure for BR can be regarded as a reflection of greater efficiency (expressed in table 7.6 as train-kilometres per member of staff employed). In the ten years to 1991 passenger travel by rail grew in Britain and rail's share of personal travel is not markedly lower in Britain than in a number of other European countries (table 6.3). Average passenger train loading in Britain is well below the average for other countries. Rail's share of freight transport is much lower in Britain than in some other European countries (table 6.4), although that can be partly explained by the shorter distances involved. Efforts are being made in some other countries to reduce public expenditure in support of railway systems.[34] Past levels of support to public transport in other countries may not provide useful guidance about the levels of support that would be justifiable in the UK for the future.

7.64 To avoid the unfavourable and misleading connotations of the term 'subsidy', we use the term 'public contribution' to refer to net public expenditure in support of either public or private transport, where such expenditure is justified on either social or environmental grounds. Where appropriate we refer to 'public capital contributions' and 'public revenue contributions'.

7.65 One type of justification for a public contribution to the costs of public transport operators is the social desirability of ensuring continued mobility for particular groups in the population such as the elderly (through the provision of bus passes which give free or reduced price travel), or the disabled (through dedicated transport services). Another justification would be ensuring continued access to shops and other services for remote rural communities. In some circumstances assistance may be given directly to potential travellers. In many cases, however, a public revenue contribution is made to the public transport operator to meet the cost of concessionary fares for particular groups. For buses and light rail outside London, public revenue contributions for this purpose amounted to about £300 million in 1991/92, more than two-fifths of total public revenue contributions; there was also a contribution of more than £100 million to London Regional Transport for this purpose, about an eighth of the total public contribution. Public contributions to BR were not categorised in this way.

7.66 A public contribution to the costs of public transport may be justified where there are benefits to road users from a reduction in congestion resulting from the availability of public transport. In these circumstances it can be argued that the cost of such assistance should in principle be met from payments by private road users, for example, from parking charges or a road pricing scheme.

Table 7.6
British Rail compared to continental railways (1992/93)

	British Rail	average of other CER railways
train-kilometres (loaded and empty) per member of staff employed	3,205	2,527
average passenger train loading (passenger-kilometres divided by loaded and empty passenger train - kilometres)	86	143
average freight train loading (freight tonne-kilometres divided by loaded and empty freight train - kilometres)	315	324
support from public funds as percentage of GDP	0.20	0.56

The other countries making up the Community of European Railways (CER) are Austria, Belgium, Denmark, France, Germany, Greece, Ireland, Italy, Luxembourg, Netherlands, Portugal, Spain and Switzerland. The CER average also includes the Northern Ireland railway system.

7.67 Our particular concern is that we believe there is also a justification for public contributions to public transport on environmental grounds. The main justification we foresee for such contributions is to achieve the targets which we propose later in this report for increasing public transport's share of personal travel, and rail and water's share of freight transport. There may also be cases where the environmental benefits, and perhaps also the saving in road infrastructure costs, as a result of carrying people by public transport over a particular route, or as a result of carrying goods by water or rail rather than road, would be substantial, but the direct cost to an operator of doing so would be significantly higher than the cost of private road transport. An alternative mode will not therefore be used unless a public contribution is provided.

7.68 BR received a public contribution of £930 million in 1993/94, which represented about a quarter of turnover. Most of it was revenue grant to Regional Railways and Network SouthEast. After deducting the contributions mentioned above which finance concessionary fares, the public contribution to London Regional Transport in 1991/92 was almost £700 million, about 70% of turnover; about two-thirds of this was capital grants. The contribution to buses and light rail outside London was about £400 million, almost a quarter of turnover, of which three-fifths was revenue support and the remainder the rebate given to bus operators on fuel duty. In addition to this rebate, public transport receives favourable tax treatment in two other ways. Oil used for diesel locomotives is subject to duty at a much lower rate than fuel for road vehicles[35]; and VAT is not levied on public transport fares, a concession which was worth £850 million in 1992/93.

7.69 The present public contributions and tax concessions have not hitherto been the subject of specific justifications put forward on environmental grounds, and may well not represent the most appropriate and cost-effective forms of assistance to public transport for the future. For example the lower rates of duty on fuel reduce the incentives for fuel efficiency. The immediate policy priority, however, is to prevent any further decline in public transport use. Existing forms of assistance should therefore be maintained until policies are in place which will meet the targets for modal share we propose later in this report. **We recommend that the present net level of support for public transport (including fuel duty rebates for public service vehicles, the low rate of fuel duty for railways and the benefit of zero-rating for VAT) should be at least maintained in real terms until superseded by measures for achieving a major increase in the role of public transport.**

7.70 There will be a transitional period in which increases in fuel duty for private road vehicles will still be taking effect, and in which it will be important to ensure that public transport provides an increasingly attractive alternative. In taking decisions about continuing policies for public contributions to the costs of public transport the emphasis should be on overcoming the obstacles identified above (7.61–7.62) by improving the attractiveness of public transport in terms of convenience, reliability and journey times (so initiating a virtuous circle); and by making it possible for operators to levy charges which are a closer reflection of full additional costs (including additional environmental costs), rather than average costs. In some cases, such as new light-rail systems, a public capital contribution may be essential before a scheme will be undertaken. In other cases finance for capital investment may be available from the private sector, and the most pressing need may be for a guarantee of future public revenue contributions so that the use of private finance becomes viable.

7.71 A danger with a public contribution to costs in any field is that it will encourage inefficiency. Future public contributions to the costs of public transport should be designed in a way that provides operators with incentives to manage the system efficiently. For example the amount of any revenue contribution should be tightly controlled and of a predetermined amount. Any capital contribution should be linked to specific improvements in the level or quality of services. Our specific recommendations relating to public contributions towards freight, urban and long-distance transport are contained in chapters 10–12.

Applying economic instruments to aircraft emissions

7.72 Like other forms of transport, air transport has brought considerable benefits. It is the most rapidly growing mode of transport. The environmental costs it imposes, and the much greater environmental costs that will be imposed if it continues to grow at the present rate, were discussed in chapter 5.

7.73 It is doubtful whether demand for air transport would be growing on the present scale if passengers had to face the costs of the damage they are causing to the environment. We believe the price mechanism should be used to ensure that the full additional environmental costs of air travel are reflected as far as possible in air fares. We discuss in chapter 12 the possibility of attracting passengers from internal and cross-channel flights to forms of high-speed surface travel which would be less environmentally damaging. Measures that increase the relative cost of air travel will encourage such transfers.

7.74 In contrast to surface carriers, airlines do not pay any tax on their purchases of fuel (although until 1985 a small tax was paid on fuel used on domestic flights). The appropriate form of economic instrument might therefore be a levy on fuel purchases. As with increases in the price of fuel for road vehicles, a levy on fuel purchases would increase the incentives for airlines to put pressure on manufacturers to improve fuel efficiency more rapidly and would also influence demand by raising the cost to travellers.

7.75 In order to be fully effective, an environmental levy on airline fuel purchases would have to be applied at EC and preferably, international level. Since 1989, taxes have been imposed on aircraft emissions in Sweden at a level which, if they were to be extended throughout Europe, would produce some £6 billion a year in revenue. On the evidence of a survey of airlines, the effect of Europe-wide taxes would be that aircraft engines would be modernised, fares would increase and there would be some shift to surface transport.[36] **We recommend that the government negotiate within the EC, and more widely, for the introduction of a levy on fuel purchases by airlines that will reflect the environmental damage caused by air transport.** As measures to improve fuel efficiency may increase emissions of nitrogen oxides (5.31) it would be important to design and apply a levy in such a way as to obtain the best overall result in environmental terms.

7.76 Airports in several countries, including France, now make charges to airlines based on the weight and noise category of their aircraft and use the income to finance sound insulation for nearby houses.

A recent government working party (the Batho Committee) recommended that consideration should be given to imposing a levy of this kind on either passengers or airlines in the UK, but DOT rejected this recommendation on the ground that 'airports already pay for measures to ameliorate noise round airports'.[37] Even if the cost of insulation grants is ultimately met by airlines and their passengers, the cost is not at present distributed in a way which makes this apparent or reflects the respective contributions they make to the noise problem. **We recommend that the government implement the Batho Committee's recommendation for a noise levy on movements at airports.** The government has recently placed a tax on passenger movements at UK airports but the Treasury stated in evidence that this was not intended to reflect the costs of environmental damage.

Effects on the economy

7.77 The primary reason for the recommended increase in fuel duty for road transport is to provide an efficient method of achieving environmental targets which are an essential part of a sustainable transport policy. In a broader context it is also consistent with the emphasis which the European Community's Fifth Environment Action Programme places on the desirability in economic terms of shifting the burden of taxation to activities which cause damage to the environment. The macroeconomic effects of the increases in fuel duty will be for the government of the day to determine. One option would be to reduce other taxes to offset the increased income from fuel duty, so that there would be a neutral effect on total government revenue; the overall effect on the economy should then also be small. If the government of the day had other reasons for wishing to deflate the economy by reducing the public sector borrowing requirement, another option would be to use the increases in fuel duty to achieve that effect while also bringing benefits to the environment.

7.78 A third option is to link the revenue received from increases in fuel duty to increased expenditure on less environmentally damaging forms of transport. Such a link would help to secure and retain public acceptance for the increases, and dispel any impression that they were simply another means for the government to raise revenue. The basic environmental justification for the increases remains the same with or without a link of this nature. However, under a policy which combines the carrot of improved public transport and cycle routes with the stick of higher fuel prices, the price increase required to achieve any given target for emissions ought to be smaller than would be required if the two elements were not linked. The concluding chapter of this report summarises briefly what additional expenditure on public transport we consider would be desirable, the extent to which it could be funded by a reduction in expenditure on the road programme, and the scale of additional public expenditure for which the increases in fuel duty could provide a source of funding.

7.79 The increases in fuel duty we have recommended will tend to produce a shift in private expenditure away from private motoring and towards other forms of expenditure, including travel by public transport. That may reduce profits and employment in the industries supplying vehicles and components. The size of the effect will depend on how rapidly they adjust to the new circumstances by making major improvements in the fuel efficiency of the vehicles they produce. In the other direction there will be a favourable impact on the public transport industries and their suppliers of capital equipment. In that public transport requires a high labour input for its operation, there will be some shift towards employment in services.

7.80 Even if other taxes are reduced to balance the increases in fuel duty, there will be a significant adverse effect on some motorists if they are heavily dependent on using a car, have low incomes, and cannot adapt to the new circumstances over the ten years in which the increases in fuel duty will be taking effect. A larger number of people will benefit, however, from the improvements in public transport we envisage. These differential effects raise issues of equity to which we return at the end of the report.

7.81 The impact of the increases in fuel duty on prices of goods will not in general be substantial. Although total transport costs have been estimated to represent as much as a tenth of the final price of goods on average (2.9) the cost of fuel represents only about a seventh of transport costs.[38] To the extent that the higher cost of fuel would exert a downward pressure on the distances over which goods are moved, that would be in accordance with our objectives for protecting the environment. There would of course be a bigger impact on bulky low-value goods, which could lead to changes in supply and transport patterns.

7.82 There would also be adverse effects on road haulage firms. However, these will be accompanied by new business opportunities to develop cross-modal and local distribution services along lines discussed later in the report. It should be emphasised again that the increases in fuel duty will be phased in over a period, so allowing time for adjustment.

7.83 A further issue is the possibility of competition from freight operators based in countries with lower fuel prices. If there are significant differences in fuel prices between European countries there will be some diversion of fuel purchases (by all international road hauliers, regardless of the country in which they are based) towards the countries in which prices are lower. It is doubtful whether this would become a major problem for the UK road haulage industry, even with the Channel Tunnel. Nevertheless, it is clearly desirable from the environmental point of view that the general level of fuel prices should rise substantially throughout the European Community, and indeed in other developed countries; and that it should not be held down by cross-border competitive pressures in Europe. **We recommend that the government press for revision of the EC Directive on fuel prices so as to ensure a sustained year-by-year increase in fuel prices across the Community.**

Conclusion

7.84 This chapter has demonstrated that there are very substantial environmental costs which are not taken into account in decisions by users of the transport system (7.20), and that there are tax rules (7.39) and differences in the way people make payments towards the cost of public and private transport (7.24) which introduce additional distortions into such decisions. These factors have made an important contribution to the present unsustainable trends in transport. Because money values cannot meaningfully be attached to some types of environmental costs (7.13), and because even for other types of environmental cost there would be enormous difficulty in assigning money values to the effects of a particular journey (7.29), we do not believe it would be practicable or appropriate to attempt to base transport policy on balancing costs and benefits at the margin. We have adopted a pragmatic approach to producing a sustainable transport policy, based on setting targets (7.30). Action to achieve those targets, using economic instruments, is likely to bring about a closer correspondence between costs and benefits (7.31). While electronic charging systems have considerable potential in the medium term, increases in fuel prices, at a faster rate than the existing government commitment implies, should be the main instrument for the time being for encouraging a less environmentally damaging use of the transport system and influencing demand for mobility (7.58). We now go on to consider the implications of price increases on this scale and the complementary measures which need to be taken in relation to vehicle technology (chapter 8), land use (chapter 9), freight transport (chapter 10), local journeys (chapter 11) and long-distance transport (chapter 12), in order to achieve environmental objectives and targets.

Chapter 8

ROAD VEHICLE TECHNOLOGY AND PERFORMANCE

8.1 Transport is dominated by the technology which has been developed around the internal combustion engine. Despite many years of engineering progress, such engines are still the subject of intensive research and development. Much of the focus is on improvements in environmental performance. This chapter deals in turn with emissions of airborne pollutants, fuel efficiency and noise. It describes relevant European Community (EC) standards and makes comparisons with the USA and Japan. It assesses the contribution technological modifications can make in bringing about further improvements in standards and achieving the environmental objectives and targets proposed in chapters 3 and 4, and how that contribution can be realised in practice. The environmental effects of using a road vehicle are determined, not only by the design of the vehicle, but by factors such as adequacy of maintenance, fuel quality and driving style; these too need to be analysed. Finally this chapter considers the longer-term options for vehicle propulsion.

8.2 As noted earlier (3.6) there is a conflict in some circumstances between improving fuel efficiency in order to reduce carbon dioxide emissions and improving air quality. In assessing the relative priority of these aims one relevant consideration is that, whereas transport is the dominant source of several airborne pollutants, it gives rise to only a quarter of UK emissions of carbon dioxide. Reductions in emissions of the main airborne pollutants therefore make more difference to the overall UK position than the same percentage reduction in carbon dioxide emissions from the transport sector. Another consideration is that adverse health effects from poor air quality would be more immediately undesirable than the consequences of climate change. We have concluded therefore that small increases in carbon dioxide emissions are acceptable in certain circumstances in the short term as a necessary by-product of achieving better air quality. Fitting a catalytic converter to a petrol car for example reduces its fuel efficiency, but is justifiable because the reduction in emissions of pollutants obtained in this way has a higher priority. In the medium term this adverse side-effect will be far outweighed by the general improvement in fuel efficiency that will flow from our recommendations.

8.3 Diesel engines represent a different situation: their superior fuel efficiency can make a major contribution to achieving targets for carbon dioxide emissions, but they are at present more polluting in some respects than petrol engines fitted with three-way catalytic converters. After discussing emissions of pollutants from petrol engines and from diesel engines, we begin our discussion of fuel efficiency by considering the overall desirability of diesel engines from the environmental point of view (8.31–8.34).

Reducing emissions affecting air quality

Petrol vehicles

8.4 At present new cars sold in the UK are required to comply with the EC stage I limits on emissions shown in table 8.1.[1] Petrol cars can meet these limits only if they are fitted with a closed loop three-way catalytic converter (which uses platinum, palladium and rhodium catalysts to remove 75–90% of the carbon monoxide, hydrocarbons and nitrogen oxides in the exhaust gases) and have fuel injection and a system to control the air/fuel ratio in the engine.

8.5 A serious limitation of catalytic converters is that they operate only after they have reached a certain temperature; when an engine has been started from cold, and emissions are highest, the exhaust gases take some time to reach this temperature, and they may drop below it again if the engine is idling.[2] A high proportion of car trips are too short for the catalytic converter to achieve efficient operation (3.11): a study of Germany, France and Britain[3] found 47–48% of trips were of less than 3 km and 20–22% of less than 1 km. This cold start problem is one of several reasons why countries in which this type of equipment was made compulsory some years ago have not achieved the expected improvements

Table 8.1
Emission limits for cars

grams per kilometre

	carbon monoxide	hydrocarbons (HCs) and nitrogen oxides (NO_x)		particulates
EC Stage I 1993				
petrol and diesel	3.16	1.13		0.18
EC Stage II 1997				
petrol	2.20	0.50		-
diesel, indirect injection	1.00	0.70		0.08
diesel, direct injection	1.00	0.90		0.10
EC Stage III 2000: German government proposal				
petrol	1.5	0.2		-
diesel	0.5	0.5		<0.04
European Parliament proposal		HCs	NO_x	
petrol	1.00	0.10	0.10	-
diesel	0.50	0.10	0.30	0.03
US Tier 1	2.94	0.58		
California				
TLEV	2.94	0.50		
LEV	2.94	0.33		
ULEV	1.76	0.28		

US standards have been adjusted to indicate what emission levels vehicles complying with those standards would achieve in the different test cycle used by the EC.

As well as setting a limit for HCs + NOx the US authorities set separate limits for NOx, total hydrocarbons and non-methane hydrocarbons.

TLEV transitional low emission vehicle
LEV low emission vehicle
ULEV ultra low emission vehicle

in air quality. However, various devices are now being developed to bring catalytic converters to the threshold temperature quickly and it is reasonable to assume that effective technology for that purpose will become generally available within the next few years.

8.6 Manufacturers establish compliance with EC limits by testing sample cars over a prescribed cycle covering 11 km. The present test cycle has been criticised for not being sufficiently representative of actual conditions. In particular:

it is carried out at a temperature between 20 and 30°C, whereas the average ambient temperature in the UK is about 10°C. The US cycle includes a test at minus 7°C;

emissions are not measured during the first 40 seconds after the engine has been started.

If the test were carried out at a temperature of 10–15°C and measurements began after 5 seconds, most petrol cars would fail the carbon monoxide limit by nearly 600%.[4] It is important that the prescribed test cycle should give an accurate picture of the cold start performance of present petrol cars, diesel cars and petrol cars fitted with the heating devices now being developed. **We recommend that the government seek amendments to the EC test cycle for cars at the earliest possible date, so that it will adequately represent typical operating conditions.**

8.7 The EC limits which came into effect in 1993 were part of a three-stage process. Agreement was reached in March 1994 on stage II, which will come into effect on 1 January 1997 and set separate limits for petrol and diesel cars. For petrol cars the stage II limit for carbon monoxide is more stringent than the US federal (tier 1) standard or the low emission vehicle (LEV) standard which will have to be met by most cars sold in the state of California from 1999 onwards. For hydrocarbons and nitrogen oxides the stage II limit is less stringent than the LEV standard, slightly more stringent than the US federal standard and similar to the California transitional low emission vehicle (TLEV) standard.

8.8 The European Commission has not yet put forward proposals for stage III, which is due to come into effect on 1 January 2000. Table 8.1 shows a proposal put forward by the German government[5] and, for comparison, the more stringent emission limits proposed by the European Parliament and the ultra low emission vehicle (ULEV) standard which will have to be met by 10% of vehicles sold in California from 2002 onwards.[6]

8.9 There is no evidence that manufacturers will have difficulty in complying with the stage II limits for petrol cars. Many new cars tested on the prescribed EC cycle already achieve the stage II limits.[7] Recent experience in the USA shows that, provided adequate notice is given, the prospect of stringent limits on emissions can be most successful in forcing the further development of known technologies. Compliance with the ULEV standard is proving much less difficult than manufacturers forecast[8]: it is now estimated that a customer in California will pay only about $200 more for a car meeting the ULEV standard than for a present-day (tier 1) car.[9] Part of the gain however has come from changes in the composition of petrol (8.16).

Diesel vehicles

8.10 In recognition of the different characteristics of diesel engines the stage II EC limits for diesel cars shown in table 8.1 differ from those for petrol cars in:

setting a limit on emissions of particulates;

setting a more stringent limit for carbon monoxide;

setting less stringent combined limits for hydrocarbons and nitrogen oxides.

As diesel cars are more fuel efficient, there is general acceptance in the European Community that a higher limit for nitrogen oxides is justifiable. There is a further concession in stage II for direct injection diesel engines, which are 12–15% more fuel efficient than indirect injection engines. In Japan on the other hand the government's view is that emissions of nitrogen oxides from diesel vehicles should be no higher than from petrol vehicles fitted with catalytic converters, and diesel vehicles of less than 2.5 tonnes have been banned in a number of areas.[10] In the USA there are only small numbers of diesel cars; separate standards have not been set for them.

8.11 The EC limits for light goods vehicles which come into effect on 1 October 1994 are shown in table 8.2.[11] For the lightest vehicles they are identical to the stage I limits for cars. Higher limits have been set for heavier vehicles: those vehicles in this category which have petrol engines are likely to require three-way catalytic converters in order to meet the limits, while those with diesel engines are likely to need two-way oxidation catalysts to reduce emissions of carbon monoxide and hydrocarbons.

8.12 Controls over emissions from heavy goods and passenger vehicles are being progressively tightened. The EC limits (table 8.3[12]) apply to engines used in heavy duty vehicles and are therefore expressed in grams per kilowatt-hour rather than grams per kilometre. The stage II limits, which

Table 8.2
EC emission limits: light goods vehicles

grams per kilometre

	carbon monoxide	hydrocarbons and nitrogen oxides	particulates
up to 1.25 tonnes gross vehicle weight	3.16	1.13	0.18
1.25 to 1.7 tonnes gross vehicle weight	6.0	1.6	0.22
1.7 to 3.5 tonnes gross vehicle weight	8.0	2.0	0.29

come into effect on 1 October 1996, are broadly as stringent as the 1994 US standards.[13] The EC has committed itself to more stringent stage III limits to take effect by 1999; the European Commission has not yet put forward proposals for that purpose, but table 8.3 includes a proposal put forward by the German government.[14]

8.13 Like diesel cars, heavy diesel vehicles can meet present emission limits without pollution control devices fitted to their exhausts. Advances in engine technology such as turbocharging, intercooling and electronic injection may mean that this continues to be true for the stage II limits but particulate traps may have to be fitted to exhausts in order to meet the stage III limits. Present designs of trap are too bulky to be fitted to smaller vehicles. Provision needs to be made for regenerating the trap by burning off the trapped particulates. The constraint on using a high engine temperature to minimise the amounts of particulates and hydrocarbons produced is that this increases the amounts of nitrogen oxides produced.

Table 8.3
EC emission limits: heavy duty engines

grams per kilowatt-hour

	carbon monoxide	hydrocarbons	nitrogen oxides	particulates
Stage I	4.9	1.23	9.0	0.68/0.40
Stage II	4.0	1.1	7.0	0.15
Stage III - German government proposal	2.0-2.8	0.6-0.77	4.9-5.0	0.1

It has proved difficult to develop a durable and effective catalyst for removing nitrogen oxides from diesel exhausts: the type of catalytic converter fitted to petrol cars cannot be used because the oxygen content of the exhaust gases is too high. One major manufacturer has yet to improve on an initial 50% reduction in nitrogen oxides, declining to 20% after 10,000 miles. In Japan this research is taken so seriously that the major vehicle and catalyst companies have co-operated to fund and staff a research institute. There is a general expectation in Germany and Japan that an effective catalyst will be introduced by the early years of the next century.[15] **We recommend that European vehicle and catalyst manufacturers pool their research on DeNOx catalysts in a programme sponsored by the European Commission.**

Fuel quality

8.14 The chemical composition of fuel has an important influence on emissions. Sulphur is present as a contaminant in petroleum. The limit on the sulphur content of diesel imposed by EC legislation is being reduced from 0.3% by weight to 0.2% from 1 October 1994, and to 0.05% from 1 October 1996. Under UK legislation, the sulphur content of petrol will be reduced from 0.1% to 0.05% from 1 January 1995.[16] A lower sulphur content not only reduces emissions of sulphur dioxide/hydrogen sulphide but can improve performance. US car manufacturers have advocated a reduction by a further order of magnitude in the sulphur content of petrol, on the ground that this would improve the performance of catalytic converters by up to 20%.[17] A lower sulphur content in diesel improves combustion and reduces the amounts of particulates produced.[18] However, a price has to be paid in energy terms: for each tonne of sulphur extracted from diesel during refining, an additional 20 tonnes of carbon dioxide are emitted from the refinery.

8.15 One method of raising the octane number of petrol (in other words, reducing its tendency to detonate or 'knock') is to increase the proportion of hydrocarbons which are aromatic compounds. Unleaded petrol usually has a higher aromatics content for this reason. Benzene is an aromatic compound, and is also formed from other aromatic compounds during combustion (see appendix A.18). Unleaded super premium petrol, which is not essential for any car engine but accounts for about 8% of petrol sales, typically has a much higher aromatics content (about 45%) than either premium leaded (about 30%) or premium unleaded (about 33%).[19] In view of the unnecessary risk to health represented by its high aromatics content, **we recommend that the government act to end the sale of unleaded super premium petrol.** The average benzene content in UK petrol is 2–3%. The EC limit is 5% by volume[20], but the German government wants it reduced to 1%. **We recommend that the government support a reduction in the permitted benzene content of petrol to 1%.**

8.16 Some other countries are going further in improving the environmental characteristics of fuels. From 1996 a new specification for petrol ('phase 2 gasoline') will be imposed in California; it is claimed that this reduces emissions of pollutants by 25–38%.[21] Aromatics will be reduced from the present US average of 32% to 22%, with a benzene content of 0.8%, half the US average. Olefins, which play a significant role in ozone formation and are the main source of 1,3-butadiene in most urban areas, will be reduced from the US average of 9.2% to 5.5%. Sulphur will be reduced to the level car manufacturers have advocated. Substances containing oxygen, such as MTBE (8.18) and ethanol, will be added to restore the octane number. This type of petrol will have a pre-tax price up to 25% higher than the present specification. Cleaner forms of diesel are also being developed. In Sweden a 'city diesel' with 0.001% sulphur and a greatly reduced aromatics content had obtained 15% of the diesel market by the end of 1992 because of a lower rate of tax.[22] In our view the higher cost of cleaner fuels is likely to be justified by their environmental benefits. **We recommend that the government collaborate with the oil industry and vehicle manufacturers to develop specifications for cleaner fuels which will contribute to achieving our targets for air quality.**

8.17 A number of additives are used to improve particular properties of petrol or diesel. For example, Shell has developed an advanced diesel that contains an ignition improver to reduce noise and smoke, an anti-corrosion additive to protect the fuel injection system, a detergent to keep the injectors clean and an anti-foam compound to reduce spills during refuelling. Detergents are also now used to keep petrol engine components free of carbon deposits. The share of the market held by petrol containing a detergent was about 40% in 1993 and is growing rapidly.

8.18 Some additives have environmental benefits. For example trials have shown that detergents significantly reduce emissions of carbon monoxide and hydrocarbons.[23] Inclusion of a detergent is now mandatory in California.[24] On the other hand there may be unexpected environmental effects. For example, methyl tertiary butyl ether (MTBE), which is added to unleaded petrol, has been found in groundwater in south-east England and East Anglia; although not particularly toxic, it poses a problem in drinking water because of its strong taste.[25]

8.19 The Commission expressed concern in its Fifteenth Report that diesel additives could be brought into widespread use without testing and assessment to determine their potential health and environmental effects, including the effects of their combustion products. Fuel additives which are introduced onto the EC market as new substances are assessed under the EC New Substances Regulation.[26] Existing substances used as fuel additives will be assessed in due course under the EC Existing Substances Regulation.[27] However, neither of these regulations requires the manufacturer of a substance to submit toxicity data for its combustion products. **We recommend that the Department of the Environment (DOE) and the Health and Safety Executive exercise their powers to require manufacturers to provide them with additional information about the combustion products of fuel additives.** It is reassuring that emissions of lead from road vehicles have now declined sharply, primarily because of the reduced lead content of leaded petrol; and are projected to fall by an order of magnitude by 2006 as use of leaded petrol declines.[28]

Alternative fuels

8.20 A number of fuels have been advocated for use in road vehicles on the ground that they are less polluting than petrol or diesel. Some of them are being or have been used on a substantial scale in various parts of the world. As table 8.4 shows, such fuels frequently have disadvantages as well as advantages in pollution terms. Some produce lower carbon dioxide emissions, but are more expensive. There is no optimum fuel in environmental terms and the choice depends on which problem is the priority at a particular time and place. In the light of an analysis we commissioned[29] we have concluded that there would not be any overall environmental advantage in widespread use of alternative fuels in the UK, and such use should not be expected in the medium term, barring large-scale government or EC intervention. What we say above about the need for adequate toxicological testing of fuel additives applies with even greater force to the toxicological safety of alternative fuels. The longer-term prospects for hydrogen or electric-powered vehicles are discussed later in this chapter.

8.21 However, there would be environmental advantages in the use of methane in the form of compressed natural gas (CNG), in substitution for diesel, in relatively large vehicles which make frequent stops in urban areas, such as buses, road sweepers or refuse collection vehicles; there are about 150,000 such vehicles in large urban fleets in the UK. Vehicles fuelled by CNG produce virtually no particulates and the hydrocarbons emitted may have as little as half the ozone-forming potential of those emitted by diesel vehicles. Emissions of nitrogen oxides are also claimed to be lower, although there is no agreed figure. Some CNG-powered vehicles have met the California ULEV standard. For urban use, total carbon dioxide emissions are similar to those for a diesel engine. Although CNG is broadly competitive with diesel in terms of cost of fuel, the greater size and weight of the tank required imposes a cost penalty. There is already an extensive distribution network for natural gas. British Gas is converting many of its own vehicles and predicts that 200,000 UK vehicles could be CNG-powered by the end of the century. Liquid petroleum gas (LPG), which consists mainly of propane and butane, also produces lower emissions than petrol or diesel, but is less attractive in safety terms and may not be available in such large quantities as natural gas. **We recommend that the government consider the case for incentives to operators of fleets of heavy vehicles in urban areas to use natural gas-powered vehicles.**

Effect of tighter emission limits

8.22 The government has said[30] that 'dramatic reductions in emissions from the UK vehicle fleet' will result from the EC legislation summarised in tables 8.1, 8.2 and 8.3. The effect of this legislation will be gradual, however, in that only part of the vehicle fleet is replaced each year, and will be partly offset by the forecast increase in traffic. We commissioned Earth Resources Research (ERR) to carry out a modelling study in order to assess the effects of the new limits in the context of this increase in traffic[31] and key data from their report are reproduced in appendix B. ERR made the following assumptions:

Table 8.4
Alternative fuels: advantages and disadvantages

fuel	net percentage changes in greenhouse gas emissions (cradle to grave)	change in air pollution impacts	other environmental and safety aspects	cost disadvantage
natural gas/methane	-21 to +5	significant net benefit but little NO_x advantage	Potent greehouse gas. Resource conservation if sourced from landfill. Compression required.	slight
liquid petroleum gas (LPG)	-30 to -10	as for natural gas	Safety of low temperature storage and handling.	slight
hydrogen (made with solar or nuclear power)	-70 to -10	major net benefit	Resource conservation if sourced from water. Major infrastructure changes. Compression or cooling required	major
ethanol	-75 to -40 (if from wood) -20 to +30 (if from corn, sugar etc)	} modest net benefit but aldehyde problem	Resource conservation. Land constraints and landscapes.	major
rape methyl ester (RME)	not known	modest net benefit but benzene and PAH problem	Resource conservation. Land constraints and landscapes.	significant
methanol	+30 to +70 (if from coal) -15 to +5 (if from natural gas)	} modest net benefit but formaldehyde problem	Toxic, soluble in water.	significant

i. stage III EC limits on emissions from cars will be roughly equivalent to the Californian ULEV standard; stage III limits for heavy duty diesels will halve emissions of particulates; and parallel changes will be made in the limits for light goods vehicles;

ii. by the end of the century the catalytic converters fitted to new cars will perform adequately from cold starts;

iii. diesel cars will obtain 20% of the UK new car market in 2005 and retain that. As an alternative it was assumed that they will increase their share further, from 20% in 2005 to 40% by 2025;

iv. traffic growth will be within the range shown in the 1989 National Road Traffic Forecasts.[32]

No allowance was made for the effect on emissions of any change in the level of congestion.

8.23 Even if traffic increases to the extent shown in the 1989 National Road Traffic Forecasts, ERR calculated that road vehicles would emit smaller totals of the major transport-related pollutants in 2025 than now. Emissions are projected to reach their lowest levels in 2007 and then rise at varying rates. By 2025 the emissions of some pollutants from road vehicles would represent substantial proportions of the 1990 totals, particularly if traffic levels match the high forecast:

emissions of **carbon monoxide** in 2025 are estimated to be about 30% of the 1990 level on the low forecast (appendix B, figure B-I) and about 37% on the high forecast. Almost all these emissions are from petrol cars (appendix B, table B.1);

emissions of **VOCs** are estimated to be about 24% of the 1990 level on the low forecast (figure B-II) and about 30% on the high forecast. Cars are the dominant source throughout, but to a lesser extent in the later years (table B.2);

emissions of **nitrogen oxides** are estimated to climb back to 42% of the 1990 level on the low forecast (figure B-III) and 56% on the high forecast. Heavy goods vehicles become the most important source (table B.3);

emissions of **particulates (PM10)** from cars are estimated to increase considerably, although this will be more than offset by reductions in emissions from heavy diesel engines. If diesel cars have obtained 40% of the market in 2025, total emissions of particulates would be half the 1990 level on the low forecast (figure B-V) and more than 60% on the high forecast (figure B-VI). In the latter case, more than 60% of emissions would come from cars (table B.5).

8.24 Sulphur dioxide, for which transport is only a minor source, shows a different pattern. Emissions will be reduced as a result of the reduction in the maximum permitted sulphur content of diesel (8.14); and are projected to reach their lowest level in 1997, just after that takes effect. From the late 1990s cars become the dominant source (table B.4). Emissions in 2025 would be almost two-thirds of the 1990 level on the low forecast (figure B-IV) and over 80% on the high forecast.

8.25 As the significance of emissions may differ according to the area, ERR also produced estimates of total emissions on urban roads. In urban areas, emissions of carbon monoxide and VOCs (tables B.1 and B.2) are projected to decline more rapidly than elsewhere (assuming cold start performance improves) and, in 2025, would represent only 14% and 20% respectively of the 1990 level even on the high forecast of traffic growth. Emissions of nitrogen oxides are also projected to decline rather more rapidly in urban areas (table B.3), but would remain a cause for concern because they would still represent a substantial proportion of 1990 emissions (45% on the high forecast or 37% on the low forecast). Emissions of particulates are projected to decline less rapidly in urban areas (table B.5). The main reason is that cars predominate on urban roads and by 2025 would produce a much higher proportion of particulates. If diesel cars obtain 40% of the market, emissions of particulates on urban roads in 2025 would be 26% below the 1990 level on the low traffic forecast and as little as 7% below on the high forecast (figure B-VII). With a 20% share for diesel cars, the corresponding figures would be 45% on the low forecast and 30% on the high forecast (table B.5).

Contribution of technology to improving air quality

8.26 The policy objective identified in the first part of this report was to achieve standards of air quality that will prevent damage to human health and the environment. Of the specific targets proposed, two related directly to air quality:

to achieve full compliance by 2005 with World Health Organization (WHO) health-based air quality guidelines for transport-related pollutants (3.44);

to establish in appropriate areas by 2005 local air quality standards based on the critical levels required to protect sensitive ecosystems (3.54).

We discuss in chapter 13 how local air quality management plans should be used to co-ordinate and give impetus to action to reduce air pollution from all sources. It is not possible to specify in advance what combinations of measures will be required to achieve particular levels of air quality in particular areas. It is clear, however, that stringent limits on emissions are a vital element in achieving air quality targets. We welcome the government's assurance that it 'will continue to press the European Commission to propose . . . the tightest feasible emission requirements for all types of road vehicle, whether fuelled by petrol or diesel'.[33]

8.27 Decisions have yet to be taken about the stage III limits due to come into effect in 1999–2000. In a tripartite initiative, the European Commission and the European automobile and oil industries are reviewing the reductions in emissions needed to achieve air quality standards and how such reductions could be brought about in the most cost-effective way. The options being considered include improved inspection and maintenance and changes in fuel characteristics as well as engine modifications and pollution control devices.

8.28 The German government has put forward proposals for stringent stage III limits (see tables 8.1 and 8.3), which seem likely to be achievable without excessive cost. **We recommend that the government give support to stage III limits for all types of vehicle at the levels represented by the German proposals and press for even more stringent limits on emissions of particulates. However if alternative proposals emerge from the tripartite initiative which produce a greater net benefit for the environment, they should be preferred.**

8.29 It seems clear that even these stringent limits on emissions from new vehicles are unlikely to be sufficient to achieve full compliance with WHO guidelines in all areas. It may be necessary to establish lower emission limits for certain categories of vehicle that operate predominantly in urban areas, as the government has accepted in its response to the Commission's Fifteenth Report[34], or for all vehicles operating within certain defined areas. Particular causes of concern are that young children on pavements may be exposed to high concentrations of pollutants, and that high concentrations occur inside vehicles (3.12). **We recommend that the government study the case for requiring filtration of the air supply to vehicle interiors.**

Reducing fuel consumption

8.30 The first part of this report also considered emissions of greenhouse gases, especially carbon dioxide, from transport (3.55–3.65), and proposed the following targets as an appropriate contribution to reducing UK emissions:

to reduce emissions of carbon dioxide from surface transport in 2020 to no more than 80% of the 1990 level (3.78);

to limit emissions of carbon dioxide from surface transport in 2000 to the 1990 level (3.79).

As it is impracticable to remove carbon dioxide from emissions, and as the amount of carbon dioxide produced is broadly proportional to the amount of fuel used, achievement of these targets depends on the extent to which total fuel consumption can be reduced. As 95% of the petroleum products used in surface transport are used in road transport (table 3.1), reductions in the amount of fuel used by road vehicles will be decisive. The year-by-year increases in the price of fuel which we have recommended (7.58) would provide the incentive for major improvements in fuel economy. The possible responses to this market signal are discussed below under the headings of replacement of petrol by diesel, changes in customer preference and behaviour, and developments in vehicle design. The overall potential of such responses is then assessed.

Replacing petrol by diesel

8.31 As diesel engines are more efficient than petrol engines (3.5), an obvious option for reducing fuel consumption is to use diesel rather than petrol in cars and other light vehicles. After allowing for the greater density of diesel and the smaller amount of energy required to produce it at the refinery, the net

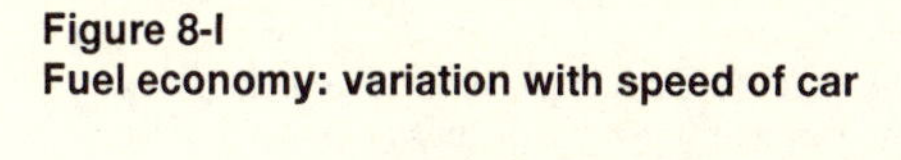

Figure 8-I
Fuel economy: variation with speed of car

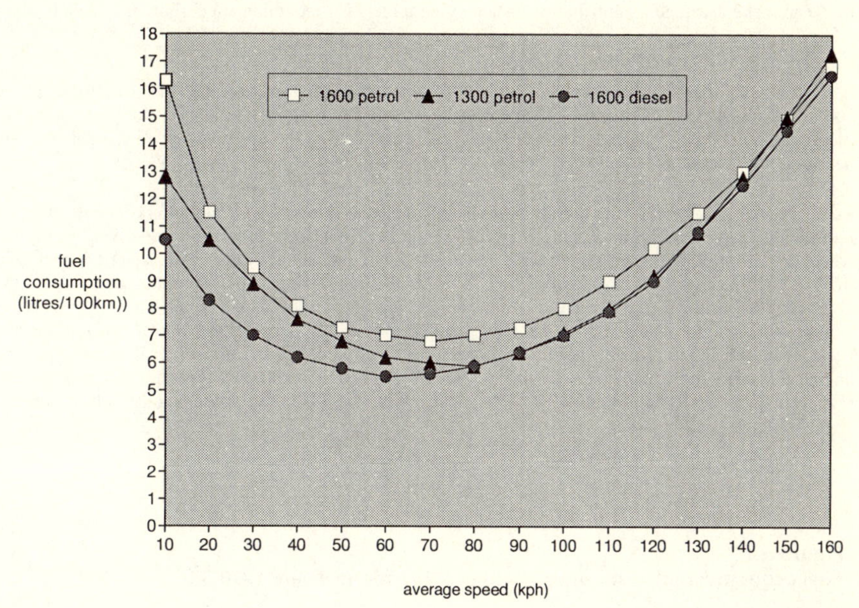

effect on carbon dioxide emissions as a result of substituting diesel for petrol is estimated to be a reduction of 20–30%. Diesel cars have a particular advantage in efficiency at slower speeds, up to about 40 mph (figure 8-I[35]). In the 1989/91 National Travel Survey the fuel consumption recorded for diesel cars of all ages was 31% lower than for petrol cars (41.4 mpg against 28.7 mpg). With changes in refinery practice, the oil industry could reduce petrol output to 20% of crude intake if necessary.[36]

8.32 In a recent report[37] DOE's Quality of Urban Air Review Group (QUARG) concluded that 'any increase in the proportion of diesel vehicles on our urban streets is to be viewed with considerable concern unless problems of particulate matter and nitrogen oxides emissions are effectively addressed.' Our conclusions on health effects in chapter 3 were consistent with the views expressed by the Commission in its Fifteenth Report: a precautionary approach should be adopted to particulates and there is a strong case for seeking reductions in emissions of nitrogen oxides. If recent reports about possible links between certain types of particulates and ill health are substantiated, it will reinforce the need to improve technology in order to achieve further reductions in emissions.

8.33 A diesel car produces far more particulates than a petrol car, and more nitrogen oxides than a petrol car fitted with a catalytic converter. On the other hand a diesel car produces less carbon monoxide than a petrol car with a catalytic converter. Whilst development of clean-up technology for petrol engines has relatively few further avenues, diesel engines in cars are only now receiving specific attention and major advances can be expected under the spur of the EC stage III emission limits. The progressive reduction in the sulphur content of diesel (8.14) should significantly reduce emissions of particulates. We have advocated switching heavy vehicles in urban fleets from diesel to natural gas (8.21), and that could bring a substantial benefit for air quality in urban areas. Our assessment of the diesel car in environmental and health terms is not as negative as QUARG's, but the precautionary principle restrains us from endorsing a major shift to diesel cars until the stringent limits on emissions recommended above are in place.

8.34 Additional pollution control devices are likely to involve a fuel penalty. Nevertheless diesel cars will retain a significant advantage in fuel efficiency over petrol cars. If the obstacles can be overcome they can make a valuable contribution to achieving targets for reductions in carbon dioxide emissions.

Customer preference and driver behaviour

8.35 Recently there has been a big swing in customer preference towards diesel cars in some European countries, presumably the result of their lower fuel costs. From an insignificant level a few years ago, diesel cars had come to account for over 20% of new cars sold in the UK by the end of 1993. In France nearly half of car sales are diesel.

8.36 Working in the opposite direction has been a tendency for the cars bought to become heavier and more highly powered. An increasing share of the market has also been taken by relatively heavy four wheel drive vehicles. The average engine size of new cars registered in the UK has risen from 1396cc in 1973 to 1502cc in 1988 and about 1540cc in 1992; and their average power output rose by about 35% between 1982 and 1992.[38] Over the same period individual models of car have tended to become significantly heavier (table 8.5[39]): the body shell has often become lighter, but a larger number of accessories and safety and anti-pollution devices are now fitted. For a time technical improvements in engine efficiency more than compensated for these trends and the average fuel consumption of new cars registered in the UK fell by about 20% between 1978 and 1987, but subsequently it has worsened slightly (figure 8-II[40]).

Figure 8-II
Fuel economy: registration-weighted average for new cars 1978-93

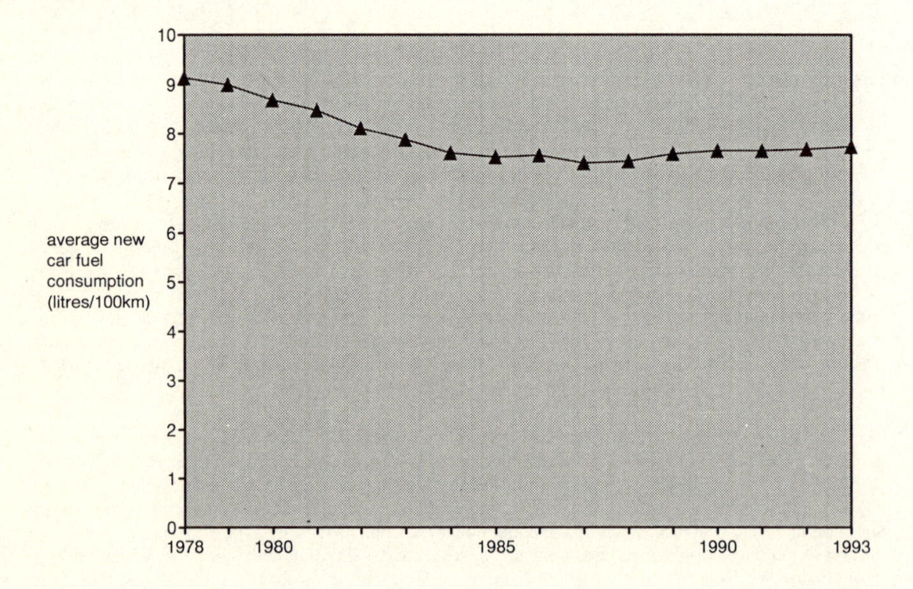

8.37 Reliable data about fuel economy need to be available to car buyers to help them make informed choices. In the USA, the Environmental Protection Agency (EPA) publishes a guide showing city fuel consumption for each model. Every vehicle must carry a label specifying its consumption as determined by the EPA, an estimate of the annual fuel cost based on 15,000 miles of operation and the range of fuel economy achieved by similar sized vehicles of other makes. In the UK, there is a requirement that fuel consumption data be available at showrooms but the Department of Transport (DOT) has recently started charging the public for the booklet. **We recommend that the government explore the possibility of introducing a system of fuel efficiency labelling for cars on US lines.** We hope that car manufacturers will increasingly emphasise safety and fuel economy in their advertising, rather than speed and acceleration.

8.38 Fuel consumption varies considerably with speed, as figure 8-I shows: it is lowest at about 40–50 mph for present designs of petrol car and 30–50 mph for diesel cars, and at low or very high speeds can be twice as high, or even higher. Fuel consumption is also affected by the frequency with which a driver brakes and accelerates. It can be almost four times as high in congested urban traffic as in free-flowing motorway traffic (table 8.6[41]). While traffic conditions are outside the driver's control, fuel consumption (and also emissions of pollutants and noise) are significantly affected by the skill and attitude of the individual driver. It is estimated that 10–15% of fuel could be saved if drivers moderated their speeds, avoided rapid acceleration and made more appropriate use of gears.[42] There are promising signs of greater environmental awareness by motorists[43] and we welcome the 'Greener Motoring Forum' launched by DOE to co-ordinate the activities of motor manufacturers, environmental groups and motoring organisations in promoting such awareness. The government should also ensure that publications such as the Highway Code and vehicle manuals encourage the adoption of environmentally responsible styles of driving and give drivers the information necessary to know what that involves. This might be linked to publicity aimed at reducing the severity of accidents by encouraging motorists to reduce speeds. We welcome the emphasis the government is now placing on the road safety benefits of lower speeds in urban areas. **We recommend that DOT conduct a major educational and advertising campaign highlighting the environmental and safety effects of driving styles.**

Technical improvements in fuel efficiency

8.39 The willingness of customers to buy cars with a lower fuel consumption will be a crucial factor in achieving our proposed targets for carbon dioxide emissions. With such a willingness, very large improvements in fuel economy could be secured through the application of known technology and minor further developments. The conclusions of a seminar held by the European Conference of Ministers of Transport in 1992 referred to reductions of 50% or more in carbon dioxide emissions if drivers were prepared to accept lower performance.[44] DOE and DOT said in their evidence that 'it is technically possible to improve the weighted average fuel efficiency of new cars by 40% in the next ten to fifteen years.' In November 1993 the French and German Environment Ministers suggested the EC should aim for a 5 litres/100 km (56 mpg) average consumption across the car fleet by 2005, close to a halving of the present figure.

Changes in design

8.40 Table 8.7[45] summarises what can be achieved through changes in design within a 10-year period. The various changes would complement each other. The most important principle is to reduce size and weight. Reducing the weight of the body reduces the power required and makes it possible to use a smaller, lighter engine. Shedding 10% of a car's weight reduces fuel consumption by about 7%. If the average engine size of new cars in the UK returned to the 1973 level, it has been estimated that the saving in fuel would rise by 2000 to the equivalent of more than 10% of 1988 petrol consumption.[46]

8.41 There has been a wide-ranging search for lighter materials for use in cars which would be acceptable in cost and other respects. High-strength low alloy steel is only 80% of the weight of ordinary steel of equivalent strength, aluminium 45% of the weight, and plastics and plastic composites about a third of the weight. Aluminium is already used for engines and engine parts. A plant has been built in Germany to produce up to 100,000 aluminium bodies a year. Other materials being investigated include metal matrices (plastic sandwiched between very thin layers of steel) and thermoplastic polymers. Ceramics offer potentially very light substitute materials for components exposed to high temperatures, as does magnesium for components not exposed to high temperatures. Increasing proportions of plastic will be used by all manufacturers on body panels and, in time, on engine and load-bearing components.

Table 8.5
Changes in weight of car models

in kg

small cars	1960	1965	1970	1975	1980	1985	1990
VW Beetle	731	780	800	820	780	780	820
Citroen 2 CV	499	510	540	560	560	600	600
Renault 4	575	575	600	695	695	665	720
-BUT-							
Mini	634	634	600	615	620	630	630

medium to large cars	1967	1977	1987
Audi 80/90	880	933	1070
BMW 1600/316	936	1050	1110
Ford Zephyr/Granada	1240	1310	1330
Fiat 125/132/Croma	1050	1180	1230
Porsche 911	1090	1160	1270
VW Golf	--	860	930
-BUT-			
Mercedes	1375	1390	1370

These changes offer scope for much lighter vehicles but need to be reconciled with the ability to recycle material.

8.42 Reducing the size of vehicles using traditional materials and designs can reduce safety for the occupants as there is less material to buffer impacts. Contrary to popular belief, however, safety can be improved when light, well-designed cars make use of substitute materials. Structural plastics used in bodywork can provide forty times more resistance to damage than steel for half the weight. Some of the lightweight materials and aerodynamic designs applied to electric vehicles to compensate for extra battery weight could usefully enter the mainstream of design. The idea that small cars lack the power to accelerate out of trouble is a myth because their power to weight ratio frequently matches larger ones. Last but not least, lighter vehicles pose less danger to other cars, cyclists and pedestrians.

8.43 A recent advance for diesel cars is idling control, which switches off the engine when the vehicle is stationary or coasting and restarts it at a touch on the throttle. The Volkswagen Golf Ecomatic uses this device, which in city centre driving can reduce fuel consumption by up to a third, and emissions of carbon monoxide and nitrogen oxides by even more. Idling control adds £1,000 to the price of a conventional diesel VW Golf, but the savings on fuel can be balanced against that.

Table 8.6
Fuel economy: effect of driving conditions

	1	2	3	4	total
	congested urban traffic	free-flow urban traffic	motorway traffic	other road traffic	total
% of mileage	2.2	15.7	32.6	49.6	66463km
% of consumption	7.3	24.0	26.1	42.6	—
average speed (km/h)	8	23	93	51	43
number of accelerations/km	24	3	0.2	0.8	1
consumption (l/100km)	25.1	12.0	6.4	7.1	7.7

8.44 Significant benefits for fuel economy could also come from the adoption of unfamiliar types of petrol engine. Lean-burn engines using a high air/fuel ratio are fuel efficient but have not been able to meet the EC limits for nitrogen oxides and hydrocarbons because the air/fuel ratio prevents a three-way catalytic converter from operating effectively. The Japanese firm Mazda has announced that a new

Table 8.7
Fuel economy: main options for modifying cars

percentages

	reduction in fuel consumption produced by each option		total reduction in fuel consumption	
	short-term (current technology) 4-5 years	medium to long-term 7-10 years	short-term	medium to long-term
improved engine efficiency	5-20	6-25		
transmission optimisation	5-15	7-20	10-30	15-50
reduced weight (and/or size)	4-15	7-19		
reduced aerodynamic drag	2-10	2-12		

catalyst, made of zeolite coated with precious metals, will enable it to launch a lean-burn car capable of meeting the strictest emission limits and giving 5–8% better fuel efficiency. Two-stroke engines are much smaller and simpler than four-stroke and can dramatically reduce fuel consumption. However they have traditionally been noisy and smoky. Recent advances have revived interest and an Australian orbital two-stroke design is now licensed to such firms as Ford, General Motors and Fiat. We consider that research on two-stroke and lean-burn designs should be given greater priority by manufacturers than hitherto.

8.45 Looking further ahead, the Rocky Mountain Institute in the US has already proposed a 'supercar' capable of 280 mpg. It would be a hybrid vehicle with no more than a 250 cc engine, a small battery pack, electric motors at each wheel and ultralight body materials. The US Government has launched an ambitious collaborative programme with manufacturers to produce a car by 2005 which is up to three times more fuel efficient than present cars and has broadly the same performance. It is envisaged that this will require a fundamental re-examination of all aspects of design, and that, irrespective of the source of power, it will be necessary to use electric motors to drive the wheels.[47]

8.46 There is less potential for improvements in fuel economy in the case of heavy duty diesel vehicles than in the case of cars. Cost pressures on operators have provided a continuing incentive for reduced fuel consumption. The benefits of further technical improvements will be balanced by the fuel penalty attached to pollution control measures or used to increase vehicle power. Even with new materials weight reduction may not be technically feasible in the case of haulage vehicles. It is unlikely therefore that an improvement in fuel economy of more than about 10% could be achieved for heavy duty diesel vehicles. Light goods vehicles will continue to benefit from the application of car technology to the lighter end of the class.

Obtaining improvements in practice

8.47 Year-by-year increases in the price of fuel will exert a powerful influence on purchasers' choice of cars and on design and marketing decisions by manufacturers. However, we do not believe that they will by themselves exert a sufficiently strong influence on purchasers to choose high levels of fuel economy in cars in preference to other features such as size, performance or price. We have therefore considered what other economic instruments might be introduced to supplement them.

8.48 The most obvious possibility would be a sharply graduated sales tax on new cars. This appeared to be effective in New Zealand following the 1974 oil crisis. A sales tax which rose from 30% on cars under 1350cc to 60% on those over 2700cc helped to cut the average engine capacity of new cars by 16% in the course of 6 years. On the other hand a tax of this nature could depress sales of new cars, with the result that older and less efficient cars would remain in use for longer. We do not therefore recommend that a graduated sales tax should be introduced, but we believe the government should keep it in mind as a possible additional instrument if the fuel efficiency of new cars fails to improve sufficiently.

8.49 The alternative which we favour is to convert annual vehicle excise duty from a flat-rate payment to a graduated one. This has the advantage that it would also influence attitudes towards existing cars and purchases of second-hand cars. At present it costs the same to license a Rolls Royce and a Mini and the government could give a quite different price signal to companies and private owners by creating two higher rates of duty, respectively three and five times the lowest rate. The change to a graduated system should be phased in over several years; and, as we have rejected the option of increasing vehicle excise duty (7.36), it should be revenue-neutral (in other words, the total amount paid in vehicle excise duty should remain the same). While engine size has traditionally been a reasonable proxy for fuel efficiency, the relationship is weakened by the availability of quite large but still efficient diesel engines; cars of the same engine capacity can vary in fuel efficiency by up to 40%. **We recommend that the annual excise duty on cars be steeply graduated, and based on the certified fuel efficiency of a car when new.** It is already a legal requirement that the manufacturer should publish certified fuel efficiencies. However, if the government were to conclude that it is impracticable to base a graduated annual excise duty on certified fuel efficiency, we consider that a graduated duty based on engine size should be introduced.

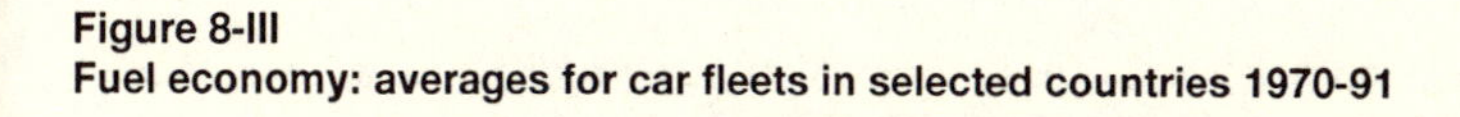

Figure 8-III
Fuel economy: averages for car fleets in selected countries 1970-91

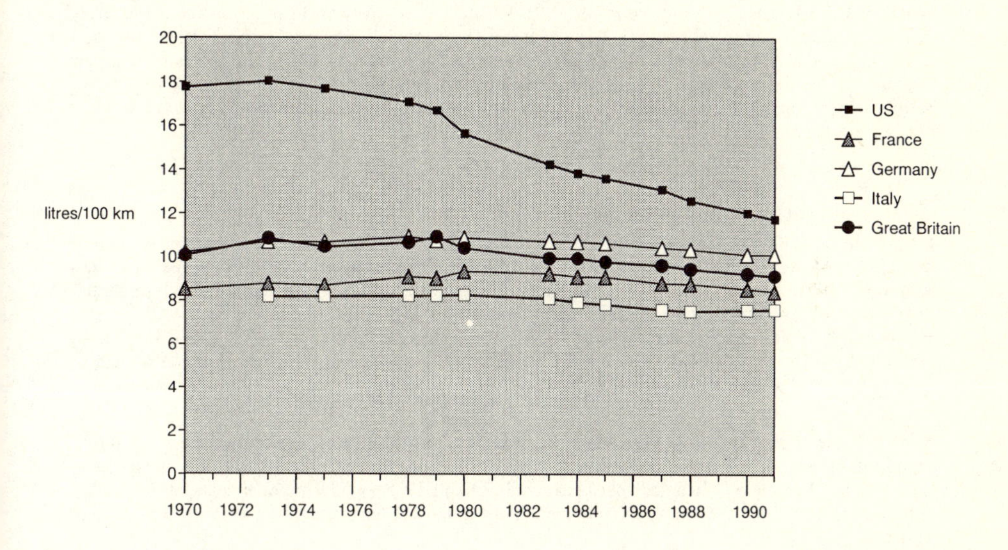

8.50 As figure 8-III[48] shows, there was a 25% improvement between 1970 and 1990 in fuel consumption per kilometre by US cars albeit from a very high starting point. This appears to be the result of the Corporate Average Fleet Economy (CAFE) standards which have been mandatory in the USA since 1976. The sales-weighted average fuel consumption of all cars sold by a manufacturer has to equal or better 8.55 litres/100 kms (27.5 miles per US gallon compared with less than 18 mpg before CAFE). Failure results in fines for the manufacturer.

8.51 The advantage of CAFE was to reduce uncertainty for manufacturers by helping them know what was required before they made expensive design decisions. At present they are reluctant to use available technology because they risk becoming uncompetitive if fuel economy remains a minor factor in buyers' decisions. Pointing out that models offering 80% better fuel economy than average have taken less than 1% of the UK market, the Society of Motor Manufacturers and Traders (SMMT) said in evidence: 'In [a low fuel price] environment a manufacturer who made the enormous investment necessary to produce a new concept of economy car, for sale in high volume, could be so out of step with the market in the price he would need to charge that he could be putting his entire business at risk.'

8.52 The Japanese achieve up to 90 mpg in urban driving with cars below 550 cc, which have a 10% share of their market, and the top-selling model has a 660 cc engine. The popularity of very small cars rested on two incentives, large tax concessions and exemption from parking permits. The first has now been removed but the second is sufficient to sustain their appeal. As well as exceptional fuel economy, these cars have the other benefit of consuming less materials in manufacture. There are some small vans operating in the UK, mainly Japanese, but the small cars have hardly been seen due to a quota system which limited total car imports and gave the Japanese a strong incentive to sell their highest value cars. The quota has now been replaced by an arrangement covering the whole EC until 1999.

8.53 Discussions have been taking place in the EC since 1991 about methods of reducing carbon dioxide emissions from road vehicles, but have been hampered by differences of view between Member States and the European Commission. In 1991–92 the UK put forward a proposal for a system of tradeable credits, but this has not made progress. Germany has proposed setting demanding fuel economy standards for three size classes and levying fines on manufacturers who sell non-compliant models. We see this approach, using direct regulation, as a second best, which might need to be given serious consideration if our preferred approach of using economic instruments fails to achieve the necessary targets.

Effect on emissions of greenhouse gases

8.54 The modelling study we commissioned from ERR (8.22) also produced projections of carbon dioxide emissions. The baseline assumption was of a 10% improvement in the average fuel efficiency of new cars between 1993 and 2005 but no change subsequently. If diesel cars do not obtain more than 20% of the new car market, it is estimated that carbon dioxide emissions from road vehicles will increase by about 40% between 1990 and 2020 on DOT's low forecast of traffic growth (figure 8-IV) and by about 76% on the high forecast (figure 8-V). There is some increase in the proportion contributed by goods vehicles, but cars remain the dominant source.

8.55 The Science Policy Research Unit at Sussex University calculated that fuel efficiency improvements of between 2.1% and 3.45% a year would be needed to limit carbon dioxide emissions to the 1989 level in 2005.[49] It noted that between 1973 and 1983 annual improvements of 2.8% were achieved, and said that there is sufficient technical potential but that economic incentives appear to be lacking.

8.56 A number of measures that would reduce carbon dioxide emissions from cars were modelled, individually and in various combinations, in ERR's study and in a study by the Transport Research Laboratory (TRL).[50] On the assumption that vehicles continue to rely on energy from fossil fuels, and traffic increases in accordance with DOT's high forecast, the TRL study concludes that, even if new technology spreads rapidly and people are persuaded to buy smaller cars, carbon dioxide emissions from cars would stabilise at about 12% *above* 1990 levels and remain at that level beyond 2020. On the other hand, if traffic increases in accordance with DOT's low forecast, carbon dioxide emissions could be reduced to 1990 levels by 2002 and remain at about 10% below 1990 levels beyond 2020. The ERR study tests the effect, singly and in combination, of major improvements in fuel efficiency, lower speed limits and no traffic growth in urban areas. With all these measures combined some reduction in carbon dioxide emissions by 2020 is achieved but, even on the basis of DOT's low traffic forecast, it would not be enough to meet our proposed target.

8.57 The Society of Motor Manufacturers and Traders also analysed the effects of different measures which could limit carbon dioxide emissions from cars to the 1990 level in 2000. Savings of 4 million tonnes of carbon (mtC) would be required. SMMT believes that, if all drivers adopted economical driving styles, carbon dioxide emissions from cars would fall by 15%, or about 3–4 mtC. It also identifies possible savings of 1 mtC from improved car maintenance, 1–2 mtC from reflecting fuel economy in vehicle price, and 0.5 mtC each from a modest improvement in new car efficiency and a shift to 30% diesel cars.

8.58 ERR have also made estimates of the total global warming potential of future emissions from cars in the UK.[51] Reductions in emissions from cars of other gases which contribute directly or indirectly to global warming would not be sufficient to offset the increase in carbon dioxide emissions stemming from

Figure 8-IV
Projected emissions of carbon dioxide to 2025: low forecast of traffic growth

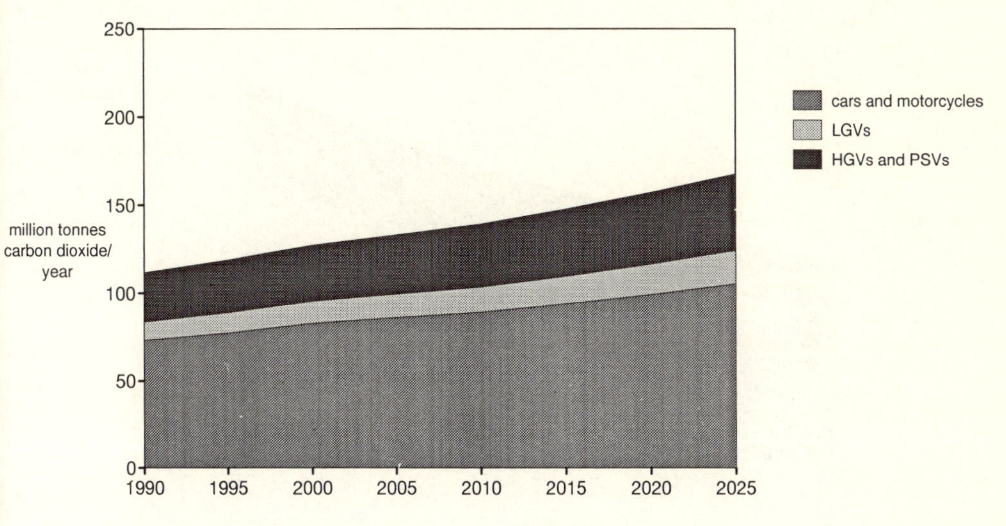

To convert to million tonnes carbon multiply by 12/44.

DOT's high traffic forecast. Under the low forecast there could be a reduction in the total global warming potential of car emissions in the first decade of the next century; but this would eventually be overtaken by further traffic growth.

Policy implications

8.59 As one would expect given the numerous assumptions required, the analyses by ERR, SMMT and TRL are not entirely consistent, but they give a broadly comparable picture. Improved design can make a major contribution to achieving the targets for reducing carbon dioxide emissions from transport. However, it is unlikely to be applied in practice unless government provides appropriate signals to influence decisions by manufacturers and the behaviour of the general public as purchasers and as drivers. The European car manufacturers' association (ACEA) has offered a 10% improvement in sales-weighted efficiency of its model ranges by 2005 but this is widely recognised to be derisory. The crucial signal will be the year-by-year increase in the price of fuel and we have also recommended a graduated system of vehicle excise duty. To reinforce these signals, we propose the following targets:

> **To increase the average fuel efficiency of new cars sold in the UK by 40% between 1990 and 2005, that of new light goods vehicles by 20%, and that of new heavy duty vehicles by 10%.**

As appendix D shows, however, even an improvement in fuel efficiency on this scale, in combination with the increase in the price of fuel, will not be sufficient to achieve the targets for reducing carbon dioxide emissions. It will be necessary to pursue other types of measure in parallel, which we discuss in later chapters.

Figure 8-V
Projected emissions of carbon dioxide to 2025: high forecast of traffic growth

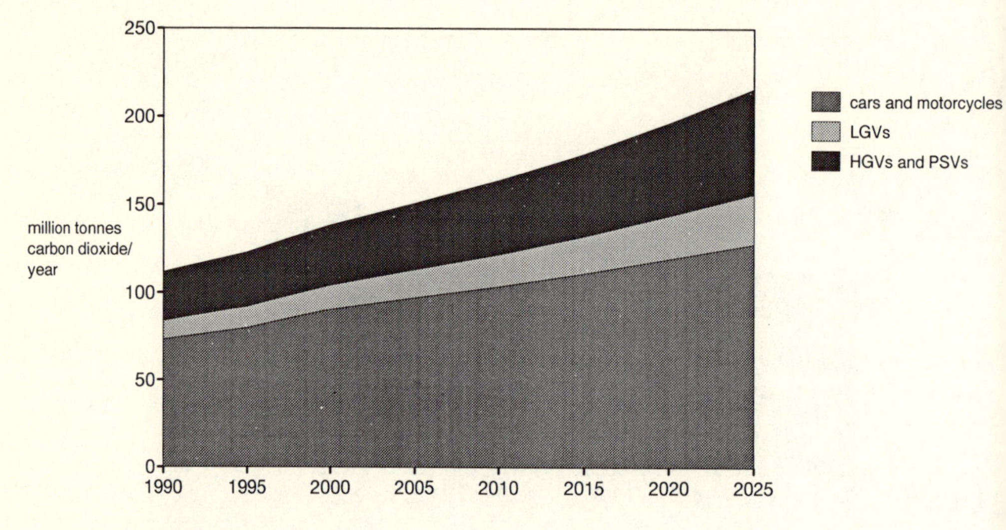

To convert to million tonnes carbon multiply by 12/44.

Reducing noise

8.60 The targets recommended earlier in this report (4.18) for noise levels from road and rail at the external walls of housing were 65 $dBL_{Aeq.16h}$ during the day and 59 $dBL_{Aeq.8h}$ at night. Reductions at source in the sound produced make an important contribution to achieving these targets. Limits imposed on new designs of road vehicles by UK legislation follow those in non-mandatory EC legislation. These limits were reduced by 2–4 dB(A) in 1988 (equivalent to a halving of sound intensity) and will be reduced by as much again in 1995 (table 8.8[52]). From 1995 any new design of car will be required not to exceed 74 dB(A) when accelerating in low gear at full throttle. The limit for the heaviest types of vehicle will be 80 dB(A). New motorcycles must not exceed 77–82 dB(A), according to engine size; these limits will be reduced to 75–80 dB(A) in 1995.[53] The European Commission put forward a proposal in 1983[54] for a directive on sound produced by trains but this has not made progress.

8.61 The sound produced by vehicles comes from several sources: the combustion process, the mechanical parts in the engine and transmission, the air intake (in heavy vehicles), the exhaust (which is silenced), the displacement of air at high speeds and the contact between tyres and the road. Modifications which reduce noise may also have other environmental benefits, for example better control of the combustion process. Improved aerodynamics can significantly reduce fuel consumption and the amount of spray. Other items of equipment may become significant sources of noise, such as warning devices during reversing, air brakes or (in stationary lorries) refrigeration units.

8.62 In principle, sound can be neutralised by generating waves which are an exact inversion of those produced by the original source ('anti-sound'). This can most easily be done for regularly repeated low frequency sounds. This technique has been successfully used to counter both noise and vibration from helicopters and aircraft propellers. In experiments exhaust noise from ship and HGV engines has

Table 8.8
Maximum sound levels for new designs of vehicles

dB(A)

	1983	1988 - 1989	1995 - 1996
cars	80	77	74
buses <150 kW	82	80	78
buses >150 kW	85	83	80
small buses, delivery lorries } <2 tonnes	81	78	75
small buses, delivery lorries } 2-3.5 tonnes	81	79	77
lorries >3.5 tonnes, <75 kW	86	81	77
lorries 75-150 kW	86	83	78
lorries >150 kW	88	84	80

For cars and for buses and goods vehicles of less than 3.5 tonnes the limit values are 1dB(A) higher if the engine is diesel.

been reduced by 40 dB. Lotus is using anti-sound to reduce engine and transmission noise inside some models of car. However this technology is too complex and costly to apply to ordinary ranges of small vehicles.

8.63 The two-stage reduction in sound levels shown in table 8.8 is very welcome, especially in the case of heavy vehicles, but represents less than a halving in perceived noise. In the long term, sound may be reduced, though not eliminated, by electric propulsion. For internal combustion engines further improvements in sound levels should be sought but the scope will be limited. Noise from road vehicles will remain intrusive where there is any significant flow of traffic. In traffic streams which include heavy vehicles there is likely to be little overall benefit from reductions in the sound produced by cars. Achievement of the targets for noise will therefore depend to a considerable extent on traffic management, the use of quieter surfacing materials and screening rather than improvements in vehicle design and performance.

The existing vehicle fleet

8.64 Any new standard applied to new vehicles will take a considerable time to have its full effect in reducing environmental damage. It will take 12–15 years before the existing fleet of cars has been completely replaced. The period is shorter for goods vehicles but even longer for buses. It will take at least five years for the proportion of UK cars fitted with three-way catalytic converters to rise from little over 10% at the end of 1993 to 50%. In 2000 it is estimated that about 30% of cars will have been manufactured to EC stage II standards, about 40% will have been manufactured to EC stage I standards, and the remaining 30% will still not have a catalytic converter.[55]

8.65 There have been some pilot schemes in the USA, financed by the vehicle manufacturing industry, to buy up older cars which are particularly polluting and scrap them. However, schemes of this kind are likely to attract many cars which are not receiving extensive use or would not have continued in use for any length of time or have even been sent for scrapping already.[56] They would not therefore be a cost-effective way of using public money to improve air quality.

8.66 A more cost-effective approach would be to provide incentives for the early purchase of vehicles complying with a new standard. Despite the general desirability of prolonging the life of existing vehicles (4.70) there will be cases in which the new standard marks such a considerable advance that it would be justifiable for government to provide such incentives. Recent EC legislation permits such incentives in the case of heavy vehicles bought before the relevant standard takes effect; they can be regarded as compensating for the higher cost of a vehicle complying with the new standard. For example DAF is manufacturing trucks which comply with the stage II HGV standard which becomes mandatory in 1996 and which are over £2000 more expensive; Germany and the Netherlands give tax rebates of about £300 and £3,000 a year respectively to firms operating these models before 1996.[57]

8.67 Another approach which has considerable attractions in some cases is the fitting of a new engine or new pollution control equipment to an existing vehicle ('retrofitting'). This can be a way of resolving a dilemma that would otherwise exist in environmental terms between contributing to an early improvement in air quality by reducing emissions and prolonging the life of vehicles. The Commission's Fifteenth Report recommended various measures to encourage retrofitting of particulate traps on heavy diesel vehicles and new or rebuilt engines in older buses. Retrofitting of three-way catalytic converters would be too complicated, but it could well be worthwhile to fit oxidation catalysts to cars and light goods vehicles which have a life expectancy of several years. Modelling predicts that retrofitting will be a very cost-effective way to reduce vehicle emissions.[58] **We recommend an immediate study to determine which categories of vehicles not designed to the latest standards justify retrofitting with catalysts or particulate traps, and whether government grants should be offered.**

8.68 We have already expressed our disappointment (in the letter reproduced as appendix E) at DOT's negative reaction to the recommendations in the Commission's Fifteenth Report about the use of economic instruments to encourage the replacement of polluting engines in old buses and the early purchase of less polluting vehicles. Developments since that report, both on vehicle standards and on the scientific understanding of air quality, have considerably strengthened the case for measures of this kind, which will bring particular benefit to urban areas. **We recommend that vehicle excise duty on heavy vehicles be graduated according to the emission limits their engines are designed to meet, and that:**

> **a reduced rate of duty should be payable for vehicles with engines meeting planned new emission limits which are being operated before those limits become mandatory for new vehicles;**
>
> **an increased rate of duty should be payable for vehicles with engines not designed to meet the emission limits currently mandatory for new vehicles, unless they have been retrofitted with effective pollution control devices.**

8.69 An issue which has attracted increasing attention is the extent to which vehicles which comply with a given standard at the time of manufacture continue to perform at the same level when in service. Surveys of cars on the road reveal a large variation in emissions: a small group of cars (about 10%) produce more than half the carbon monoxide emitted.[59] There is considerable variation even among cars fitted with catalytic converters. In a German government survey half of such cars failed to meet the relevant emission limits, particularly for carbon monoxide: cars with high mileages and those being driven at high speeds performed very badly.[60] UK experience shows that catalysts are being damaged or failing at a higher rate than had been foreseen. They are vulnerable to grounding, drowning, bump starts, running out of petrol and overheating; one manufacturer has suggested that a car will, on average, need a new catalyst every three years.[61] Detailed tests show that, even in normal use, emissions from petrol engines with catalytic converters show considerable variation and a tendency to deteriorate

Figure 8-VI
Deterioration of vehicle emissions in use

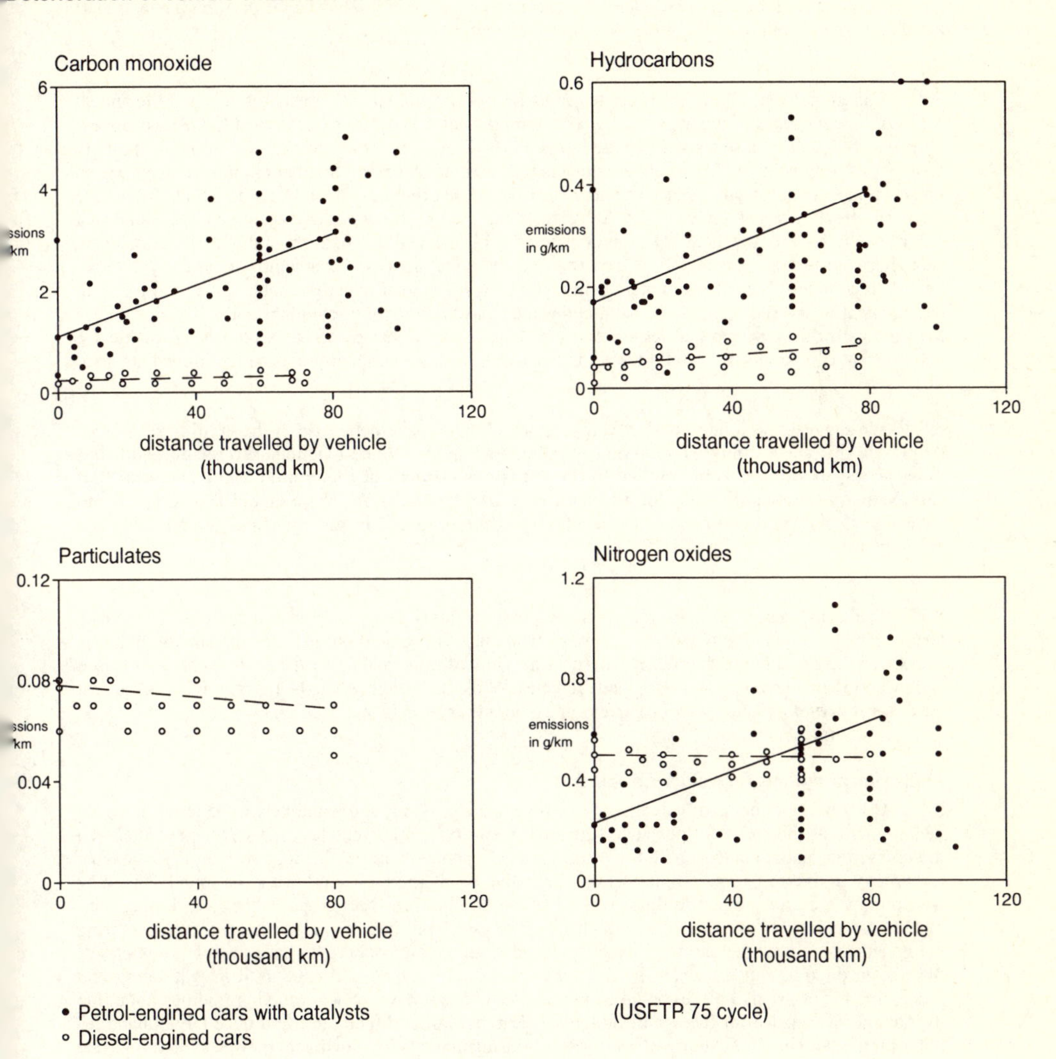

(figure 8-VI[62]). They also show that diesel engines give a more reliable performance in this respect. A Dutch government survey of cars on the road found that a third of cars with catalytic converters and 80% of other petrol cars needed tuning, compared with only 6% of diesel cars.[63]

8.70 EC requirements for type approval of cars include testing of emissions after 80,000 km of specified driving cycles.[64] Some other countries impose more stringent requirements about durability and continuing operation. In the USA manufacturers must certify that the pollution control equipment in individual cars will last 10 years or 100,000 miles. Sweden requires manufacturers to ensure that pollution control equipment can keep car emissions below prescribed limits for five years or 80,000 kms, while HGVs have to achieve eight years or a distance ranging from 200,000 to 500,000 kms depending on their class. From this year US federal standards require each car to have a diagnostic system to alert the driver to any failure of the pollution control equipment.

8.71 Emissions from UK vehicles are tested as part of the annual MOT tests for cars and the annual roadworthiness checks for goods vehicles, which implement the requirements of the EC Roadworthiness Directive.[65] Because these tests are applied to vehicles of all ages the criteria are not stringent and only grossly polluting vehicles fail. There are powers to prosecute owners of highly polluting vehicles but no prosecutions have ever taken place.[66] There is evidence that the present MOT requirements have little practical effect: a recent survey of 25,000 cars on the road by the Royal Automobile Club found that a third would have failed the MOT emissions test.[67] This has led to debate about the merits of roadside spot checks or remote sensing of vehicle exhausts, and about the case for new powers to carry out spot checks. It is doubtful whether remote sensing technology is yet sufficiently reliable to be used for legal enforcement on a large scale. Some local authorities may wish to take action against polluting vehicles but, although this may have a deterrent effect, it is unlikely that resources would be available for enforcement campaigns that would have a noticeable impact on air quality outside the immediate area.

8.72 We suggest that a more cost-effective approach is to make better use of the existing MOT test. At present this checks only the emissions produced while the engine is idling. **We recommend that more stringent standards be applied in the emissions element of the annual MOT test, and that this element become obligatory for all cars a year after registration.** There should also be a publicity campaign to inform drivers about the benefits of accurate tuning in minimising both emissions and running costs.

8.73 The Batho report on noise recommended that the MOT test should also include metered noise testing. The government has said this is impractical, and has pointed out that the test already includes a check on the condition of the silencer. In two states in Australia random roadside noise checks are used and owners have to remedy non-compliant vehicles. While this is unlikely to be justified for cars it might make sense for other categories of vehicle, for example, motorcycles.

Longer-term options for vehicle propulsion

8.74 Oil is likely to be readily available for some decades (7.50) and is unlikely to be replaced in the medium term as the source of power for the great majority of road vehicles (8.20). It is nevertheless a finite resource and reliance on it for transportation cannot therefore be regarded as a sustainable technology in the long term. Eventually an alternative or alternatives will have to be found. This may well prompt a closer interest in those fuels that can be obtained from crops (table 8.4), despite their higher cost and, in some cases, other drawbacks. The use of biofuels is one possible way of countering the greenhouse effect: the carbon dioxide removed from the atmosphere by the growing crops offsets the carbon dioxide emitted when the fuels are burned. However the energy used in cultivating and processing the crops also has to be taken into account. There have been conflicting findings about the overall energy balance for the use of biofuels.[68] The area of land required to produce large quantities of biofuels would be very extensive. The most promising markets for bioethanol may be as an oxygenate added to petrol or as a replacement for diesel in special situations, rather than as a fuel for general use on its own.

8.75 There is experience of using ethanol fermented from sugar cane on a large scale in Brazil, where at its peak in 1989 it provided 28% of vehicle fuel. However, the falling price of oil made it uncompetitive and the government had to provide a subsidy of £2 billion a year to ensure its continued use. Because of concern over crop surpluses produced under the Common Agricultural Policy, the European Commission is working on a Bioethanol Directive which would almost eliminate excise duties on biofuels, at an estimated cost of £2.8 billion a year to EC governments. The aim would be to replace 5% of petrol and diesel consumption with bioethanol fermented from crops such as wheat, sugar beet, and artichoke.[69] Environmental groups have opposed this proposal on the ground that it might encourage the use of unsustainable agricultural practices and increase damage to habitats.[70]

8.76 Many firms worldwide are working to develop electric cars and other types of electric vehicle. The immediate attraction of electric power is that it does not produce emissions at the point of use. One stimulus has been a requirement placed on car manufacturers by Californian legislation that 2% of the cars they put on sale in the state from 1998 onwards (an estimated total of 40,000 cars a year), and 10% from 2003 onwards, should be 'zero emission vehicles' (ZEVs). The French take a particular interest, partly because their electricity is mainly nuclear and hydro. In a programme at La Rochelle, the utility EdF and Peugeot Citroen have equipped the town with recharging posts and provided 50 cars for rent. There are plans to provide Tours with several hundred electrically powered versions of popular Peugeot models.

8.77 At present, most UK electricity is obtained from burning fossil fuels. Electric propulsion could have an overall advantage in terms of air quality but, because of energy losses during generation and transmission of electricity, its net effect could be to increase the amount of carbon dioxide emitted per vehicle-kilometre. There are nevertheless strong reasons for regarding electric propulsion as a desirable direction for development in the longer term. An electric motor requires less maintenance than an internal combustion engine and, with few moving parts, is more reliable. Another significant advantage is the ability to recapture kinetic energy by using regenerative braking. There is a considerable reduction in noise. Last but not least, this form of propulsion opens up a wide range of options for future sources of energy for vehicles.

8.78 So far the problems of storing electric power for use by vehicles have proved difficult to overcome. They relate primarily to the weight, cost and life of the storage medium and the time required for recharging. Additional problems from the environmental point of view are that the materials suitable for use in electric batteries are often toxic or scarce and considerable quantities of energy may be needed in manufacture. Some of the materials that have been used would create a hazard if significant amounts were released in a road accident. We conclude that there is no overall benefit for the environment in the widespread use in the UK of electrically powered cars and heavy vehicles of the types at present available.

8.79 Lack of progress in battery development has encouraged investigation of other technologies. An alternative way of storing energy is a flywheel spinning in a vacuum at up to 200,000 revolutions a minute. This is much lighter than a bank of batteries and can be recharged within seconds. However a prototype car has yet to be built.[71] Another possible storage technology is a capacitor in which a massive electric charge is placed on the surface of carbon floss.[72]

8.80 Hybrid vehicles offer an alternative to complete dependence on batteries. A gas turbine or small internal combustion engine drives a generator which in turn produces electrical energy to power the transmission system or charge the storage system. Such a vehicle can run on electric power within urban areas; outside urban areas, it can operate like a conventional vehicle and recharge its storage system at the same time.

8.81 In the long term the preferred technology for electric propulsion may be fuel cells. These are based on an electrocatalytic process in which hydrogen, or a hydrogen-rich fuel such as hydrocarbons or aliphatic alcohols, reacts with oxygen to produce electricity, heat and water. The major technical problems in developing fuel cells are the limited life of the stack in which the reaction takes place and the weight and cost of all the principal components. If hydrogen was used, there would also be safety problems to overcome. The efficiency of fuel cells is likely to be double that of internal combustion engines, and they produce 40% less carbon dioxide and only minimal amounts of nitrogen oxides and other pollutants. Their ability to use a range of feedstocks as fuel may become important as oil and gas reserves decline. The vehicle industry is showing renewed interest in the long-term potential of this technology. **We recommend that the government increase its support for fuel cell research.**

8.82 Hydrogen can also be used as a very clean fuel in internal combustion engines (table 8.4). However, the amount of energy required to obtain hydrogen (for example by decomposing water) makes this an unattractive option in energy terms. Eventually photovoltaic cells may be developed to the point where they can provide renewable energy to obtain hydrogen at low cost. By then fuel cells are likely to provide a more advantageous way of using hydrogen to propel vehicles.

8.83 There are some purposes for which electric propulsion is particularly suitable. In the case of heavy duty vehicles which operate over relatively short distances in urban areas, the weight and bulk of electric propulsion systems may not be such a crucial consideration and they may well eventually be more attractive than natural gas-powered vehicles as a replacement for diesel vehicles. Trials now taking place in various parts of the world include a shuttle service of electric buses between the railway station and the town centre in Oxford.

8.84 The most promising direction for applying electric power to private vehicles may be neighbourhood electric vehicles (NEVs), intermediate between a cycle and a car. They would have one to three seats and limited performance which would be adequate for use within a town or city. There are, however, safety concerns if they were to share road space with conventional vehicles; the first NEV in the UK, the Sinclair C5, failed on these grounds. If NEVs are to gain popularity, they may need access to dedicated cycle/NEV lanes that avoid high risk areas. General policies of traffic calming and lower urban speed limits will enhance their prospects. NEVs are often envisaged as an addition to a household's car or cars. If they were to become popular on that basis, the environmental benefits would be much less than if they were to replace household cars.

8.85 Electric and hybrid vehicles could have a valuable role in improving local air quality in accordance with air quality management plans. The biggest benefit in terms of emissions would come from buses and other heavy vehicles although NEVs could have a contribution to make. **We recommend that the government encourage the development of electric power for public transport or fleet vehicles which operate in urban areas with frequent stops and for small private vehicles for neighbourhood use.**

The roles of government and citizens

8.86 To summarise the message of this chapter, improvements in vehicle design have a considerable contribution to make in achieving the targets for noise and emissions of pollutants and carbon dioxide proposed in chapters 3 and 4. In general the most rapid and effective progress is likely to come from innovations by vehicle and component manufacturers. For some difficult problems of special importance, such as developing a DeNOx catalyst for diesel engines, collaborative research programmes are desirable sponsored by the European Commission. Fuel cells have considerable long-term potential for propelling vehicles and research into their use for that purpose should receive full support from the government, as should the development of other forms of electric propulsion systems for the most promising applications.

8.87 Government must use an appropriate combination of direct regulation and economic instruments to force the pace of technological development and foster markets for new products. In the case of noise levels and the emission of pollutants direct regulation in the form of EC legislation should continue to be the primary method used to reduce the environmental impact of vehicles. Direct regulation should extend beyond compliance with limits for new vehicles to include much more effective enforcement of environmental standards applying to the existing fleet.

8.88 The use of economic instruments in the form of increased fuel duty and graduated vehicle excise duty should be the primary method for obtaining reductions in carbon dioxide emissions through increased fuel efficiency. In addition, economic instruments can play an important part in promoting the early adoption of less polluting designs of commercial vehicle and unfamiliar technologies with environmental advantages such as use of natural gas as fuel.

8.89 The attitudes of the public as purchasers of cars and as car drivers will be crucial. Economic instruments, especially the much higher fuel duty, will influence those attitudes. However, there needs to be a more deep-rooted cultural change in the way people view car speed and performance. We discuss what government can do in relation to speed limits and other traffic management measures in chapters

11 and 12. Fuel efficiency deteriorates fairly quickly as car speeds rise above 55 mph (figure 8-I) and emissions of nitrogen oxides (and, in cars fitted with catalytic converters, carbon monoxide) increase. The risk of severe accidents also increases. Over and above these specific consequences of high speed, cars will continue to be more environmentally damaging than need be the case so long as they are predominantly designed for long-distance travel at speeds above the legal speed limit in the UK and used to a large extent for relatively short journeys in urban areas for which their fuel efficiency is low and their emissions of pollutants even larger than at high speeds.

Chapter 9

TRANSPORT AND LAND USE PLANNING

9.1 Relationships between land use and transport are of considerable significance for the environment. Transport systems have direct land requirements and major effects on habitats and landscapes (considered in chapter 4); they also interact with land use in more subtle ways and in the long term. Transport is an important factor in the evolution of settlements; and patterns of development in turn influence the demand for movement.

9.2 Land use and travel patterns have reinforced each other in the emergence of a more dispersed and highly mobile society. Decentralisation of population and employment to suburbs and, since the 1950s, to smaller free-standing settlements in non-metropolitan areas, has been permitted by rapidly increasing personal mobility. This outward spread of people and jobs has been accompanied by an increasing physical separation of homes, jobs and other facilities, reinforced by the concentration of shops, schools, hospitals and other services into fewer, larger units.[1]

9.3 Interactions between land use and transport are neither simple nor readily quantifiable. But since the growth of traffic will have to be restrained to meet environmental targets, it is a cause for great concern that modern land use patterns are making car and lorry use a matter of necessity rather than choice. Much of the evidence submitted to the Commission reflected this concern, together with a growing recognition that land use and transport planning should be integrated at all levels of policy making, not merely in terms of the formal and statutory responsibilities, but in ensuring that the signals made and decisions taken are consistent with each other.

9.4 In this chapter we look at both sides of the dynamic relationship between land use and transport and its implications for the development of a sustainable transport system. We begin by considering how the cost of travel and the provision of transport infrastructure (particularly roads) influence activities, development pressures and patterns of land use. We then summarise the evidence concerning the effects of the density and distribution of development on travel behaviour. The remainder of the chapter is devoted to examining recent changes in land use planning policies at the national level and recommending further measures that we believe the government should take to prevent land use and transport policies coming into conflict in a wasteful and environmentally damaging way.

The influence of transport on land use

9.5 In this section we consider how the provision of transport infrastructure influences land use. The cost of travel — measured in terms of both money and time ('generalised cost') — affects the mobility of individuals, the accessibility of locations and, potentially, the development of land. New transport infrastructure tends to reduce the generalised cost of travel (usually by reducing journey times) and in consequence may give rise to development pressures reflecting new patterns of accessibility. We have already recommended increasing the monetary costs of travel so that they better reflect environmental costs which have previously been excluded. We therefore go on to consider briefly the possible longer-term effects of such changes on land use.

Effects of transport infrastructure

9.6 Historical evidence (on the influence of the railways in the 19th century and on development around ports and airports for example) demonstrates the power of transport infrastructure to shape patterns of land use.[2,3] More recently the car and lorry have not only enhanced the centrifugal tendency initiated by other forms of transport, but have spread the means of access rather than concentrating it in corridors. This has enabled manufacturing activities, professional services, warehousing and distribution, and housing development to be much more widely dispersed.

9.7 Provision of transport facilities alters the way in which activity is distributed within the existing built environment, as well as leading to new development pressures and changes in land values and use.

For example, the motorway network has enabled 'just-in-time' goods deliveries to be substituted for traditional stockholding in many industries. Similarly, large numbers of people use motorways to travel to work over longer distances than would have otherwise been practicable. The question of whether new roads generate additional traffic has recently been studied by the Standing Advisory Committee on Trunk Road Assessment (SACTRA). Although SACTRA's findings have not yet been published, press reports suggest it has concluded that there is such an effect.[4] We discussed earlier some of the implications for the methodology of highway investment appraisal of the effects of new roads in both redistributing existing travel requirements and in generating new trips as people adjust to fresh patterns of accessibility.

9.8 Because of the long time-scales involved and other factors influencing development pressures, the effects of specific infrastructure projects are not easy to measure or predict. As a report by ECOTEC on the potential of land use planning to reduce the growth of transport emissions points out: 'The limited number of studies of the impact of transport infrastructure projects on development in the UK, and the evidence that the nature of the impacts depends on a series of contextual factors, warns against simple or general conclusions'.[5] Nevertheless, broad patterns of development indicate a permissive, if not a causal, relationship between transport systems and land use. Though there are differences of opinion about the strength of this relationship, many (including the AA and the Institution of Civil Engineers) argued in evidence to the Royal Commission that the effects are substantial, especially in the medium to long term. In the light of its review, the ECOTEC report concluded that road, and in some cases commuter rail, schemes can significantly affect development pressures in two circumstances: where they facilitate movement between an area with a buoyant land market but development constraints, to another with lower land prices but more development opportunities; and where they significantly change the relative accessibility of parts of an urban area.

9.9 One of the few detailed studies of such effects shows how the original stretch of the M40 motorway from London to Oxford had a dynamic effect on the use of near-by land.[6] The construction of the motorway divided previously open country and created a new boundary for the development of neighbouring settlements. It also altered patterns of access to the settlements and made the motorway junctions into attractive locations for activities which depend on easy vehicle access. As a result, development which was not in accordance with the approved development plan took place on previously undeveloped land. Its nature and intensity (superstore, multi-screen cinema, home furnishings store, and headquarters office block, all with associated parking) are radically different from previous development in the area (housing, a hotel, and open space). Traffic generated by the development has added to highway problems and brought forward the need for further improvements to the road network.

9.10 New roads around urban areas create particularly strong development pressure. Modelling carried out by the International Study Group on Land Use/Transport Interaction suggests that 'provision of a fast outer ring road encourages more decentralisation of employment, particularly service and retail.'[7] A prime example is the London orbital motorway, the M25. Between 1989 and 1991, an estimated 26 million square feet of offices (enough for 160,000 workers) were completed in the districts surrounding the motorway[8]; and in 1993 some 50 million square feet of office space were awaiting construction or planning permission.[9] Sites close to the motorway have also proved extremely attractive for retail and leisure facilities. Such development generates traffic and undermines the strategic purpose of the road; the traffic generated will cause greater congestion than had been expected, and thereby reduce the achieved time savings to less than had been assumed when calculating whether road construction was justified. This means that more vehicles (or vehicles travelling a longer distance) receive benefits from the road, but the benefits per vehicle are smaller. Using the social cost-benefit principles applied to the current procedures for evaluating road schemes, the net result of these two effects in congested conditions is to reduce the total estimated benefits.

9.11 Development pressures may be difficult to resist even when they run counter to existing planning policies (the planning system in England and Wales is outlined in box 9A). The Institution of Civil Engineers told us that powerful developers have often been able to achieve amendments to local plans either directly, by negotiating with the local authority, or by having the local authority's decision overturned on appeal. It remains to be seen how this situation will be affected by the requirement, under the Planning and Compensation Act 1991, for planning decisions to accord with the development plan, unless material considerations indicate otherwise. In any case we consider the present situation, in which trunk roads are effectively superimposed on existing development plans, to be unsatisfactory and make appropriate recommendations below.

9.12 Bypasses may create development pressures on a smaller scale. There may be cases where it is sensible to plan for new development at or near to the intersections on bypasses but, in general, this will tend to clog the bypass with local traffic, reducing its effectiveness; it is better avoided. Some bypasses have attracted very considerable development. Purley Way, part of the A23 trunk road, which bypasses Croydon town centre, has become one of the biggest centres in the country for out-of-town shopping.[10] There are now six retail parks in the industrial area bordering the road, with a total of over 1 million square feet of retail space and parking provision for over 5,600 cars.

9.13 We conclude from the available evidence that provision of new transport infrastructure, especially new trunk roads and motorways, can affect land uses and hence further demand for travel as well as the environment and should be taken into account in land use planning.

Effects of travel restraints on land use

9.14 Just as increasing mobility has had a profound influence on patterns of development, restraints on travel are likely to lead to modifications in land use patterns. The nature of the modifications is difficult to predict because people can respond to price increases or other changes without permanently altering their travel behaviour: social and behavioural influences on travel patterns are still relatively poorly understood.

9.15 We have already discussed price elasticities of demand for fuel and for travel (7.53ff). Evidence suggests that land use patterns are unlikely to respond in the short to medium term to increases in the cost of travel. But in the longer term (say more than ten years), they might respond to sustained increases in real fuel prices, as people adjust locations and lifestyles.[11,12,13] If so, there might be some slowing of recent land use trends, with urban and suburban centres becoming increasingly attractive, though a simple reversal of decentralisation is unlikely. These effects remain speculative, but the evidence suggests that the price mechanism alone is unlikely to change land use trends in efficient and equitable ways. The land use planning system will have an important role in enabling travel patterns to be adjusted in response to higher travel costs. In any scenario in which travel restraints are great enough to have a longer-term influence on development pressures, the land use planning system must itself be a policy instrument in reducing the demand for travel.

9.16 Other forms of restraint might have different effects on land use. For example, there are indications that road pricing would have significant short to medium term influence on travel patterns in specific areas.[14,15] A study for the Department of Transport (DOT) on the effects of road pricing on land use patterns in London and the South-East is as yet unpublished. High parking charges and reductions in highway capacity might, in the absence of attractive alternative means of access to urban centres and effective planning policies, increase pressure on land uses in areas where such restrictions do not apply. Strong planning policies are therefore needed to prevent undesirable outcomes.

Land use patterns and travel behaviour

9.17 It is widely held that patterns of land use influence travel behaviour and that development density, the mixing of land uses and the location of development in relation to transport networks are related to the frequency, distance, speed and mode of travel. The strength and significance of these interactions is a matter of some dispute, but there is a broad consensus that land use planning policies should seek to minimise the need for travel and encourage the use of less polluting forms of transport. Planning policy will be an important way of maintaining access and choice in a future which is less heavily dependent on the private car.

9.18 If residential areas and other facilities are close together, there is greater potential for trips to be shorter, for journeys to be more readily combined and for more trips to be made on foot or cycle. The separation of activities is related to development densities and to the extent to which different land uses are clustered together or dispersed, confined to city centres or integrated in residential areas. Higher densities permit greater choice for less travel, but with some sacrifice of space. If some jobs and services are integrated in residential areas, access to them may involve less travel.

9.19 The pattern of land uses affects the need for travel, that is, the minimum amount of travel required to carry out a given set of activities. People may, however, choose to travel further than the minimum, particularly if travel is cheap and convenient; evidence suggests that when mobility is high, travel is not very sensitive to patterns of land use.[16] The pollution caused by transport depends on mode and speed as well as distance travelled. Shorter journeys in congested conditions may be as polluting as longer journeys elsewhere.

9.20 Attempts to identify 'ideal' land use patterns are unlikely to be successful; there are too many variables, too many other considerations and always a need to be flexible and responsive to local circumstances. Some land use patterns are clearly transport intensive, however, and some are more appropriate for a future less dependent on cars. For example, development within existing towns and cities is likely to require less car-based transport than dispersed, low-density or piecemeal development at or beyond the urban periphery. Similarly, development which helps towns, or well-defined areas within larger cities, to be relatively self-contained has the potential to generate less travel than dormitory suburbs or green field business parks which by their very nature require car transport for commuting and other travel.

Development patterns and land use

9.21 Current knowledge of the effects of land use on demand for travel derives mainly from models and from empirical comparisons of different land use patterns. Neither of these approaches has provided unambiguous information about the relationship between land use and transport, nor are they ever likely to do so, given the complexity of land use and transport patterns, and the social and economic processes in which they are embedded. Nevertheless, they provide some insight into development patterns and planning policies that could provide access to a range of jobs and services without the need for excessive mobility.

9.22 Many studies have explored the relationship of urban density to travel patterns. Most confirm that as urban density increases, travel (and transport energy use) fall.[17] One international comparison suggested that fuel consumption for transport rises sharply as densities fall below about 30 persons per hectare.[18] In the UK, National Travel Survey data show that distance travelled by all modes, and by car, varies with density; travel demand rises quickly as densities fall below 15 persons per hectare and falls sharply as density increases above 50 persons per hectare.[19] For comparison, densities of new private sector developments are typically around 25 dwellings per hectare; population densities in British cities are typically between about 30 and 50 persons per hectare. There is some evidence that the total amount of time spent on travel by all modes taken together is roughly the same, on average, in areas of widely different densities. What seems to happen is that in denser areas there is greater use of public transport, at slower speeds, for shorter journeys, and in less dense areas there is more car use, at higher speeds, for longer journeys.[20] But some research suggests that income and car ownership are more important determinants of travel demand than population density, and the extent to which density acts as an independent variable is disputed.[21] One way of reducing the distances that people need to travel is by increasing densities but this may not be sufficient to reduce travel demand.

9.23 The mixing of residential and other land uses to reduce the need to travel has also been frequently advocated. Some models and empirical work suggest that in large urban areas, dispersal of employment and services (especially if these are clustered at particular locations) would lead to lower travel demand[22]; other studies find that mixing of land uses has little effect on travel, might encourage car use at the expense of public transport or may lead to greater fuel consumption than centralisation if people are very mobile.[23] In some industries, integration might be compatible with modern trends towards smaller units but where there are large economies of scale, as in retailing and certain sorts of manufacturing, the trend has been towards fewer, larger outlets. In part, such economies of scale have been achieved at the expense of the environment.

9.24 The lack of consensus results to some degree from different assumptions about mobility, which are related to social, economic and policy factors. These affect both the strength of the land use/transport relationship and also the relative efficiency of different land use patterns. The extent and quality of public transport networks are also important factors. In some studies, land use has very little effect on travel behaviour; in others, differences in travel or fuel consumption vary by as much as a factor of three between land use patterns.[24]

9.25 The compact centralised city emerges as one potentially transport-efficient pattern of urban development. A recent study of alternative development patterns for a hypothetical city and its surrounding region found that the compact centralised city had some advantages over a pattern of development dispersed in the hinterland, reducing travel demand through shorter trip lengths by 9–14% over 25 years.[25] In contrast, further use of the same model to simulate alternative development patterns for an archetypal town produced no significant variations in fuel use between the alternatives.[26]

9.26 The former findings are broadly reinforced by simulations in the ECOTEC study which concluded that concentrated, relatively centralised patterns of growth of free-standing towns were likely to be more efficient from an emissions perspective than lower-density, more peripheral patterns. Simulations found that the most effective land use and transport strategy in a metropolitan sub-region was a combination of extended public transport provision, limited additional highway capacity and the regeneration of existing centres. After a 20-year period, this combination produced a 16% reduction in peak hour air pollution from private vehicles and minimised journey times for public and private transport users, albeit marginally. The differences between patterns are modest — and given the uncertainties may be within the error of the estimates — but this is not surprising given the assumption that people will be as mobile as at present. Overall, the findings suggest that urban regeneration could have transport and environmental benefits; but measures to control traffic and traffic congestion would seem to be a prerequisite, both to reduce the demand for travel by car and to contribute to higher quality urban environments.

9.27 Smaller towns and 'urban villages' within existing cities can also be relatively efficient in transport terms. One study of Oxfordshire settlements concluded that important factors in transport efficiency were 'the availability of good local services and facilities, local employment, good public transport, and a high proportion of walk trips'.[27] However, a pattern of smaller urban areas may generate more travel than compact centralised cities because of movements between towns. We were told by the County Council that this happens in Kent, where 18 towns of over 15,000 population are in fairly close proximity. The more mobile people are, the larger, or more isolated, any urban area would have to be to achieve thresholds for economic provision of services and to approach self-containment.

Land use, location and transport mode

9.28 Much of the evidence we considered suggested that land use policies could influence not only the need for travel but also the choice of mode. Shorter journeys are more likely to be made on foot or by bicycle; concentration of homes and facilities maximises access to public transport routes and encourages high load factors; dispersed patterns of development are difficult to serve by conventional public transport; and some patterns of development effectively require the use of cars. Similarly, the redevelopment of former railway or waterfront land forecloses the possibility of using those modes for the carriage of goods.

9.29 Land use configurations may at least in part account for significant variations in travel patterns in different locations. For example, a comparative study of Milton Keynes and Almere (Netherlands), new towns designed in about 1970, showed that in Almere, 64% of local shopping trips were made on foot or by bicycle, but in Milton Keynes only 34%.[28] Although the study was not able to isolate the relative contributions of land use patterns, travel facilities and cultural influences, it suggests that the cause lies partly with the layout of Milton Keynes which was designed on a low-density (20 houses per hectare) gridiron pattern.

9.30 Factors other than distance are significant. The degree of emphasis on safety, amenity and priority for cyclists and pedestrians may have an important effect on modal choice; this view was supported in much of the evidence submitted to the Commission. As discussed later in the report, surveys show that reducing the volume and speed of traffic in urban areas is a high priority for many people. There appears to be a growing belief that attractive and viable towns are not compatible with dominance by the motor car and people appear to be prepared to accept the need for change. The planning system should be used to complement other measures to encourage walking and cycling and create an attractive built environment. For the most part, these policies are a matter for implementation built on existing good practice, rather than further research.

9.31 The location of specific developments has an important influence on mode of travel. Unlike many aspects of land use policies, this can have rapid and striking effects. Many facilities are now sited to be accessible by car and lorry and may be virtually inaccessible by any other means. They may also reduce choice in surrounding areas and the viability of facilities which can be reached on foot or by public transport. Dudley, for example, is the worst affected of several towns in the West Midlands in which the quantity and quality of retailing has been adversely affected by the Merry Hill retail centre.[29] Whilst such developments might ease congestion in urban centres, they may also act as significant traffic generators.

9.32 Effects are striking. ECOTEC reports that in a major Midlands office centre the proportion of train journeys to work falls off from 11% for an office next to the station to 0.4% for an office a quarter of a mile away, while one survey suggested that 93% of employees in out-of-town office sites travelled to work by car.[30] A comparison of supermarket sites in London showed that of journeys to a free-standing outer-London site, 95% were by car and 3% on foot; to an inner-London site, 33% of journeys were by car and 50% on foot.[31] Two-thirds of employees' journeys to work and 85–90% of visitor trips to Merry Hill are by car.[32]

9.33 There is similar evidence from abroad. In the Netherlands, when the Ministry of Physical Planning was moved to a location adjacent to the railway station in the Hague, the proportion of employees using public transport increased from 34% to 77%. Research suggests that the relocation of two Amsterdam city hospitals to a new site outside the city led to an increase of 116% in the total number of car kilometres travelled to and from the hospital. Another Dutch finding is that, in general, the transfer of offices from the city to the perimeter raises car use by 10–40% and reduces travel by public transport and bicycle by 4–8%.[33] One study of the impact of the Bay Area Rapid Transit system in California showed that people who had moved to clusters of housing development very close to the stations made up to five times as much use of rail as people living further away, although the study concluded that if 'transit-based housing' of this form were to give significant environmental benefits it would have to be accompanied by growth of employment opportunities which were also clustered close to stations, and some increase in motoring and parking costs.[34] Thus whilst greater internalisation of environmental costs could have a direct influence on the attractiveness of such developments, planning has a key role to play.

Densities, mixed land uses and amenities

9.34 Some of the land use measures which might reduce the need to travel and encourage the use of public transport, walking and cycling may appear to conflict with amenity. Mixing of land uses may bring into closer proximity activities which in the past it has been thought desirable to separate. Higher densities may mean less space for individual households and possibly a loss of open spaces in urban areas. Concern has been expressed about this possibility and the Department of the Environment has commissioned a study on the intensification of development in urban areas.

9.35 Denser urban development need not be of low quality. Advances in pollution control may enable activities to be brought closer together than has previously been acceptable. Moreover, quite moderate densities could (potentially) achieve many transport-related objectives. We also note that large areas of land in urban areas are now required to accommodate cars — Friends of the Earth suggested up to 20% of urban land area. If dependence on the car were reduced, some of this could be released so as to reduce the potential conflict between need for urban green space and a more compact pattern of development.

9.36 We conclude, from a large amount of often confusing evidence, that there is no single pattern of land uses that will reduce the need to travel and so reduce the effects of transport on the environment. Nevertheless some important principles emerge. A range of settlement patterns, including compact, centralised cities and small to moderate-sized towns or urban villages with a good mix of employment and services, has the potential to offer efficiency, access and choice. The size of settlement and precise arrangement of land uses could never be planned solely on grounds of transport efficiency but avoidance of obviously travel-intensive development patterns would be a significant improvement on the present situation. Land use planning has a role here but there are two important qualifications. Bringing different land uses into closer proximity is a necessary but not a sufficient condition for reducing travel demand; and the environmental and social costs of mobility may have to be internalised first, before it becomes 'economic' to provide more local facilities.

Implications for planning policy

9.37 The evidence on land use and transport interaction points to the need for a full consideration, at all levels of decision making, both of the land use implications of transport infrastructure provision, and of the travel and consequent environmental implications of land use policies. This consideration should cover the direct and indirect effects and the medium to long term as well as the more immediate future. It will translate into specific policies which local planning authorities should apply. In our view these

BOX 9A THE PLANNING SYSTEM IN ENGLAND AND WALES

The planning and control of development in England and Wales is operated largely by local planning authorities within a framework set by central government. For these purposes, 'development' is defined in section 55 of the Town and Country Planning Act 1990 as 'the carrying out of building, engineering, mining or other operations in, on, over or under land, or the making of any material change in the use of any buildings or other land.' The Act excludes some operations or uses of land from this definition. National policy guidance is issued by the Department of the Environment and Welsh Office, mainly in a series of Planning Policy Guidance (PPG) notes. There is no statutory regional planning system but, in England, the Department of the Environment issues Regional Planning Guidance (RPG) notes which are based broadly on advice prepared by groups of local authorities. (The equivalent in Wales is Strategic Planning Advice which is being prepared by the Welsh Office in response to advice submitted by local authorities.) This advice informs the structure plans prepared by county councils which cover broad land use issues and provide a framework for the local plans in which district councils set out more detailed development policies for their areas. In areas served by single-tier authorities (at present, London and metropolitan districts) two-part unitary development plans fulfil the purposes of structure and local plans.

The process of drawing up RPGs is intended to allow the interactions between land use planning and transport infrastructure to be examined so that the guidance itself can promote both their integration and co-ordination, and strategies to reduce the need to travel. All types of development plan should include land use policies and proposals relating to the development of the transport network and to the management of traffic. They should also include all schemes in the government's trunk road programme. Planning decisions must accord with the development plan unless material considerations indicate otherwise.

should include the promotion of development which does not rely on car access (hence no new out-of-town superstores, retail centres and business parks unless they bring demonstrable environmental benefits); the location of developments which generate high travel demand where they can be reached on foot, bicycle or public transport (planning policies in the Netherlands attempt to do this and to ensure that the location of other types of business matches their transport needs[35]); the encouragement of a wide range of facilities at local level so that journeys can be shorter and made on foot or bicycle; traffic management (for instance by limiting parking provision); the encouragement of housing development which enables people to live near their work; the siting of freight depots where they can be served by rail or water; the encouragement of lively and attractive town centres; and the adoption of measures to foster walking, cycling and public transport.

9.38 There are encouraging signs that such policies are being adopted and that, more generally, the significance of the links between land use, transport and the environment is being recognised at all levels of policy making. Agenda 21 acknowledges its importance, while the European Commission identifies land use planning as one of five key transport measures, and transport itself as a major target area, in its Fifth Environment Action Programme. In Britain the 1990 White Paper on the Environment explicitly acknowledged a role for land use planning in influencing the demand for transport and this was again endorsed in the UK Strategy for Sustainable Development. The White Paper also contained a commitment to review all Planning Policy Guidance Notes (PPGs) to ensure that they reflect the current environmental agenda. Of particular relevance to land use and transport are the revised PPG 6, on town centres and retail development, published in July 1993 and PPG 13, on transport, published jointly by the Environment and Transport Departments in March 1994.

9.39 As yet, there are no equivalent policy statements for Scotland but similar guidance is under preparation by the Scottish Office, following the commitment in the 1990 White Paper. A new series of National Planning Policy Guidelines (NPPGs) and Planning Advice Notes is being issued. The Scottish Office told us that, where appropriate, published NPPGs address transport issues. The Welsh Office, too, is preparing planning guidance on transport.

9.40 PPG 6 places renewed emphasis on the economic, social and environmental importance of town centres, and on access for everyone to a wide range of opportunities. Local planning authorities must consider the effects of new retail developments on the vitality and viability of existing town centres, and ensure that development is sited where it is likely to be accessible by a choice of means of transport, and to encourage economy in fuel consumption.

9.41 PPG 13 advises how local authorities should integrate transport and land use planning. It emphasises the need to reduce growth in the number and length of motorised journeys, encourage more environmentally friendly means of travel and reduce reliance on the private car. The specific measures it advocates are broadly consistent with those which we endorsed above. It rightly recognises that, although the number of new developments each year is relatively small, they contribute to patterns which will endure long into the future and therefore have potential to influence — for good or ill — the long-term effectiveness of policies to reduce the environmental effects of transport.

9.42 PPG 13 is an important statement of intent. Its effectiveness will be judged by the policies and decisions which follow it and other planning guidance. Ministerial pronouncements have been equivocal, stressing, for example, that the government is 'not against all out-of-centre developments.'[36] It will be important to demonstrate consistent commitment to the new policies over the long term, especially in appeal decisions taken by the Secretary of State.

9.43 Planning policies are a necessary instrument for reducing travel demand but in isolation they will not suffice. The effectiveness of the new measures will depend crucially on the transport policy context as a whole and on the ability of local planning authorities to achieve the necessary shifts in policy, when it is not always clear that they have the powers to do so. To move towards more sustainable land use patterns will not be easy, and it will be necessary for the Department of the Environment to support consistently the aims of PPG 13. For this reason we welcome the Department's plan of producing a good practice guide to demonstrate practical ways of implementing appropriate policies. **The government also plans to monitor the effectiveness of PPG 13. We welcome this and recommend both that the results be made public and that any shortcomings which are revealed be corrected without delay.** Further long-term work may also be needed to understand fully the effects of planning decisions on travel behaviour.

9.44 It will also be necessary to address specific limitations in the wider policy framework which will otherwise frustrate the intentions of integrating land use and transport policies. For example, there are currently no targets relating to traffic growth or emissions from the transport sector to provide a context within which local planning authorities might assess the contribution of land use measures. (How should they judge, for example, whether the travel and emissions implications of particular policies or proposals are acceptable?) Furthermore, the authorities, although they are being urged to relate development to public transport, have limited control over public transport provision. Nor can they control the routes of trunk roads: PPG 13 sees 'little need or scope to vary the [trunk road] programme or alter priorities within it, in connection with local planning priorities'.

9.45 Clearly planning guidance notes, even on broad subjects such as transport, are not necessarily the most appropriate medium in which to address wider questions of policy. Measures which we recommend elsewhere concerning targets, public transport and the roads programme address some of the limitations outlined above and if adopted would help to ensure the effectiveness of locational policies. But there is still a need for greater consistency of purpose between policies, and for environmental objectives and targets adopted at the national level to be translated into a local framework. **We recommend that the government explore, in conjunction with the local authority associations, the potential to define and adopt targets for specific planning areas: these should be addressed in the context of local air quality plans (13.28ff) and might relate to emissions and traffic growth as well as factors such as cycle routes built or bus lanes designated.**

9.46 Individual local planning authorities would also be assisted by regional planning guidance consistent with the aims of PPG 13. The high mobility of suppliers, employees and customers increases the range of locations acceptable to developers and weakens the resolve of local planning authorities to refuse planning permission for traffic-generating facilities. This is a particular problem in the metropolitan areas and London where there is no longer a single strategic planning authority. PPG 13 rightly warns of the destructive potential of competitive provision by neighbouring authorities (citing the example of parking). Its remedy is to include strategic policies in RPGs and structure plans. The requirement for planning decisions to accord with the development plan, unless material considerations indicate otherwise, goes some way to check the trend to exploit different approaches by neighbouring authorities but we doubt whether RPGs and structure plans alone will be sufficient to ensure a consistent approach. Later in this report we recommend ways of achieving more consistent approaches between authorities.

9.47 It has been a long-standing criticism that proposals for new trunk road schemes are often imposed on development plans without sufficient regard to the needs of the local area or the development pressures which they may generate. One topical example is the Department of Transport's proposal for link roads in the south-west sector of the M25, which in the opinion of the South East Regional Planning Conference 'are not compatible with the regional strategy — indeed they are in conflict with it.'[37] Surrey County Council, which opposes the M25 schemes and therefore omitted them from its structure plan, is reported to have been directed by the Department of the Environment to include them.[38] Stressing that trunk roads are primarily for long distance through traffic, PPG 13 now states that decisions on the trunk road programme will 'take into account' the overall strategy set out in Regional Planning Guidance. Since that guidance is issued by the Secretary of State, this may not mark much of a change in practice. Thus there appears to be a contradiction between the government's urging of local planning authorities to reduce the need to travel, and the way the trunk road programme, still committed to serving traffic growth, is brought into structure plans. The regional planning conferences do not appear to contain a mechanism to analyse interactions between land use planning and transport infrastructure (including the trunk road programme) at anything other than a rather late stage and a level of generality which has effectively enshrined DOT's national traffic forecasts and policies for interurban traffic. Hence land use pressures created by trunk routes are left to local authorities to address in their structure and local plans.

9.48 We consider that this is thoroughly unsatisfactory. Proposals for new or improved long-distance routes should not be imposed from above but should flow from the development plans and should be open to public scrutiny as those plans are developed. Since routes of this nature necessarily affect several local authorities, overall need for them should be considered in a forum which is able to take broad interests into account. **We therefore recommend that the overall need for a new or improved long-distance route, the possibilities of providing improvements by road or rail, and alternatives based on managing the level of demand or giving different priorities to short and long-distance users of the route, should be jointly assessed by the regional planning conferences and the government's newly integrated regional offices.** Later in this report we recommend strengthening the powers and status of regional planning conferences: this additional task would itself both build on and contribute to that strengthening. The assessment would take fully into account how best to protect the environment from damage at an acceptable cost in accordance with the concept of Best Practicable Environmental Option. A high degree of public involvement would be essential. In addition, as the route was incorporated in structure plans, the examination in public of the plan could provide a further opportunity for public discussion of the need for development of the route.

9.49 This new procedure would integrate broad land use planning and transport issues, openly, at a regional level, but the need for integration is at least as strong at a more local, detailed level. Local authorities and developers are required to obtain planning permission for their road schemes and we see no good reason why the government's proposals should be treated differently. The Department of Transport told us in oral evidence that its procedures provide the same sort of safeguards as the planning legislation. It argued that trunk roads and motorways form a national system and local authorities would not therefore be competent to assess proposals. Yet, according to the government's present plans, the national road programme will in the future be concentrated on widening existing motorways and providing bypasses rather than building completely new routes.

9.50 For decisions on road schemes to take full account of land use factors, this separate treatment must end. **If a system of trunk roads is retained, we recommend that all trunk road schemes be considered initially as an intrinsic part of local authority structure plans and integrated fully into the development control system.** The effect of this recommendation would be that all proposed road construction would be treated like other major development.

9.51 If a need for a new road was identified, the proposals would be developed in co-operation with the appropriate strategic planning authorities so that they reflected, and were reflected in, development plans and were not imposed on them. When a scheme was to be built, an application for planning permission would be submitted to the local planning authority. Permission would normally be granted if the scheme accorded with the development plans. If permission were refused, an appeal against the decision could be lodged with the Department of the Environment in the usual way. Individual local authorities would thus be unable to obstruct schemes for which there was a national or regional need but they could influence the details of a scheme. Their involvement in the decision would encourage co-operation between developer and planning authorities throughout the design process. Compulsory purchase orders would follow the grant of planning permission (when necessary).

Resource implications

9.52 Land use policies will not lead people to use cars less if there is no attractive alternative. PPG 13 acknowledges that good quality public transport, and town centres which are safe and attractive to cyclists, pedestrians and public transport users, are essential if habits are to be changed. The government should therefore ensure that these new priorities are observed in its annual resource allocations for local authority transport expenditure and that the necessary resources are made available. We discuss these issues further in subsequent chapters.

9.53 We believe that the priority must now be to implement the contents of PPG 13 without delay. This will place no small burden on local authorities. Their professional staff will need to develop skills which bridge the traditional gap between transport and land use planning. In particular, there will be a need for readily accessible methodologies for assessment of the travel implications of land use policies and location decisions as well as the land use implications of transport policies. **We recommend that the government supplement PPG 13 and other recent guidance with more specific advice on these issues.** There may also be a need for new types of training in planning education to enable these tasks to be accomplished.

Strengthening the planning system

9.54 Our study has suggested several other ways in which the planning system could be adapted to take better account of environmental factors.

9.55 In many cases policy decisions rather than 'immutable trends' have shaped land use and transport patterns. Policy makers and service providers appear to have been too ready to seek road-based solutions to transport demands without sufficient systematic exploration of alternatives. Strategic environmental assessment provides an opportunity for such exploration, and a means to identify the Best Practicable Environmental Option (BPEO) in transport policy. Extending the latter concept, explored by the Royal Commission in its Twelfth Report, means that when transport policies are developed or transport decisions taken, attention must be paid to assessing how best the environment (land, air and water) can be protected from damage, at an acceptable cost (see box 9B). It will include assessment of demand management options, using the land use planning system or other instruments. A key element in the BPEO approach is the imaginative search for alternative options. In this context, these should include non-transport options, and options which do not involve the construction of new infrastructure, or which make use of alternative modes, possibly in combination with demand management methods. That the government has stressed the link between land use plans and transport policies and programmes is a helpful development in this respect although we wish to see similar links established at regional, national and international levels. Clear, quantified targets must be established to provide a framework for environmental appraisals.

9.56 **We recommend that decision-making at all levels of transport policy be based on the identification and pursuit of the best practicable environmental option.** It should be an integral part of planning procedures and should inform regional, structure and local planning as well as decisions on applications for planning permission. The long-term aim of this approach would be to provide people with access to the goods, services and activities they desire, without unsustainable environmental degradation. In this sense, the planning system is an instrument for designing flexibility and personal choice into developments.

9.57 There is a risk that the investigation of a large number of different ways of achieving objectives might create more widespread planning blight. Decisions should therefore be taken as quickly as is consistent with adequate exploration of options.

9.58 It seems somewhat anomalous that, when a new development is approved, developers do not bear any of the external costs of the traffic that the development will induce, (although they are now sometimes charged for the provision of additional road capacity). **We recommend that the government consider ways of making developers responsible for a charge reflecting the external costs of traffic induced.** This would be especially useful as an interim measure before a full external cost system was applied to road users directly, since it would be much easier to administer. It should apply not only

BOX 9B THE PRINCIPLES OF BPEO APPLIED TO TRANSPORT

The concept of Best Practicable Environmental Option (BPEO) was defined in the Royal Commission's Twelfth Report as 'the outcome of a systematic consultative and decision-making procedure which emphasises the protection and conservation of the environment across land, air and water. The BPEO procedure establishes, for a given set of objectives, the option that provides the most benefit or least damage to the environment as a whole, at acceptable cost, in the long term as well as in the short term.'

BPEO was first put forward as an approach to the control of industrial pollution but it also has a role to play in strategic planning and in matters of national and global importance. It can help in the development of transport policies and in finding solutions to transport problems. The following hypothetical example expands on material in the Twelfth Report to illustrate how the BPEO approach might be used in addressing the problems of traffic congestion and pollution in a town.

Step 1 : Define the objective

The objective must be defined in terms which do not prejudge the means by which it is to be achieved. It would therefore be inappropriate to set as the objective 'to improve car access to the town centre' or 'to exclude cars from the town centre.' For this illustration the objective might be: **to make the town centre a more attractive and accessible place.** At the same time as the objectives are set, a statement of the constraints imposed on the decision-making must be formulated, whether the constraints are legal, technical, social or economic. The implications of alternative proposals should also be considered, including the 'do nothing' option.

Step 2 : Generate options

All feasible ways of achieving the objective should be identified and the aim should be to find those which are both practicable and environmentally acceptable. In this context, 'practicable' implies that the option must be in accordance with current technical knowledge and must not have disproportionate financial implications. The search should be as wide-ranging and imaginative as possible and should, in this context, include non-transport options. As in this illustration, complex objectives are likely to call for combinations of options to achieve them. The options in this case are likely to include:

land use plans to situate facilities where they do not require car access and to discourage developments which are accessible only by car;

improvements to public transport and to cycling and pedestrian infrastructure;

car and heavy goods vehicle restraint;

measures to stagger the rush hour (eg different starting times for schools and other destinations susceptible to local authority influence, greater use of teleworking);

development of delivery services for shopping (and perhaps placing orders by telephone or computer);

encouragement of entertainment and cultural facilities and perhaps residential development in the town centre in order to enhance its liveliness out of business hours;

construction of new infrastructure.

Step 3 : Evaluate the options

The advantages and disadvantages of each option for the environment should be evaluated, using both quantitative and qualitative methods as appropriate.

Some effects (for example on atmospheric emissions, noise generation, water pollution (eg from run-off), accidents) should be capable of fairly accurate estimation. Others will be less readily quantified (eg how far the attractiveness of the town centre has been increased, how sympathetic any new development is) but even here there is increasing evidence which would permit some assessment of the effectiveness of options (eg how much a pedestrian scheme might enhance business opportunities).

Step 4 : Summarise and present the evaluation

The results of the evaluation should be presented concisely and objectively, and in a format which can highlight the advantages and disadvantages of each option. The results of different measurements and forecasts should not be combined if to do so would obscure information which is important to the decision. The cumulative effects of a series of complementary options should, however, be spelled out.

Step 5 : Select the preferred option(s)

The choice will depend on the weight given to the environmental impacts and associated risks, and to the costs involved. Decision-makers should be able to demonstrate that the preferred option(s) do(es) not involve unacceptable consequences for the environment.

Step·6 : Review the preferred option(s)

The proposed option(s) must be scrutinised closely to ensure that no pollution risk (including regional and global ones) has been overlooked. It is good practice to have the scrutiny made by individuals who are independent of the original team.

Step 7 : Implement and monitor

The achieved performance should be monitored against the desired targets, especially those for environmental quality. This is intended to establish whether the original assumptions were correct and to provide feedback for future developments.

Throughout steps 1 to 7: Maintain an audit trail. The basis for all choices or decisions through each stage should be recorded. This will include the assumptions used, details of evaluation procedures, the reliability and origins of the data, the affiliations of those involved in the analytical work and a record of those taking the decisions.

Public involvement and consultation should be an integral part of the decision-making and implementation procedures. It will be especially important in steps 1, 2, 4 and 5.

to private developers, but also to public sector decisions on the location of hospitals, recreation centres, schools, military establishments and other facilities.

Environmental assessment

9.59 Environmental assessment of certain projects is required by EC Directive 85/337.[39] In the UK, consideration of environmental assessments normally takes place as part of the process of deciding applications for planning permission. Road construction by the Secretary of State is excepted: in this case the assessment is carried out under section 105A of the Highways Act 1980 or its Scottish or Northern Irish equivalents.[40]

9.60 The construction of motorways is always subject to environmental assessment. Other road schemes need be assessed only if they are likely to have significant effects on the environment. The Department of Transport told us that, in practice, environmental assessment is carried out for most schemes in the trunk road programme. For local authority roads, the government advises that, outside urban areas, new roads or road improvements over 10 km in length (or 1 km for roads passing through a national park, or through or within 100 metres of a SSSI, a national nature reserve or a conservation area) may require environmental assessment.[41] We consider that these criteria may underestimate the likelihood of environmental harm's being inflicted by roads which fall outside these categories and **we recommend that all road construction proposals be subject to environmental assessment.** This does not mean that minor schemes must have appraisals running to several volumes in length. It would be appropriate for a less extensive assessment to be made of small schemes (unless they put particularly sensitive areas at risk).

9.61 The fact that environmental assessments have been carried out does not mean that satisfactory standards have necessarily been achieved. There has been widespread criticism, endorsed by the Standing Advisory Committee on Trunk Road Assessment, of the nature and quality of environmental assessments of trunk road proposals. SACTRA found, for instance, that 'where a series of individual schemes has been generated by a Route Identification Study or . . . when a series of schemes taken altogether amount *de facto* to an improved route, detailed environmental assessments are always made but only of the individual component schemes. Corresponding assessments are not made at the aggregate level in relation to the overall route and its regional impact.'[42] Others argued in evidence to the Royal Commission that environmental assessment has been made at too late a stage, when fundamental alternatives cannot be evaluated, and with inadequate opportunities for public involvement in decisions, and that the issues are effectively marginalised by excessive reliance on cost-benefit analysis. There has been dissatisfaction with the quality of environmental assessments which have been carried out, with examples of basic data omitted, some effects unconsidered or unquantified, inadequate discussion of effects in the text of environmental impact statements and unsupported comments.[43,44,45]

9.62 Following SACTRA's recommendations, the government has published revised guidance on environmental assessment.[46] This takes a broader view than the previous guidance and goes some way towards the three stage assessment procedure recommended by SACTRA. It also acknowledges that 'in some cases assessment may need to cover the combined and cumulative impacts of several schemes'[47] because this may result in a better choice of alignment and design in both environmental and traffic terms. It stresses that environmental assessments should be conducted by people with sufficient relevant expertise. These are welcome changes. There are signs that the quality of recent assessments has improved[48] and we hope that the new guidance will reinforce that trend. The government now makes stage 1 and 2 assessments available when trunk road schemes are put to public consultation and we welcome its decision to monitor the effectiveness of the new guidance and amend it as necessary.

9.63 Environmental appraisal is only one part of the process of appraising new transport infrastructure. It has been widely argued in evidence that differences in appraisal methods between road and rail schemes have put rail schemes at a serious disadvantage in comparison with road schemes. For example, environmental costs have not been taken satisfactorily into account in either rail or road schemes, rail safety is determined by standards which imply a markedly higher valuation of human life than is used in the case of road safety, and the treatment of time savings is not on an equal basis and has a distorting effect on the decision-making process. In such circumstances it is difficult, if not impossible, for decision-makers to select the Best Practicable Environmental Option.

9.64 We consider that a consistent overall transport policy will require that broadly similar criteria and methodologies are applied to roads and public transport, to policies on investment and operations, and to local and long-distance networks. All these aspects interact so intensively and continually that to single out any one for special evaluation leads to uneven and distorted treatment. **For these reasons, we recommend that the role of the present Standing Advisory Committee on Trunk Road Assessment (SACTRA) be broadened to include the development of appropriate appraisal techniques to ensure consistency of evaluation methods as between the different transport modes.** This might be associated with a change of title to the Standing Advisory Committee for Transport Assessment. Alongside this, we consider that it should report not only to the Secretary of State for Transport, as at present, but also to the Secretary of State for the Environment, with appropriate provision for liaison with professional staff in the two Departments. It should, however, remain essentially a technical and methodological body, not a political one. Its first task under this new brief should relate to the methodologies for forecasting and evaluating shifts in demand and investment priorities that are implied by the recommendations in this report.

Other measures

9.65 The environmental appraisal of development plans which local authorities are now required to conduct provides a vehicle for assessing the environmental implications of strategic and local land use and transport policies. We comment on the appraisal of central government policies later in the report. There is further scope to enhance the effectiveness of the land use planning and development control systems in reducing the need to travel and promoting environmentally friendly modes of transport. **To this end, all significant applications for planning permission should contain an analysis of the transport implications, including pedestrian, cycling and public transport access, and freight movements in the case of industrial developments, and we so recommend.**

9.66 The Town and Country Planning (Use Classes) Order specifies when a change of use of a building or land does not require planning permission. Since there are circumstances in which such a change of use could have considerable effects on the volume of traffic going to and from the development, **we recommend that the order be reviewed to ensure that new uses which generate appreciably higher levels of traffic cannot take place without a fresh grant of planning permission.**

9.67 The enhanced status of development plans should provide a greater degree of stability than hitherto in the strategic framework for land use planning and development control. Nevertheless, as we noted earlier, the legislation allows planning permission to be given which is contrary to existing plans. At present, developers have the right to appeal against unfavourable decisions by local planning authorities but third parties have no equivalent right. We have already alluded to the pressure which can be put on local authorities to accept controversial developments in their area on the grounds that, if the development is refused in one place, it will be relocated elsewhere. Major developers invest considerable resources in obtaining planning permission where it will most benefit them. Stronger procedural mechanisms are needed to inhibit such pressure from bringing environmentally damaging results and **we recommend that a right of appeal by third parties to the Secretary of State be introduced against decisions by local planning authorities which are contrary to up-to-date development plans.** The possibility of such appeals would inhibit developers from playing off district against district and be a firm check on the decisions which were taken. Rules about costs have been increasingly used in recent years to inhibit unjustifiable appeals by developers and we feel it should be possible to develop equivalent rules to restrain vexatious or frivolous appeals by third parties. Similarly, the Secretary of State must be prepared to use call-in powers to ensure that wider national strategic interests are considered. This will be especially important if local government reorganisation fails to retain effective strategic institutions.

Conclusions

9.68 Land use planning is an important component of policies designed to reduce the environmental effects of transport. Its role should be enhanced both in relation to the planning of transport infrastructure, especially new and improved roads, and in the extent to which travel and environmental implications are considered when land use decisions are made.

9.69 It is frequently suggested that land use planning policies can influence travel demand only in the long term. For the most part land use patterns change incrementally and the average rate of turnover of the built environment is small — of the order of one to two per cent per annum.[49] In areas of rapid growth, however, land use patterns can evolve quickly, and certain kinds of development, for example major regional shopping centres, can significantly affect travel patterns within only a few years.

Even when effects are long term, the policies adopted in Regional Planning Guidance and Structure Plans will influence patterns of growth and development over several decades: such considerations strengthen the case for integrated land use and transport planning.

9.70 In some circumstances transport systems have a dynamic influence on land use. These effects, and the further travel demands which they generate, have not been systematically taken into account, and local planning authorities have been left to deal with development pressures arising from road schemes. This is not satisfactory. Proposals for new or improved trunk roads should flow from development plans and should be open to public scrutiny as land use and transport policies are formulated. In the case of trunk routes, integration may most effectively be achieved at regional level and all such routes should be subject to strategic environmental assessment which includes consideration of alternative solutions as well as less direct and longer-term effects.

9.71 Land use patterns influence travel behaviour, though it is difficult to separate their effects from those of income, car ownership, and the availability and quality of different modes of transport. Development patterns tending to increase the need for travel include dispersed, low-density and/or piecemeal growth at or beyond the periphery of urban areas, for example dormitory suburbs and villages, business parks and many out-of-town facilities. Such developments are permitted by high mobility and reinforce car-dependency. Development patterns with the potential to generate less movement are those which add to the relative self-containment of towns or 'urban villages'. Possibilities range from city centre regeneration to mixed development in moderate sized urban areas or free-standing towns. A high quality built environment need not be travel-intensive.

9.72 Land use policies also have the potential to influence the choice of mode for different journeys. Policies designed to promote safety, amenity and priority for pedestrians and cyclists have both a land use and transport dimension and exemplify, at the local scale, the need for integrated land use and transport planning. In considering the location of particular facilities, such as offices, schools and hospitals, there is a need to give serious and explicit attention to access by public and non-motorised transport.

9.73 Though these findings are broadly supported by the available evidence, there is sufficient uncertainty to make the search for 'ideal' development patterns unproductive: in any case, land use planning policies must start with the physical infrastructure that already exists and the choice of development patterns and location policies must be flexible in response to local circumstances. Nevertheless, all planning policies and applications should be assessed for their probable effects on travel behaviour and the environment. While they may not always be overriding, such factors should always be considered.

9.74 The significance of interactions between land use and transport has been explicitly recognised in recent government policy statements. We welcome this recognition and broadly endorse the measures advocated in new planning policy guidance on transport (PPG 13). But it is important to recognise that these are not sufficient. In isolation, land use planning is a relatively blunt instrument for changing travel behaviour, critical though it may be in the long term. Appropriate policies may be difficult to implement in competitive environments; even when they are adopted, simply bringing land uses into closer proximity cannot make people travel less. The problems of traffic growth and congestion must, in the first instance, be tackled by more direct means. Unless the measures advocated in PPG 13 are consistently upheld on appeal, and unless they are implemented within a coherent transport policy framework combining the measures recommended elsewhere in this report, they are unlikely to achieve the intended effects on the amount and mode of travel. Nevertheless, we firmly believe that taking proper account of the interactions between land use and transport is crucial if essential access is to be available within the framework of an environmentally sustainable transport system.

Future policies towards transport

Chapter 10

FREIGHT TRANSPORT

Introduction to chapters 10–13

10.1 In the light of the findings in earlier sections of this report about the effects of the recent massive growth in private road use and the further very large growth that has been forecast, and about the consequences of taking land to provide new infrastructure, we have concluded that an important objective for a sustainable transport policy must be:

> **TO INCREASE THE PROPORTIONS OF PERSONAL TRAVEL AND FREIGHT TRANSPORT BY ENVIRONMENTALLY LESS DAMAGING MODES AND TO MAKE THE BEST USE OF EXISTING INFRASTRUCTURE.**

10.2 A sustainable transport policy will necessarily involve a new approach in all sectors of transport. In this third part of the report we consider in turn the three major sectors of surface transport, beginning with freight in this chapter and local journeys in chapter 11. In chapter 12 we consider the future of the long-distance transport system in the light of the conclusions reached in this chapter about freight and an analysis of longer-distance personal travel. We discuss the implications of the 1989 National Road Traffic Forecasts and the main factors that will influence future demand for movement; the potential for transferring traffic to modes that are less damaging to the environment; and further action that should be taken to limit the environmental effects of particular transport modes. Our first purpose is to indicate how the general objective stated above can be pursued through quantified targets for changing significantly the balance between transport modes. Our second purpose is to identify the kinds of new policies for each sector which will help achieve our overall set of objectives for a sustainable transport policy.

10.3 In chapter 13 we consider whether the present and planned organisation of government and the main transport services in Britain is capable of devising and delivering a coherent strategy for a sustainable transport system and, if not, what changes need to be made. We also consider how some EC policies and legislation may need to be reformed in order to make possible the achievement of the Community's own stated objective of a sustainable transport policy for Europe.

Future demand for freight transport

10.4 Freight transport has grown very rapidly, although less rapidly than personal travel (figure 2-III). The 1989 National Road Traffic Forecasts foresaw light goods vehicle traffic rising in line with growth in gross domestic product (GDP) to 2025 and heavy goods vehicle (HGV) traffic growing more slowly but still showing 140% growth under the high forecast by 2025. That is also the industry's view of the future.[1] The forecasts assumed that, contrary to the objective stated above, there will be a further increase in the proportion of freight which is transported by road.

10.5 Freight transport is considerably affected by developments in the rest of the economy. For example, the decline of coal as a fuel has been reflected not only in the weight of goods carried but also in rail's share of freight. Conversely, the increase since the early 1970s in the proportion of freight moved by water results from the growth of the North Sea oil industry. The more recent growth in freight transport has been in other types of goods, for which 90% of the movements are by road.[2]

10.6 One reason for the dominance of road transport is its flexibility to carry goods in large or small consignments when and where they are wanted. The main factor in the overall growth of freight transport has, however, been the increase in the average length of freight trips by road (2.5). This has largely resulted from reductions in the cost of road transport made possible by improvements in the interurban road network, increases in the permitted weight and length of HGVs, increases in their power and improvements in their aerodynamic design.

Table 10.1
Just-in-time logistics: cost and energy use

An industrial plant is assumed to have 5 suppliers, each 50 km away; a round trip of all 5 suppliers is assumed to be 180 km.

	efficient non-JIT logistics	JIT	JIT with consolidation of loads
weekly pattern of deliveries	5 return trips of 100 km, each with a 15 tonne load	25 return trips of 100 km, each with a 3 tonne load	5 round trips of 180 km, each with a 15 tonne load (3 tonnes from each supplier)
size of vehicle	38 tonne	12 tonne	38 tonne
cost per km	£0.6175	£0.3875	£0.6175
annual distance travelled	26,000 km	130,000 km	46,800 km
annual cost	£16,055	£50,375	£28,899
annual cost as multiple of non-JIT logistics	1	3.1	1.8
fuel used annually	2,080 gallons	4,362 gallons	3,744 gallons
energy use as multiple of non-JIT logistics	1	2.1	1.8

10.7 New patterns of freight transport have emerged, such as long hauls from a small number of distribution depots serving the whole of Britain, and 'just in time' (JIT) logistics. JIT involves precise planning of deliveries to match production or sales needs. It has adverse consequences for the environment because many more deliveries have to be made (though probably by smaller vehicles) than in a system depending on higher stockholding; vehicles may not be filled to capacity, and more fuel is used. One study, summarised in table 10.1[3], found that energy consumption for transport was twice as high in JIT as in conventional logistics.

10.8 In Japan, JIT is associated with the clustering of component suppliers within a short distance of major industrial plants, thus limiting the distances travelled. That reflects the structure of Japanese industry and the often poor road system. Road links in the UK are generally good and it is unlikely that

Figure 10-I
Distribution costs as percentage of turnover 1971-1991/92

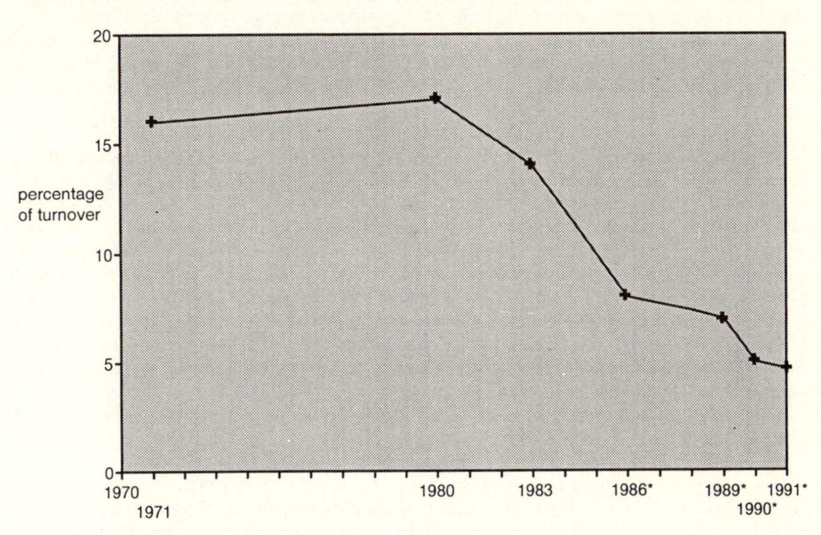

* - data are for financial years 1986/87, 1989/90, 1990/91 and 1991/92

similar clustering would emerge.[4] Nevertheless, there is scope to limit the extra traffic generated by JIT logistics. The Nissan car plant in Sunderland found that it was receiving JIT deliveries from 22 suppliers in the Midlands, about 300 km away. It was able to reduce considerably the distances travelled (by 80% or 3.9 million miles a year) by taking responsibility for collecting consolidated loads of components from the individual suppliers.[5] 'Consolidated JIT' has environmental advantages over the normal form of JIT, but still uses significantly more energy than conventional logistics (see table 10.1).

10.9 The example in table 10.1 also shows that transport costs can be over 200% higher in JIT, and 80% higher in consolidated JIT, than in conventional logistics. The higher transport costs may be outweighed by the financial advantages of eliminating warehousing and releasing capital that would otherwise be tied up in stocks. Transport costs represent only about a third of distribution costs on average. Expressed as a percentage of sales income, distribution costs have themselves fallen sharply since 1980 (figure 10-I[6]).

10.10 Growth in the international transport of freight by road has been even more rapid. Roll-on roll-off traffic to and from the continent has doubled since 1982[7], and traffic to and from Ireland has also increased sharply.[8] Production and distribution of the most everyday products may now involve series of long-distance movements across national boundaries. The completion of the Single Market and the enlargement of the European Union will not only give a further stimulus to the movement of goods across frontiers and make it easier for firms to centralise production and distribution but will also have a direct effect in liberalising road haulage and enabling hauliers to reduce their charges for international freight by perhaps a quarter.[9]

10.11 It was concluded earlier that the amounts paid in road taxation by UK operators of HGVs are substantially less than the costs their operations impose by way of environmental damage and requirements for infrastructure (7.19). The effect of this underpricing has been to distort competition in favour of road freight and encourage the progressive growth in the distances over which goods are moved. The increase in fuel duty already recommended (7.58) would not be sufficient, at the upper end of the estimated range of environmental costs, to bridge the gap shown in table 7.3 between revenue from road taxation and quantified infrastructure and environmental costs. Moreover, the calculations do not include unquantified costs such as the environmental damage caused by the use of land for transport infrastructure. There is therefore a risk that, even after the full increase in fuel duty has taken effect, use of road freight will continue to be boosted by a degree of underpricing.

10.12 A further consideration is the position of UK freight operators in relation to firms operating from elsewhere in Europe. We discussed earlier the effect of differing fuel prices in European countries (7.83). The vehicle excise duty paid by UK firms for the heaviest vehicles, which are most commonly used in international traffic, was until recently the second highest of any EC country. It has now become the highest, as a result of the German government's decision to introduce a charge *(vignette)* on vehicles of all nationalities using the German motorway system and reduce the annual duty paid by German operators. As a result of negotiations over several years, agreement has been reached on minimum levels of annual duty for HGVs in the EC, but these are well below present levels of duty in the UK. **We recommend that the government continue to press for EC legislation which will standardise the annual duty paid by HGV operators at a sufficiently high level to ensure that the total amounts paid in road-specific taxes by operators in each country fully reflect external and infrastructure costs.** As well as being beneficial in environmental terms, EC legislation on these lines is necessary to protect the competitive position of UK freight operators.

10.13 The effect on demand for freight transport from the doubling of the price of fuel which we have recommended will vary with the type of traffic. On average, fuel represents about a seventh of transport costs (7.81). The importance of freight costs as a proportion of sales income varies considerably, from up to 80% for bulk materials such as aggregates to very small percentages for high-value goods. For bulk materials a large increase in fuel prices may induce a transfer from road to a more energy-efficient mode. For many other products large increases in fuel prices will make it less attractive to carry goods over long distances, exert a downward pressure on trip length and encourage operators to make fuller use of the capacity of vehicles. However, even a doubling of the price of fuel may not by itself produce a significant transfer of freight generally to other modes, because of the other advantages of road transport or of JIT based on road transport. Additional measures will therefore be necessary. These include more effective regulation of HGVs and restrictions on the use of lorries in unsuitable locations. Steps to improve the attraction of other modes will also be essential.

The scope for mode switching

10.14 The environmental case for mode switching where practicable lies in the very different energy efficiencies, emissions, noise levels and accident rates of the main modes of freight transport. Primary energy consumption and emissions for each mode are shown in table 10.2.[10] Pipelines are the least polluting. Of the remainder, water transport has the lowest primary energy consumption, although rail produces lower emissions of some pollutants per tonne-kilometre. Road has a much higher primary energy consumption than rail, and the primary energy consumption for air transport is much higher still. It has been claimed that a freight train is 5–10 dB quieter than road vehicles carrying the same tonnage at the same speed[11], but the extent to which road, rail or air cause noise nuisance and degrade the quality of life depends on their proximity to large numbers of people. A German study which compared road and rail found the number of injuries per billion tonne-kilometres was 248 for road freight and 10 for rail freight; and estimated that the space required for road freight was almost three times as great (0.007 m^2 per tonne-kilometre per year compared with 0.0025 m^2 per tonne-kilometre per year).[12] The better safety record of rail also makes it preferable for moving hazardous substances. It is clear that, where an alternative is available, moving freight by road takes more space, uses more energy, produces more pollution and is more likely to lead to an accident.

10.15 Licensing powers which could be used to direct freight to another mode existed in the 1968 Transport Act but were repealed in 1980. This form of regulation is no longer, in our view, a realistic way of changing modal share. There has been a general trend away from it across Europe; and it is

Table 10.2
Freight transport modes: energy use and emissions

	rail	water transport	road	pipeline	air
specific primary energy consumption (kJ/tonne-km)	677	423	2,890	168	15,839
specific total emissions (g/tonne-km)					
carbon dioxide	41	30	207	10	1,206
methane	0.06	0.04	0.3	0.02	2.0
volatile organic compounds	0.08	0.1	1.1	0.02	3.0
nitrogen oxides	0.2	0.4	3.6	0.02	5.5
carbon monoxide	0.05	0.12	2.4	0.0	1.4

incompatible with EC legislation, under which HGVs from any Member State will be able to operate without restriction in any other State from 1998. Careful study needs to be given to the potential for moving a higher proportion of freight by environmentally less damaging modes and other means sought of realising that potential. The majority of freight will inevitably go by road because its origin and/or its destination are remote from navigable waterways and railway lines, and transhipment during the trip would not be worthwhile. However, a sustainable transport policy must look a considerable distance ahead; some constraints which are of major importance in the short term can be overcome in the medium term, as the building of the Channel Tunnel illustrates.

Pipelines

10.16 Use of pipelines grew rapidly during the 1960s and 1970s, from 0.2 billion tonne-kilometres in 1958 to 9.8 billion tonne-kilometres in 1978, but has subsequently grown more slowly and stood at 11.6 billion tonne-kilometres in 1993.[13] Transport statistics do not include either the considerable volumes of water and gas moved by pipeline or the movement over short distances of slurries and granular material such as coal, ores, clay, chalk, sand, gravel and farm waste. It is technically possible to move larger solids by hydraulic or pneumatic pipeline, either in free flow or packed in capsules, but this method has not been competitive in cost terms.

10.17 A new pipeline requires heavy initial investment and the successful completion of statutory procedures. It will not be economic unless there is a long-term assurance of high-volume continuous flows, a condition most often fulfilled in the oil industry. Pipelines are more likely to be an alternative to rail or coastal shipping than to road transport. They probably carry a high proportion of the materials for which they are suitable under present circumstances but higher fuel prices may improve their competitive position.

Waterborne transport

10.18 Waterborne transport tripled between 1952 and 1983, from 20 billion tonne-kilometres to 60 billion tonne-kilometres, but fell back to 55 billion tonne-kilometres in 1992.[14] Of that, 53 billion tonne-kilometres was seaborne, and 90% of the remainder used inland waterways in order to reach the sea. Seaborne transport has risen because of the growth of the North Sea oil industry. It is characterised by a high average trip length (620 km).

10.19 Because of its slow speed, waterborne transport is best suited to cargoes of bulk materials. It is desirable that maximum use should be made of freight-carrying waterways for cargoes of that nature, especially by firms which are, or could be, located on their banks. We agree, however, with the conclusion of two recent reports prepared by consultants for the Department of Transport (DOT)[15] that measures taken for that purpose through the land use planning system and in other ways are unlikely to have more than a marginal effect on the share of freight carried by inland waterways. Freight is now transported on only some 1,600 km of inland waterways in Britain, mostly in the vicinity of major estuaries (plate IX). The other navigable inland waterways are used predominantly for leisure. There are no prospects of usefully extending the waterway network or of adapting narrow canals for freight use.

10.20 Attempts have been made to develop types of vessel that could provide direct links between inland waterways in Britain and destinations on the continent. The London Rivers Association pointed out in evidence that use of barge-carrying ships (ocean-going mother ships which carry smaller feeder vessels) or ocean-going barges would make possible voyages from wharves on the Thames to cities such as Basel, Liège or Paris or to the river Danube. A British consortium has developed a design in which two barges would be rapidly combined to form an ocean-going freighter. Greater use of inland waterways in these circumstances could replace some of the road and rail traffic to and from ports, and lessen the pressure for new road capacity to serve ports.

10.21 The potential of coastwise shipping is well illustrated by Japan (table 6.4), where about 45% of domestic freight (245 billion tonne-kilometres, more than the total of freight transport by all modes in the UK) is transported in this way. The maximum lorry weight of 25 tonnes (6.49) and the poor road links between different regions of Japan help explain why shipping is so well used. In the UK, coastwise shipping can compete for bulk freight if both origin and destination are reasonably close to a seaport or estuary. Long-distance movement of aggregates is expected to increase, especially to south-east England.

Transport by water would be the preferred mode in environmental terms (4.66) but new infrastructure may be needed. The aggregates industry has suggested that as many as 15 to 20 new wharves will be needed by 2011 on the Thames alone to receive aggregates.[16] The part which the planning system can play is covered below (10.37).

10.22 The immediate scope for switching from road to water is limited. Nevertheless, we believe that a combination of consistent government policies, innovative use of technology and commercial enterprise could give water a more important role, not only for bulk materials but also for container-size loads, using existing ports which are under-used at present. The commercial prospects for waterborne transport will be improved by the increase in the price of fuel but, during the transition and perhaps beyond it, there may be valid reasons to provide a public contribution towards operating costs in some cases. The tariff rebate subsidy paid by the Scottish Office for shipping services in remote areas is an example; it is less expensive to contribute to the cost of water transport than to repair and upgrade minor roads to take regular HGV traffic. There will continue to be situations where it is worthwhile to pay a public contribution to attract or retain waterborne traffic because the environmental and infrastructure costs thus avoided would be greater. Provision was made in s.36 of the Transport Act 1981 for grants towards the cost of freight facilities on inland waterways where such facilities would be in the interests of a locality or its inhabitants, but up to 1991 only seven grants had been given.[17] The present power (s.140 of the Railways Act 1993) is framed more widely and permits grant to be given if the Secretary of State is satisfied that it is in the public interest for the relevant goods to be carried by inland waterways. We discuss future use of these powers below, in the context of the similar powers applying to railways. **In addition we recommend that, where environmental benefits could be obtained by transferring freight from road to coastwise shipping, but the inadequacy of port facilities is preventing such a transfer, the government pay grant towards the cost of providing the facilities required. We also recommend that the government make a public contribution towards the operating costs of companies carrying freight by water, at least for a transitional period, where that will achieve a demonstrable environmental benefit.**

Rail

10.23 Freight transport by rail in 1993 was about two-fifths of what it had been in 1952 and half what it had been in 1961 (figure 2-III). The freight activities of British Rail (BR) were reorganised to serve niche markets. In recent years, revenue has fallen more sharply than volume; higher-value freight has been lost and competition from road transport has held down prices. Nearly half by weight of the freight carried by rail in 1993 was coal handled by Trainload Freight, which expected its business to decline by a quarter within 18 months because of the reduced use of coal in power stations. Attempts to put Trainload Freight on a more viable basis in advance of the restructuring of the industry in April 1994 led to large increases in charges and the loss of some other major customers[18], although BR claims to have retained most of the traffic affected. Royal Mail letters have been handled by Rail Express Systems, although the growing private sector parcels and courier services use road transport exclusively, as does the publicly owned Parcelforce.[19] On-shore movements of shipborne containers, and other international and inter-modal freight, have been handled by Railfreight Distribution, which lost £62 million on a turnover of £160 million in 1993/94 (and even larger amounts in previous years). Between 1992/93 and 1993/94 some 20 million tonnes (16%) of freight was lost to rail.

10.24 Although the rail network is only half the size it was at the beginning of the century, it remains extensive and still passes close to most significant sources and destinations of freight in mainland Britain. Of 16,500 km of lines, 2,200 km are solely for freight. There was a reduction of about 4% between 1982 and 1992, affecting mainly freight lines.[20] There is considerable spare capacity for freight within the existing network. Moreover, capacity could be considerably increased by the removal of bottlenecks and by resignalling.

10.25 There have been proposals in the UK and elsewhere to create new freight-only lines. In France this has involved dedicating to freight tracks which have now been bypassed by the construction of new high-speed lines for passenger traffic. The private company operating the north-south autoroute through the Rhône Valley is said to be interested in acquiring the parallel railway line and carrying lorries on rail trucks; this will reduce the cost of maintaining the autoroute and avoid the need to increase its capacity. Netherlands Railways plan a new freight line extending into Germany to carry traffic between Rotterdam and the Ruhr.[21] The taking of land for the construction of new railway lines raises similar environmental issues to the taking of land for new roads. In order to avoid a conflict with our objective of halting losses of land in areas of conservation, cultural, scenic or amenity value (4.60), it would have to be shown that a proposed new railway line represented the best practicable environmental option. We believe the emphasis should be on making the best use of the existing rail network, with local improvements if necessary; and that this will in general be sufficient for increased levels of freight traffic for the foreseeable future. In the one case where a capacity problem can be foreseen in the UK after the turn of the century, on the lines between the Channel Tunnel and London, overall capacity will be raised considerably by the planned new high-speed line for passenger traffic. It has to recognised, however, that more intensive use of the rail network will involve running a larger number of freight trains at night. That underlines the importance of our proposed targets for noise (4.18), and measures to reduce noise from railway lines.

10.26 Use of rail for freight requires either a private siding or a conveniently located railhead. Following drastic reductions in recent decades there were only 60 freight depots remaining in 1993 and some 1,100 private sidings. In France the number of private sidings is ten times higher, or five times higher in relation to the length of track.[22]

10.27 For regular freight movements in complete trainloads, rail is competitive with road on price, even down to distances of a few miles if the flow is large enough. Only a limited range of freight falls into that category. For consignments the size of a single lorryload, the thresholds at which rail can compete vary from 150 to 600 km depending on the number of road feeder movements (none, one or two) at the ends of the rail haul, although increases in fuel prices should reduce these thresholds somewhat. Only a very small proportion of freight within Britain is carried over a distance of 600 km.

10.28 The Channel Tunnel has opened up an important new market for rail: freight traffic between Britain and other European countries is growing particularly rapidly and the trips are longer on average than domestic ones. BR expects the present low level of cross-Channel rail freight to triple by 1999.[23] Railfreight Distribution has identified nine inter-modal terminals to provide a hub-and-spoke distribution system served by the tunnel. Four of these are now in operation: at Willesden (London), Landor Street (Birmingham), Trafford Park (Manchester) and Mossend (Glasgow). The other proposed sites will be kept under review as demand builds up. Railfreight Distribution is a partner in two joint ventures which are promoting inter-modal services: Allied Continental Intermodal (ACI)[24] (with SNCF and Intercontainer) is aimed at larger distributors, while Combined Transport Ltd (CTL) (with French and German inter-modal companies, the Road Haulage Association and 28 UK transport companies)[25] is directed mainly at small and medium-sized distributors.

10.29 As well as through freight trains, the Channel Tunnel will also be used by shuttle trains carrying lorries from one end of the tunnel to the other (plate VIII). Facilitating the growth of through freight trains would clearly be the higher priority in environmental terms. There has been concern that this will be hindered by the British loading gauge (the height and width of tunnels, bridges etc), which is too small to take continental rolling stock. The cost of enlarging the loading gauge was estimated to be £3–4 billion. A study funded by Eurotunnel, local authorities and the European Commission has concluded that for only £70 million an 800 km route from the Channel Tunnel to Scotland could be upgraded by 1997 to carry road trailers on railway wagons 'piggy-back' and that by this means 400,000 trailers a year could be taken off the roads by 2000. That would be 9% (56 million vehicle-kilometres) of current HGV traffic between the UK and the continent.

10.30 The attractiveness of rail as an element in an inter-modal system can be greatly increased by the development and application of more efficient technology for transferring loads between modes or for transhipment (transfers which involve dividing up or combining loads). The simplest form of transfer is to load a lorry or semi-trailer onto a railway wagon. To overcome the limitations of the British loading gauge, road trailers can be fitted with extra wheels for use on rails, or placed on special railway bogies. There are techniques for doing this without using separate lifting equipment. It is disappointing that these new techniques have not so far been taken up on any significant scale in the UK. A private sector company which tried to take advantage of them, Charterail, went into liquidation after a short time.

10.31 An essential condition for widespread use of inter-modal transport is international standardisation of dimensions. This is a field where the standards-setting bodies (BSI, CEN and ISO) could play a valuable role. In order to give more impetus to inter-modal transport in Europe, and facilitate the free operation of the market in transport services, we consider there should be EC legislation on the subject. **We recommend that the government press the European Commission to draft and introduce legislation prescribing a range of standard dimensions for containers and bodies for use in inter-modal transport and covering other matters which are essential to ensure technical compatibility.**

10.32 The decision to permit 44 tonne lorries for trips to or from rail depots has only a modest commercial benefit for rail operators. BR sees advantages for international inter-modal services but expects little effect on the domestic market. Along with the Association of Metropolitan Authorities and the Association of District Councils, BR voiced fears that weak enforcement of the restrictions on use, and an impression that they will eventually be removed, could undermine the purpose of the decision.

10.33 The recent restructuring of the railway industry and the plans for its privatisation are outlined elsewhere (13.48–13.56). Considerable concern has been voiced about the possible adverse effect on rail freight. There has already been a rapid rundown in some parts of BR's former business. Prior to April 1994 a high proportion of the fixed costs on shared lines was allocated to passenger traffic and only the directly attributable marginal costs to freight traffic. The allocation of costs is unlikely to be as favourable to freight in future, and the fixed costs which have to be met are now substantially higher because of changes in accounting methods. Particular fears expressed to us were that under-used freight facilities and sites suitable for future freight facilities will be sold off and that private sector firms will not be sufficiently interested in exploiting the market potential described above.

10.34 On the other hand, finance for investment in suitable projects could be more readily available from the private sector and private sector firms will be in a better position to use rail in innovative ways as part of a comprehensive logistics service involving other modes. The EC Directive on third party access to national railway systems[26] opens the way for such companies to operate trains under their own control throughout Europe. Although there are exciting possibilities, there are also many uncertainties and practical problems are bound to arise. It is essential that the government and the European Commission both have a strong commitment to increase the use of rail freight and take early action to demonstrate that commitment. **We recommend that the government:**

i. **ensure that sales of BR freight subsidiaries to the private sector take place on terms which create the most favourable prospects for increasing rail's share of the freight market;**

ii. **authorise Railtrack to start work immediately on the modifications required to open a Channel Tunnel to Scotland route to piggy-back traffic, in advance of the general upgrading of the West Coast Main Line.**

10.35 The government offers two types of grant to encourage the transfer of freight from road to rail. The first is a new track access grant intended to help operators to meet the charges levied by Railtrack for the use of the tracks. Up to 100% grant is payable if the Secretary of State is satisfied that this is necessary to secure the carriage of goods by rail, and that social or environmental benefits are likely to result. The second (freight facilities grant) has existed for some years and offers assistance towards the capital costs of rolling stock owned by users and facilities such as sidings. Demand hitherto has been low but the eligibility conditions were relaxed from April 1994. Financial provision for the two grants has been increased and now averages £14 million a year for the next three years.

10.36 At the moment DOT takes the position that it has provided the framework for a competitive rail freight industry, through the grant schemes and other measures, and it is up to the private sector to make the most of the opportunities that exist. We believe that, in view of the benefits of transporting freight by rail, the government should accept a greater responsibility than this for pursuing the environmental objective identified at the beginning of this chapter, and more particularly for achieving specific targets for switching freight from road to rail, of the kind proposed at the end of this chapter. This will require vigorous marketing of the grant schemes and careful evaluation of their effectiveness. It is also important that the reference in the legislation to 'the public interest' should be interpreted to cover the full range of environmental objectives. The same considerations apply equally to the grants for facilities on inland waterways mentioned above. **We recommend that the government keep the grant schemes for freight transport under review, and make further resources available for them if the demand exists, and if necessary modify them or adopt other measures in order to stimulate the switching of freight from road to rail or water.**

The role of land use planning

10.37 The land use planning system has an important role to play in increasing the use of environmentally less damaging modes for freight. The government has recently given guidance to local planning authorities to 'encourage the carriage of freight by rail or water rather than by road wherever it can provide a viable alternative' and 'designate sites for distribution and warehousing, particularly of bulk goods, which ... are served or with the potential to be served from wharves, harbours or railway sidings'.[27] This does not deal with the problem of the many distribution centres and industrial sites which have been built in recent decades with only road access. To cater for these, local planning authorities need to identify and safeguard suitable sites for additional depots which also enjoy good road access. **We recommend that suitable sites for new wharves and rail depots are identified and safeguarded by planning authorities.**

Reducing the environmental impact of road freight

10.38 Even with a large-scale shift to rail, road will remain the dominant mode of freight transport. It is therefore important to explore all possible ways of limiting its environmental effects. Measures to reduce emissions of pollutants and the technical scope for improving fuel efficiency have been discussed already in chapter 8. The use of natural gas as a less polluting fuel for vehicle fleets in urban areas (8.21), and eventually the introduction of electric vehicles (8.83–8.85), would primarily affect buses and service vehicles. Measures to reduce noise have been discussed in chapters 4 and 8. In general the environmental effects of HGVs are greatest when they are in towns and villages. More people are exposed to their noise, vibration and emissions; their weight may damage pavements as well as roads; there may be damage to buildings; and, quite apart from the risk of serious accidents, the size of HGVs intimidates many

Figure 10-II
Weights of goods vehicles 1982-92

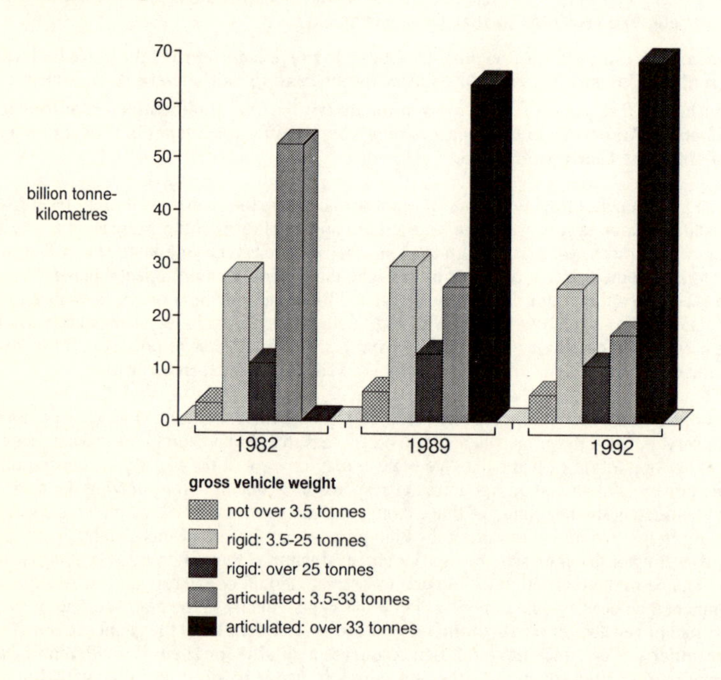

pedestrians and cyclists. The remainder of this chapter considers further ways of limiting the environmental effects of HGVs.

10.39 Despite the growth in freight transport by road, HGVs are no more numerous now than 10 years ago; in 1993 some 412,000 HGVs were in use, compared with a peak of 478,000 in 1989. When in 1983 the maximum vehicle weight was raised from 32.5 to 38 tonnes, there was a rapid shift to the heavier lorries (see figure 10-II[28]). Use of mid-range vehicles has declined and the largest articulated vehicles (over 33 tonnes) now account for 59% of all HGV tonne-kilometres. Light goods vehicles up to 3.5 tonnes have increased rapidly from about 1.6 million in 1982 to 2.2 million in 1993 and according to DOT's 1989 forecasts are likely to increase their numbers and mileage more than any other class of vehicle. However, they carry only about 5% of road freight. There is minimal scope for mode switching from light goods vehicles and policies for reducing their environmental effects must therefore be based on restraining demand and on tighter limits for emissions and noise.

10.40 Although the road haulage industry cites the absence of sustained growth in HGV numbers as evidence of increased efficiency, the reality is more complex. There has been only a slight increase in average payload since 1983. Excluding empty running, the average payload of 38 tonne vehicles in 1992 was 16.4 tonnes, some 20% less than the 20 tonne maximum payload of 32.5 tonne vehicles.[29] For HGVs generally, capacity utilisation has been drifting downwards and is now only 58%: empty running has declined slightly, from 33% of mileage in 1980 to 29% in 1993, but this has been offset by an increase in unused capacity on other trips.

10.41 The energy needed to move a given amount of freight has risen from 2.7 MJ/tonne-kilometre in 1978 to 3.3 MJ/tonne-kilometre in 1988.[30] The many technical developments which are improving HGV fuel efficiency have been outweighed by less efficient logistical practices such as JIT. A comparison of diesel consumption by road transport with HGV tonne-kilometres confirms this. Between 1989 and 1992 the recession caused HGV tonne-kilometres to decline by 8% but diesel consumption rose by 9.5%. Part of the rise in consumption came from greater use of diesel by cars but, as cars and taxis still use only about 15% of diesel, HGV operation will have been the major factor. On environmental grounds we believe the aim should be to achieve a steady increase in capacity utilisation.

Licensing of operators

10.42 The Transport Act 1968 created a system of licensing ('quality licensing') with the potential to hold operators to high safety and environmental standards. Anyone operating HGVs is required to hold an operator's licence (O-licence), granted by the Traffic Commissioner to those of good repute. Because entry to the road haulage industry is nevertheless relatively easy, there are many small firms, a proportion of which (in the UK, as in other countries) disregards legal requirements. Since 1984, local authorities have had the right to object to O-licences if a depot is or would be damaging to the local environment. Between 1986 and 1992 Surrey County Council objected to 202 applications (7% of the total) and 97% of the objections were upheld. This right of objection provides an important instrument for local authorities to use in protecting the environment in their areas, and it would be most regrettable if it were to be removed or limited under the guise of deregulation.

10.43 Although quality licensing and other related legislation have raised environmental and safety standards much of the potential has been wasted by inadequate enforcement. The efforts of police and DOT examiners are let down by the leniency often shown by the Traffic Commissioners, as was acknowledged in an internal report for DOT (the Palmer Report). In 1993/94 only 4% of new O-licence applications and 0.6% of renewals were refused, despite widespread breaches affecting mechanical standards, weight limits, driver's hours, speed restrictions and payment of vehicle excise duty. Deliberate overloading is rife in the less reputable parts of the haulage industry. In checks by DOT examiners and local authority trading standards officers at least 20% of HGVs weighed are overloaded and 6.6% of vehicles weighed in 1993/94 justified immediate immobilisation. Speeding is widespread: on dual carriageways and other A roads between 50% and 76% of HGVs break speed limits, 38 tonne lorries being the worst offenders. The road haulage industry recognises all this and complains that operators who persistently break regulations undercut good ones and push standards down across the industry. A National Audit Office report on regulation of heavy lorries and the Public Accounts Committee have both criticised DOT for lack of enforcement.

10.44 As long ago as 1978 a committee of inquiry recommended tough measures such as impounding vehicles to drive out illegal operators. No effective action was taken, and the worst operators remain unlicensed, ignore regulations with impunity and are rarely brought to justice. There are now thought to be 6–12,000 illegal operators compared to 127,000 O-licence holders. In 1991 the Road Haulage Association drafted an amendment to the Road Traffic Bill to make vehicles operated by someone without an O-licence liable to seizure. It was put forward by the Opposition in the House of Lords but withdrawn when the government promised more effective action. In 1993 a DOT working party finally proposed impoundment powers as a last resort for the Vehicle Inspectorate but Ministers rejected this advice and chose instead to ask magistrates to impose higher fines. Before that decision the South Eastern and Metropolitan Traffic Commissioner said: 'Illegal operators are like quicksilver — they disappear. It's no good imposing penalties on them because they disappear.' **We recommend immediate legislation to allow impoundment of illegal operators' vehicles.**

10.45 Illegal working by HGV operators and drivers increases the dangers associated with such large vehicles and leads to unfair competition against legal operators and against other modes. In France it has been estimated that observance of all legal obligations would increase average road haulage rates by a third. Stronger enforcement of the O-licence system and other existing laws would restore the commercial balance in relation to other modes, reduce the energy used in speeding, and reduce emissions, noise and the damage to infrastructure caused by overloading. It would ease the pressure on drivers to remain behind the wheel beyond their permitted hours. DOT has commissioned a study of fatigue and accidents, but work in the USA, Netherlands, Scandinavia and the UK has already pointed to a probable link. Proper regulation would benefit all responsible HGV operators, their drivers, other road users and those who live around depots. At present there appears to be a contradiction between the government's policies to promote road safety and the laxity of the licensing system.

10.46 Two measures are in prospect which should reduce speeds: a DOT proposal to set the motorway speed limit for HGVs over 12 tonnes at 55 mph from 1 January 1996 and an EC Directive which involves phased introduction of speed limiters at 56 mph on HGVs over 12 tonnes by the same date. These will have fuel efficiency, emission and safety benefits. By increasing journey times, they will also make rail freight that much more attractive where there is direct competition. We remain concerned about two categories of HGV: those over 12 tonnes and first used before 1988, which may have deteriorated mechanically but are not covered by the new Directive; and those between 7.5 and 12 tonnes. In both cases speed limiter settings will remain at 60 mph, that is, above the proposed motorway speed limit and the existing 50 mph speed limit on dual carriageways. **We recommend that all HGVs over 7.5 tonnes have their speed limiters set at 56 mph as from 1996.**

Restrictions on access

10.47 Each year local authorities introduce some 2,000 traffic management measures which involve access restrictions for HGVs. They normally consult the public, business and the police before taking a decision. Most restrictions in the UK are by weight (mainly 7.5 or 17 tonnes) or time, with exemptions for essential access. The majority of restrictions are in urban areas but they are also used to exclude vehicles, unsuitable by weight or size, from narrow or steep rural roads or weak bridges. For many rural local authorities this avoids the need for excessive expenditure on bridge strengthening. Loading and lorry parking bans are also employed and air brake silencers can be made a condition when exemptions are granted. During construction projects local authorities can define the routes or modes which contractors may use. There were plans in the past for strategic national lorry routes but little was achieved because the network of routes was expanded to include all roads regularly used by lorries, regardless of suitability. The Armitage Report in 1981 offered no encouragement to bans or systems of defined routes at more than local level. As a result most restrictions are reactive and unco-ordinated.

10.48 There is far less use in the UK than elsewhere in Europe of bans applying at specified times. Many continental countries forbid the use of certain classes of HGV outside the working week; some, such as Switzerland, use 3.5 tonnes as the threshold. Switzerland has now decided by a referendum that all lorries using the Alps as a transit route (some 350,000 a year) will have to be carried by rail piggy-back. The switch to rail must be completed within 10 years. France bans HGVs over 7.5 tonnes and all vehicles carrying dangerous goods at weekends, on public holidays and for part of the day preceding public holidays. Germany and Italy also ban HGVs over 7.5 tonnes on Sundays. In many urban areas on the

Table 10.3
Banning goods vehicles within the M25: estimates by Wood Inquiry

Ban at:	percentage increase in vehicle-kilometres	percentage change in total operating costs	percentage change in fuel used
7.5 tonnes	+20%	+4%	-8%
16.26 tonnes	+5%	-1%	-9%
24.39 tonnes	+1%	-1%	-8%

continent, HGVs over 17 tonnes are banned at all times and there is no doubt that logistical arrangements have adapted successfully.

10.49 Table 10.3 shows the conclusions of the Wood Inquiry, set up by the Greater London Council (GLC) in 1983, about the effects of a ban on HGVs within the M25 orbital motorway. The surprisingly small increases in mileages estimated to result from such a ban reflect the extent to which HGV capacity is under-used. Following the abolition of the GLC, the London Boroughs Transport Committee implemented a night-time and weekend ban from 1986, under which fines of up to £1,000 can be levied. It applies to all HGVs which have not been granted exemptions and to date there have been 3,440 convictions. HGV traffic in London has fallen by about 5% a year since the early 1980s but the relative contributions of the M25, a decline in manufacturing industry and the ban are disputed. We believe the ban has brought worthwhile benefits and **we recommend that the government maintain the effectiveness of restrictions on access by HGVs to Greater London**.

10.50 Local authorities are subject to conflicting pressures from the public and from business on the question of lorry bans. The dilemma is that isolated bans may damage the commercial success and employment prospects of an area if firms relocate to neighbouring areas where restrictions are not in force. Several outer London boroughs withdrew from the lorry ban on these grounds. This highlights the need for consistent national policies over restrictions on access. **We recommend that all urban authorities adopt a presumption against access for HGVs over 17 tonnes and follow a uniformly strict approach to the granting of exemptions.** This is the critical measure required to encourage transhipment depots and the use of smaller and less intrusive freight vehicles, which we discuss below. There is a danger that restrictions in urban areas could stimulate out-of-town developments and thereby encourage higher traffic levels; that reinforces what we say elsewhere in this report about the necessity of integrating land use and transport policies.

Use of smaller vehicles

10.51 Transhipment is already used extensively by distributors for deliveries to shops where small consignments or physical problems of access make it impractical to use large lorries. Greater use of transhipment and smaller delivery vehicles would contribute to retaining and reviving both traditional town centres and corner shops. Rail freight prospects would benefit because most rail depots and stations are already well inside urban areas, and are therefore better placed than the edge-of-town sites which would be appropriate for road transhipment depots.

10.52 Critics of transhipment emphasise the higher operating costs, the possibility of damage and loss, increased vehicle numbers and the need for specialist vehicles for many products. Transhipment can raise city centre delivery costs by up to 100% but, as delivery to retailers contributes only a very small percentage of the final cost of most retail products, prices are little affected. The specialist vehicle sector is not large. Refrigerated or insulated vehicles, for example, were 2.8% of HGVs in 1991. Although some companies, such as Marks and Spencer, emphasised to us their desire to keep control of vehicles in which chilled food is moved, this could be done without the use of very heavy vehicles. One haulage company with a wide range of customers has overcome these constraints with a Metro Swop system in which several small bodies, specialised when necessary, are carried on drawbar trailers and transferred directly to and from small vehicles (plate VII). None of the disadvantages ascribed to transhipment is insuperable and we were impressed by evidence from the Royal Mail on the extent of their transhipment operations, both within and between modes. The freight industry is noted for the speed with which it adapts, and we have confidence in its ability to develop efficient transhipment techniques.

10.53 A switch to more, but smaller, freight vehicles in towns could reduce energy efficiency and increase emissions in some circumstances, though table 10.3 offers reassurance about fuel use overall. Nevertheless it would be preferable on balance to the present situation. Moreover we envisage an eventual move to urban freight vehicles powered by natural gas or electricity.

10.54 According to the Freight Transport Association, over 50% of transport operators now make deliveries outside the hours of 7 am to 5 pm in order to avoid congestion. Night-time delivery can cut costs and some emissions but other environmental effects, notably noise, are suffered by those in near-by residential areas. This may give rise to complaints, which can lead to the imposition of night-time bans.

10.55 Some see the prospect of night-time deliveries as an opportunity for the use of non-polluting electric or hybrid vehicles, quiet enough to avoid disturbance and small enough to avoid vibration. The full realisation of this idea will not be possible in the short term because several prerequisites are lacking. There may need to be automated reception arrangements to minimise costly night staffing. Second, the level of noise made when loading and unloading has to be acceptable. Third, other stages of the distribution chain would have to evolve to complement night-time deliveries. This could mean more trunk hauls at night and more night-time noise at transhipment depots. A move towards 24-hour activity would squeeze more capacity out of transport infrastructure but the social implications of increased night-time working are not insignificant. On balance we feel increased night-time deliveries in non-residential areas are acceptable but more caution is needed elsewhere. It might be sensible at a future date to link permission to deliver at night to the use of electric or hybrid vehicles, providing an added incentive to their development.

Use of larger lorries

10.56 Some organisations claim that there would be economic and environmental advantages if the UK permitted heavier HGVs. An EC derogation will expire on 1 January 1999, at which point five-axled, 40 tonne lorries will in any case be permitted in the UK. There is a proposal from the European Commission for maximum weights to be harmonised at 44 tonnes on six axles for lorries operating nationally and internationally in all Member States. If this were accepted, it would apply to the UK from 1999.

10.57 From an environmental point of view, the possibility of higher gross vehicle weights offers both benefits and disadvantages. Individually, heavier HGVs tend to be more fuel-efficient, but give rise to more noise and vibration and more damage to buildings and pavements, and cause greater damage when involved in accidents. Better technology could perhaps prevent extra noise from 44 tonne lorries, but could equally be deployed to reduce the noise from existing sizes of lorry. The Freight Transport Association suggests that a 44 tonne limit would reduce the size of the lorry fleet by 9,000 (about 2%), miles travelled by 480 million and diesel consumed by 44 million gallons, and that emissions of all types would fall. A report by the Transport Research Laboratory suggests that there would be a 0.7% reduction in HGV numbers and a 3.7% reduction in road wear. This arises because, while the individual lorries would impose slightly more wear, the reduced vehicle numbers should reduce the amount of wear per tonne carried. Some freight operations would become more efficient. Oil companies, for example, move a dense product with few opportunities for load matching and there is considerable empty running on return trips: 30 tonne loads could be carried by a 44 tonne lorry instead of 25 tonnes under present regulations. Esso believes it could save 35,000 vehicle movements a year, a 20% reduction in mileage. For the oil industry as a whole it estimates that 12.5 to 15.5 million fewer vehicle-miles would be run each year. For many loads, however, the volume to weight ratio is too high to benefit from higher gross weights: a vehicle is physically full well before the weight limit is reached.

10.58 While some environmental benefits are possible there could be damaging effects. The reduction in costs, estimated at 6–10%, would encourage longer hauls and more use of transport instead of stockholding, and would attract more freight from rail and water, offsetting any reduction in vehicle numbers. The 1983 increase in permitted weight from 32.5 to 38 tonnes can be seen as a test case for these arguments. In the 1984 National Road Traffic Forecasts, DOT assumed that the larger lorries would constrain growth in HGV vehicle-kilometres. In fact, there has been only a slight increase in average payload and capacity utilisation has drifted downwards (10.40). In oral evidence to the House of Lords European Communities Committee, the General Manager of TNT Europe said: 'I do not think we saw any benefit from the 32 to 38 tonnes increase. It may have led to a reduction of a few vehicles on the road but the main benefit went to the manufacturers.'[31]

10.59 Since 1983 there has been a programme to strengthen bridges to carry the maximum drive-axle weight of 10.5 tonnes imposed by 38 tonne lorries and the 11.5 tonne drive-axle weight which 40 tonne lorries will impose. The County Surveyors' Society has recently estimated the cost of this programme as £2 billion across the UK. In DOT's view, no additional bridge strengthening would be required by a move to 44 tonne lorries with six axles and a maximum drive-axle weight of 10.5 tonnes. This is some reassurance but it is difficult to see how the cost and disruption of the existing bridge strengthening programme is justified.

10.60 The trend towards heavier and heavier goods vehicles is inconsistent with the objective of reducing the dominance of lorries in order to improve the quality of life (4.32). We would therefore prefer to see the UK oppose the European Commission's proposal for a 44 tonne limit, despite the advantages in fuel efficiency. Whether or not the limit is raised to 44 tonnes, it is desirable on environmental grounds that the heaviest lorries should be permitted only on suitable roads. The government already advises local authorities that it should be unnecessary to strengthen bridges on roads which few lorries use or for which alternative routes exist. **We recommend that the programme of strengthening bridges to carry 40 tonne lorries should be scaled down to the minimum necessary to provide a basic network giving such lorries access to main distribution centres.**

Conclusions

10.61 Much of the volume and nature of freight transport is unavoidable in a consumer society. Nevertheless a further increase in the environmental impact of freight transport by road, on the scale implied by DOT's 1989 forecasts, would not be acceptable. There are a number of important measures that should be taken to reduce the environmental effects of freight vehicles. Any substantial reduction in the overall environmental impact of freight transport however must depend, first, on finding ways of maintaining and enhancing economic well-being which do not depend on a continual increase in the length of trips; and, second, on transferring as much freight transport as possible to modes which are less damaging to the environment.

10.62 The objectives identified at the beginning of this chapter were to increase the proportion of freight carried by modes which are environmentally less damaging and to make the best use of existing infrastructure. In view of the evidence that rail and water are preferable to road transport on environmental grounds (10.14), and have considerable capacity for additional freight traffic (10.22, 10.24), we believe specific targets must be set to guide government action aimed at increasing the proportions of freight carried by rail and water, which in 1993 were 6.5% and 25% of tonne-kilometres respectively.

10.63 We were impressed by the target adopted by the Dutch government for increasing rail transport of freight in the Netherlands, where it has a smaller share of the market than in Britain and has likewise been declining (table 6.4). The target in that case was to increase goods lifted by rail to 50 million tonnes in 2010, 150% more than in 1987.[32] As the environmental effects of freight transport are related to distances travelled as well as tonnages lifted, however, it seems more appropriate to use tonne-kilometres

Table 10.4
Movement of goods: projected modal shares to 2020

percentages

	road	rail	water	pipeline
1952	35	42	23	0.2
1972	62	15	20	2.5
1993	63	6.5	25	5.5
2000	55	10	30	5.0
2020	45	20	30	5.0

in defining targets. We also prefer to express such targets in terms of modal share rather than in absolute terms. On that basis, and taking into account the unique opportunity created by the Channel Tunnel, and the likelihood that forms of freight transferred to rail will involve longer distances than much traditional rail freight in Britain, it seems possible to aim for a larger percentage increase than in the Dutch target. The target we propose for rail is:

> **To increase the proportion of tonne-kilometres carried by rail from 6.5% in 1993 to 10% by 2000 and 20% by 2010.**

10.64 There is less scope for increasing freight transport by water, which is already very substantial, and much of which is dependent on the scale of operations of the offshore oil industry. Nevertheless there

Table 10.5
Movement of goods: projected position to 2020 on different growth scenarios

billion tonne-kilometres

	road		rail		water		pipeline		total	
total demand growing each decade at	10%	20%	10%	20%	10%	20%	10%	20%	10%	20%
1993	135		14		52		11		212	
2000	125	133	23	24	68	73	11	12	227	242
2020	123	157	55	70	82	104	14	17	274	348

would be considerable environmental benefits from transferring more freight to water. The target we propose is:

> **To increase the proportion of tonne-kilometres carried by water from 25% in 1993 to 30% by 2000, and at least maintain that share thereafter.**

We recognise that these targets for rail and water transport are challenging. On the assumption that pipelines retain broadly their present share and domestic airfreight does not become significant, the effect of achieving these targets would be that freight transport in 2000 and 2020 would divide between modes in the way shown in table 10.4.

10.65 If the overall demand for freight transport continues to keep pace with the growth in GDP, it might be expected to grow at about 20% a decade. Table 10.5 shows the extent of freight transport by each mode in 2000 and 2020 on that assumption and on the basis of the modal shares shown in table 10.4. It also shows the extent of freight transport in 2000 and 2020 on the alternative assumption that overall demand for freight transport will grow at 10% a decade. The transfer of freight to rail on the scale envisaged will bring tangible benefits. On the hypothetical assumption that all the transferred freight would otherwise have been carried by articulated lorries of more than 35 tonnes for at least 200 km, the annual mileage travelled by lorries of that size on trips of that distance would be reduced by almost four-fifths. It will be apparent from table 10.5 however that, if overall demand were to grow at 20% a decade, even achieving our challenging targets for modal share would not be sufficient to prevent a further increase of almost a quarter between now and 2020 in freight transport by HGVs on roads. If, on the other hand, demand were to grow at 10% a decade, achievement of the targets for modal share would bring about a small reduction in freight transport by road.

10.66 The effect on road traffic will depend on whether there are changes in the sizes of vehicles and on the extent to which their capacity is utilised. During the 1980s any downward pressure on the weight of loads from JIT logistics and changes in the nature of goods was balanced by increases in the capacity of lorries and pressures for more efficient utilisation, and the average payload rose slightly. If there were to be little further change in the period up to 2020, the trend in vehicle-kilometres would broadly parallel the trend in tonne-kilometres shown in table 10.5. However, a side-effect of increasing use of rail for large loads and transhipment depots for local deliveries may be that smaller vehicles become relatively more important in road transport.

10.67 We believe a sustainable transport policy would be based on growth of no more than 10% a decade in overall demand for freight transport over the next 30 years, with a levelling off of overall demand in the longer term. That need not entail a reduction in economic activity. Instead higher fuel prices would operate gradually, not only to encourage a shift from road to other modes, but also to modify manufacturing and distribution patterns over a number of years by shortening the average length of trips. Until these modifications took effect there would be some increase in the freight costs faced by industry and passed on to customers. However, the increase in fuel prices would be introduced over a period; it would contribute to the efficient functioning of the economy by reflecting external costs which had not previously been taken into account; and other effects on the economy would not be seriously damaging (7.77–7.83). The overall effect would be a considerable benefit for the environment, especially in the longer term.

Chapter 11

LOCAL JOURNEYS

11.1 Most journeys are local. Although the number and average length of journeys has been increasing, almost a third are still of less than a mile and almost three-quarters are of less than 5 miles.[1] Journeys of under 25 miles (our primary focus in this chapter) account for over 55% of the average distance travelled per person each year and 96% of the total number of journeys. More details are given in table 11.1. The growth has been in car use: there have been reductions in bus travel and cycling.

11.2 Traffic congestion and pollution have spread from town and city centres into surrounding areas. The government has acknowledged that increased road capacity in urban areas is not the solution to these problems and that measures will have to be taken to manage demand and encourage the provision of acceptable alternatives to the private car.

11.3 Changes in the ways local journeys are made can help to reduce emissions of greenhouse gases. The high proportion of personal travel accounted for by local journeys lends itself particularly well to a switch from cars to less environmentally damaging modes which produce either no additional greenhouse gases (walking, cycling) or relatively smaller emissions (various forms of public transport).

Table 11.1
Why people make local journeys (1989-91)

	1 to under 2 miles	2 to under 5 miles	5 to under 10 miles	10 to under 25 miles
	percentages of journeys			
total leisure, including	26	27	28	32
social/entertainment	*19*	*23*	*24*	*26*
holiday/other	*7*	*4*	*4*	*6*
other personal business	21	21	18	14
commuting	16	21	27	29
shopping	23	21	17	14
business	2	4	5	8
education	12	6	5	3
all purposes	100	100	100	100
average number of journeys per person each year	194.8	279.0	151.2	100.9
average number of miles travelled per person each year	240	826	998	1,477

11.4 Some of the worst air quality problems arise in urban areas and, while improved technology will be a necessary component in their solution, additional measures will be essential. Policies towards local journeys can contribute to achieving the targets for improving air quality, and reducing noise nuisance from transport identified earlier (3.44 and 4.18 respectively).

11.5 We have already stated the objective that transport policies should embrace ways of improving the quality of life by reducing the dominance of motor traffic. Many local authorities have traffic management and parking control programmes in urban areas and are increasingly carrying out programmes of traffic calming. In this chapter we explore how the different aspects of policy towards local journeys relate to one another. We also consider what targets should be set for reducing the dominance of motor traffic, and how their achievement would contribute to the targets identified in earlier chapters.

Transport in rural areas

11.6 Most of the discussion in this chapter concentrates on urban journeys. The ill effects of transport are in general experienced less acutely in deeply rural areas than in towns or the neighbouring countryside but there are exceptions — ozone concentrations tend to be higher in rural areas (3.14f). Car dependency is even higher in rural than in urban areas and low population densities make it difficult or impossible to provide adequate substitutes by means of conventional public transport services.

11.7 In considering rural transport, attention generally focuses on the needs for transport of the 30% of households in the shire counties which are without a car.[2] There no longer appear to be major legal obstacles to the development of alternative transport services and many schemes have been devised to improve access to essential services.[3] They include taxibuses, community buses, post buses, social car schemes and, in less remote areas, dial-a-ride services. Such services are often dependent on voluntary help and on financial assistance from local authorities, the Rural Development Commission and other bodies. Non-transport solutions have also been sought. Some planners, not without criticism, have sought to concentrate population into larger settlements or 'key villages' in the hope that they might be able to support essential services.[4,5] The development of telecommunications may enable some services to be provided remotely, thus reducing somewhat the need to travel, and may also create pockets of rural employment (for example, in telecottages). In some circumstances these could be developed in buildings alongside rural railways, thereby possibly also stimulating greater use of the rail network.[6] Measures such as these might slightly reduce the need for car travel in rural areas. In suitable locations they could be combined with measures (such as better conditions for cyclists and pedestrians) which are more often introduced in urban areas.

11.8 As the countryside becomes an increasingly popular recreational destination, ways are needed of reducing the effects of cars at the most visited destinations. Again, many of the approaches applied in urban areas can be adapted to the needs of the countryside. Traffic management schemes can help to limit the flows of vehicles in unsuitable localities; park and walk arrangements can prevent beauty spots from being overwhelmed by car traffic; suitable country lanes could be open only to residents' cars and otherwise reserved for pedestrians; and means could be devised of slowing traffic on some country roads. Some experiments have been carried out in this country, if not wholly successfully; we are impressed by the ways in which popular destinations in the French Alps are being protected from the effects of previously unrestrained car access.

11.9 Perhaps the most intractable problems arise in the less isolated rural areas, especially in the vicinity of large conurbations. Here, road improvements, population dispersal and changes in land use have combined to produce large flows of traffic with consequently high levels of pollution and pressures for the construction of new infrastructure. (Traffic levels on major roads in Surrey, to take an extreme example, are already twice the national average.[7]) The scale and complexity of the problems in these areas will require the full range of solutions which are advocated in this report.

Alternatives to car travel

11.10 Earlier chapters outlined the complex inter-relationships between increased personal incomes, lower costs of private motoring, the convenience and attractiveness of the private car, and trends in land use which have stimulated car travel. In addition, one analysis of public expenditure between 1979/80 and 1990/91 has suggested that more capital expenditure was allocated to the road system than to public transport and, for most of that period, the largest element of public transport expenditure was used to maintain existing services, rather than improve them through capital expenditure.[8] Improved roads may help bus and coach services but they also represent powerful stimulants to greater car use which it will be difficult to counter in the short term.

11.11 We have already discussed how measures such as price increases might reinforce appropriate land use policies and changes in urban form to offer long-term potential for reducing the length of local journeys and minimising the demand for car use. To be effective, these approaches must be supported by a variety of transport-related measures which will improve the urban environment and help to reduce preference for cars for local journeys. These measures are familiar to practitioners and to the public although in the past they have not been used in this country on a wide enough scale. Many of them are cheap compared with road building. And experience on the continent shows that, if they are followed consistently over several years, a more civilised environment can be created (as in Copenhagen, described later in this chapter).

11.12 In order to encourage a shift away from the car, we recommend the following targets:

> **To reduce the proportion of urban journeys undertaken by car from 50% in the London area to 45% by 2000 and 35% by 2020, and from 65% in other urban areas to 60% by 2000 and 50% by 2020.**

These targets are expressed in terms of journeys so as to focus on the environmental benefits which could be attained by switching from car use for the many journeys short enough to be made easily on foot or by bicycle. Journeys in London are reported in the London Area Travel Survey. The percentage of journeys outside London has been based on National Travel Survey data for urban areas of over 25,000 inhabitants. Improvements in public transport in urban areas will need to make an important contribution both to these targets and to the target of carrying 20% of passenger-kilometres by public transport by 2005, and 30% by 2020 (12.54) but other measures will be required, including a greater emphasis on restraining car traffic.

Walking

11.13 There is wide international experience of the value of pedestrianisation schemes in improving the attractiveness and commercial success of central areas and in removing traffic pollution from areas frequented by large numbers of people. Well-conceived schemes are popular among the public.[9] A review of experience of pedestrianised areas in Germany[10] showed that nearly all experienced a substantial increase in the numbers of pedestrians using them, and a review[11] of 26 published studies of pedestrianisation and traffic calming in both Germany and Britain showed that there was generally a positive effect on retailing, with shops inside pedestrianised areas being more successful than those outside. There can, however, be a reduction in turnover during a transition period of 1–2 years, and the effects can be unfavourable for fringe shops just outside the developed area.

11.14 There are examples of badly designed pedestrian areas which encourage crime and are shunned by most people at night. When pedestrian schemes are imaginatively designed, however, they generally have beneficial effects on the vitality of city centres. In some cities, pedestrianisation has been linked to new development, allowing more people to live centrally. This adds a further dimension to the cities' attractiveness by helping to relieve transport problems and ensure that pedestrian areas are not deserted at night. We saw an interesting example in York but other cities are implementing similar policies. City centre managers can also help to create a clean, welcoming environment which encourages people into the city instead of an out-of-town shopping centre.

11.15 Where full pedestrianisation is not appropriate, more should be done to keep footpaths attractive to potential users and ensure that they can be used as through routes to neighbourhood shops and other common destinations. This could include better siting of street furniture, quicker repair of damaged surfaces and provision of seats. Narrow pavements are too often largely or wholly obstructed by cars parked across them. This is inconvenient for all pedestrians but a real danger to the blind and to anyone forced into the road. Parking on pavements can be prohibited locally but should be made generally illegal throughout the country. Where there are recurrent infringements, physical measures should be used to make the prohibition self-enforcing. In residential areas, traffic calming can be beneficial. At road junctions controlled by traffic lights, there should be pedestrian phases (automatic where flows are high enough, manually triggered in other cases) which change rapidly in the pedestrian's favour and allow a comfortable crossing time. Pelican crossings should have longer green phases so that the less nimble can cross more safely at them. The introduction of Puffin crossings seems to be a long-overdue step in the right direction. Measures involving traffic signals are likely to be especially important wherever motor traffic is fast moving.

11.16 In some areas it will be appropriate to construct attractive, adequately sign-posted pedestrian routes which minimise the risks both of personal assault and of conflict with vehicles and cyclists. The National Travel Survey records that between 1975/76 and 1989/91, the distance walked per year by children aged 5 to 15 fell sharply. It attributed this change partly to the increasing tendency for children to be taken to school by car though other factors are also important (2.20). Because of the importance of developing environmentally desirable lifestyles at an early age, **we recommend that local authorities should provide networks of safe pedestrian routes, especially those which will enable children to walk to school when they live close enough to do so.**

11.17 Local authorities will have a major role in implementing these measures but they should not be left to do so in isolation. We see the promotion of journeys on foot as an integral part of a national strategy to which central government and others should contribute. The voluntary sector could help plan pedestrian routes and in rural areas parish councils have a role to play. Central government should also adopt policies which make it more attractive for people to make short journeys on foot, rather than by car. The government should promote a framework which supports local initiatives, ensuring especially that its guidance to local authorities (eg on Transport Policies and Programme (TPP) submissions) and its financial controls provide for the desirable level of facilities. Where it has direct responsibility for traffic management in urban areas (eg on trunk roads) it should be particularly vigilant in improving the lot of the pedestrian. Given adequate financial support, major improvements to pedestrian facilities could be made fairly quickly (5–10 years), even taking account of the periods needed for full public consultation on the improvements. These improvements should enable good localised progress to be made towards improving air quality, reducing noise levels and diminishing the dominance of motor traffic.

Cycling

11.18 Greater use of cycles is an important alternative to reliance on the car. In congested urban areas, cycling is often the fastest means of transport.[12] While cycle ownership has grown, however, cycling has declined. Between 1972/73 and 1989/91, the percentage of households owning at least one cycle grew from 25% to 36% but the percentage using a cycle fell from 46% to 30%. Between 1975/76 and 1989/91, cycle mileage per person per year fell by 18% (from 51 to 41 miles[13]). In Britain, only about 2% of journeys are now by bicycle.[14]

11.19 This situation is not immutable. Cycling is much more common in some other countries than in Britain. The Netherlands has a long tradition of encouraging cycling and achieves high levels of use. On average 30% of trips to work there are made by bicycle. In Delft the figure is 43% and in medium-sized cities the proportion often rises to 50%.[15] Towns in Germany show wide variations in cycle use but the average for all trips is much higher than in the UK. In some German towns, cycle use reaches almost 30%, whilst in the big cities of Hamburg, Munich, Cologne and Frankfurt the bicycle has a share of about 10% of all trips and comes close to the national average of 11%.[16]

11.20 The reduced use of cycles in Britain has been attributed to fear of accidents, exposure to traffic fumes, social attitudes and the wider availability of cars and public transport.[17] Fears for personal safety, the quality of cycling routes (where these exist) and theft also play a role. We have already recommended (4.41) a target for reducing accidents involving cyclists. Not all the other disadvantages of cycling can be entirely overcome but experience on the continent and in some towns in Britain suggests they are not so significant as to prevent a much greater use of cycles. We believe that, with suitable encouragement, cycling could take a larger share of local journeys.

11.21 In order to develop innovatory schemes for safer cycling, the Department of Transport helped during the 1980s to implement and monitor cycling projects in Exeter, Kempston (Bedford), Nottingham and Stockton-on-Tees. The schemes have had somewhat modest success. They have tended to attract cyclists from busier routes and have proved popular, particularly where they have led to shorter journey

BOX 11A **CYCLING IN THE NETHERLANDS**

In the Netherlands, a high level of cycling provision has been built up gradually by supportive national and local policies over two decades. Delft is one of several Dutch towns to have capitalised on its natural advantages and created an ambitious network of cycle routes. Although in the early 1980s it increased the share of cycling by only 2%, this was from an already high level (40%). It is claimed that the 90,000 inhabitants of Delft cycle in total some 425,000 km each day. Compared with the same distance travelled by car (consuming perhaps 8,000 gallons of fuel), the environmental and health benefits are enormous.

The Delft scheme demonstrates the value of separating cycle paths from major traffic flows, and of providing segregated crossings of main roads. Delft's policy of developing a fine-scale network with a hierarchy of routes at neighbourhood, district and city-wide levels helps cyclists to find a safe route for virtually any journey and recognises the importance they attach to taking the shortest route to their destination. There have been no fatal accidents to cyclists in the last 5 years and cycling accidents are mostly minor. Delft's success has been achieved by relatively modest expenditure (some £2 million per annum (at 1994 rates of exchange) between 1982–86). The scheme was initially 80% financed by the national government as a demonstration project and subsequently funded equally by national and local government in the same way as road maintenance.

times. Users have generally considered the routes to be safer than the alternatives. Although cycling declined nationally over the period of the studies, the decline was either less marked or negligible on most of the cycle routes, with increases in some parts of Nottingham. Care is needed in interpreting the effects of the schemes since studies of them are based on surveys which took place over limited periods. Major cycle schemes in The Hague and Tilburg in the 1970s also had mixed results. A recent study of a cycle route in Southampton showed an increase in cycle traffic on the route of 28% over 3 years, compared with a 14% fall elsewhere in the Southampton area.[18]

11.22 In contrast to some continental models (see box 11A), none of the British schemes has created a dense, comprehensive network of segregated routes and the schemes have been developed in a social and political context which has regarded car travel as the norm. The lesson is that the promotion of bicycles as a means of transport must not be confined to the construction of simple cycle paths. To give much greater encouragement to cycling **we recommend that comprehensive networks of safe cycle routes which do not involve the use of heavily trafficked roads should be built up in all urban areas**. These should include: secure storage for cycles at common destinations, including railway stations, public transport interchanges, shops and offices; priority over cars for cycles at light-controlled junctions; care in the design of other junctions (especially where cycle routes cross busy roads) and new bridges or underpasses if necessary; cycle hire or loan schemes (several continental railway companies hire cycles from stations); facilities at workplaces to change and shower; and partnerships with public transport. Cycles represent a convenient means of access to bus and railway stations and more encouragement of 'bike and ride' could significantly increase their catchment areas. In addition, it would be preferable to create cycle lanes, not simply by painting a white line along the edge of a road, but by providing physical separation of the cycle lane from the rest of the highway in order to dissuade motorists (moving or parked) from encroaching on cyclists' space.

11.23 Cycling is sometimes permitted in pedestrian areas. This can provide safe and convenient connections for cyclists in or through town centres. On the other hand, large numbers of cyclists or rash behaviour by a minority can compromise the freedom of movement, safety and amenity of pedestrians. A short study for the Transport Research Laboratory found no real factors that justify the exclusion of cyclists from pedestrian areas and indicated that cycling can be more widely permitted without detriment to pedestrians.[19] By suitable design and regulation, which would differ from place to place, some pedestrian areas should be capable of providing for the needs of cyclists as well as pedestrians. Elsewhere, (especially where there are large numbers both of cyclists and pedestrians) segregation would be necessary in order to prevent accidents.

11.24 The cost of improved cycling facilities is modest in comparison with other transport expenditure. For example, the Exeter cycle route, which included a new river bridge, cost £450,000 in 1988. The cost of implementing the 1,000 mile strategic Cycle Route Network proposed for London has been estimated at £3.5m a year for ten years, or some 4% of the London Boroughs' transport supplementary grant allocation. It is important to ensure that these relatively modest sums are consistently available for improvements to cycling infrastructure but, until very recently, the rules for the local authority financial settlement have been weighted against such measures (and other non-road-based ones, such as traffic restraint). In particular, there has been no suitable methodology for assessing the benefits (including the health and environmental benefits) of cycle schemes which are not associated with other new road construction or which do not have a specific safety benefit. This illustrates how small a role cycles were considered to play. The Department of Transport has taken steps to change the basis of assessment for the 1995/96 TPP bids to give more positive weight to environmentally friendly proposals. In June, the Minister for Roads and Traffic issued a pro-cycling statement advocating some of the measures we endorse and designed to effect a change of perceptions.[20] The crucial test, however, will be the Department's willingness to support proposals for cycling infrastructure when it allocates resources. **We recommend, first, that the government make available to local authorities the relatively modest resources required to support a 10-year programme to create high-quality cycling facilities and, second, that the new appraisal method for cycling schemes be reviewed in the light of the 1995/96 TPP round.**

11.25 Cycling is a potentially satisfactory way of completing the opening and closing stages of longer journeys made by train or coach. It is therefore important to ensure that cycles can be carried on trains and long-distance coaches. Consideration should be given to ways of enabling cycles to be carried on local buses and trams, as in certain other countries. We deplore the decisions by British Rail to exclude bicycles from certain trains and to levy fares at rates which seem designed to discourage rather than encourage cyclists. **We recommend that British Rail and private rail operators should be required to provide adequate space for cycles on all passenger services. It will be crucial to ensure that new and substantially overhauled rolling stock is designed to meet this requirement. We also recommend that Railtrack provide secure facilities for cycle parking at all stations.**

11.26 We conclude that, on health and environmental grounds, there should be a long-term programme to encourage much greater use of cycles by both adults and children. This should include measures to enable children to cycle to school. Hitherto the government has vacillated in its approach. In order to ensure that the promotion of cycling receives the necessary priority, this long-term programme should be based on a national target for increased cycle use. This should be set in the light of current levels of cycle use in other northern European countries, the higher levels of cycle use in the past in the UK and the contribution that reduced car use will make to reducing environmental damage. We propose the following target:

> **To increase cycle use to 10% of all urban journeys by 2005, compared to 2.5% now, and seek further increases thereafter on the basis of targets to be set by the government.**

Local authorities should also set targets for their own areas.

Public transport

11.27 As reliance on the car has increased, the share of journeys undertaken by public transport has fallen. Many of the reasons have been touched on in earlier chapters. Car journeys, especially short ones, tend to be regarded as free. Public transport often fails to match modern lifestyles and patterns of activity. On the whole its perceived quality is poor and does not accord with the self-image of many people. Fear of crime in public places, especially at night, is growing.

11.28 Environmentally, however, public transport has significant advantages over private transport, provided that load factors are adequate. It is usually more energy efficient and is much more economical in use of road space. It can reduce the number of sources of air and noise pollution. If public transport were sufficiently widely available, planning authorities would not need to insist on high levels of parking provision when old property is converted into flats or new dwellings are built on constricted sites; hence there could be more conversions of property which has little off-street land. Such development could in turn increase the use of public transport. When public transport has been upgraded, particularly where trams have been reintroduced, improvements have sometimes also been made to the physical environment of the town itself, as members of the Commission saw in Grenoble. This complementarity of policies is important if town centres are to remain destinations which people wish to reach. In a few places, notably London, waterborne transport might offer a useful service if it was carefully planned and fully integrated with land-based services. Passenger Transport Authorities (PTAs) should explore the feasibility of introducing such services.

11.29 People are unlikely to turn to public transport, unless it is provided at reasonable cost by clean, comfortable vehicles and unless services are regular, predictable and reliable. Public concern on these points was expressed at an open meeting held by the Royal Commission in Gateshead. The best operators go to considerable lengths to clean vehicles and buildings and rid them of graffiti. More might be done to increase staffing levels on vehicles and stations in order to provide reassurance to passengers, especially at quiet times. And as far as possible, all services should be made accessible for people with reduced mobility.

11.30 Information about services is important. The ideal would be reliable and frequent services that render real-time information displays superfluous. This may be achievable only in heavily populated areas but, at least in peak periods, some rail systems in Britain and others abroad give cause for optimism. Heavy investment will be needed before most services in Britain which have the potential could be brought to this level but it is an objective which should be aimed at.

11.31 In the meantime, customers could be attracted to public transport if fuller information on services were made consistently available. This is especially desirable when there are several operators and services change frequently. Electronic passenger information systems, which are now being introduced, can reduce uncertainty by making route, waiting time and timetable information more accessible. Computerised rail timetables and touch screen facilities can simplify journey planning. Local authorities should work closely with transport operators (as do Passenger Transport Executives (PTEs) in the metropolitan areas) to ensure that comprehensive and up-to-date information about services is available in forms which people find useful for journey planning. The extended PTAs and PTEs we recommend later (13.24) would be best placed to discharge this duty in their new areas. Electronic indicator boards showing waiting times should be extended to all urban railway and light-rail systems. Provision of comparable information at bus stops is still at the pilot stage and would be expensive. If evaluation of the pilot projects shows sufficient user benefits, government grants should be made available for the progressive introduction of electronic indicators in urban areas.

11.32 Experience of travelcards, particularly in London, demonstrates how effectively the use of public transport can be encouraged by the availability of a ticket valid for any combination of public transport modes within an overall zone. It is essential to spread this concept further and to widen its appeal. Modern technology is beginning to provide the means whereby this might be done even within an overall transport framework which comprises many independent providers. The use of smart cards could cope with the inherent accountancy problems and could encourage other providers, such as taxi companies, car park operators and car and cycle hire companies, to participate. **We recommend the government ensure the availability of tickets valid for all public transport modes and services in a particular area, taking full advantage of new technology.** A range of tickets should be devised, as with present travelcards, to meet passengers' needs. This policy should be developed by the extended PTAs and PTEs we recommend.

11.33 Park and ride schemes have been widely seen as a means of encouraging mode switching from cars to buses and trains. They are most effective when combined with other measures to dissuade people from bringing cars into towns. In such cases, they have proved popular and, as in Oxford, have demonstrated their ability to complement restraint of road space and parking and to prevent traffic growth within the city. Nevertheless they can have disadvantages. They can generate additional car traffic to the park and ride facility (for instance from journeys which would otherwise have been made entirely by public transport, or not made at all), reduce amenity in the neighbouring area and, unless countervailing measures are taken, do little to reduce overall car traffic in the city itself, because of suppressed demand. They may be better suited to some road networks than others. On balance we consider that carefully designed park and ride schemes, located and operated so as to attract car users without reducing the number of people making the whole journey by public transport, should form a significant element in local transport strategies. There may also be potential for using existing car parking capacity, which is under-used during weekdays, for park and ride services for city-bound commuters.

Buses

11.34 Poor services and down-market image are powerful disincentives to bus use. Institutional changes are recommended later to remedy the defects associated with deregulation (13.41ff) but buses are major contributors to air pollution blackspots in urban centres and radical improvements are needed if they are to play a full part in reducing the adverse effects of motor traffic. The Commission made appropriate recommendations in its Fifteenth Report (see also appendix E). More investment is needed in quieter, less polluting, more comfortable vehicles designed to match passengers' wishes; journey times must be shortened (for example by giving buses priority over other traffic (11.37)); services made more reliable; more comprehensive networks and, in some places, more flexible services developed. Fear of crime, especially at night and between bus stop and home, must be tackled. These deficiencies are not irremediable and where services have been improved passenger usage has increased.

11.35 The growth in patronage has been considerable in some continental cities. In Amsterdam patronage grew by 41% between 1970 and 1992; in Hanover by 30% between 1970 and 1980; in Stuttgart by 29% between 1979 and 1991; in Vienna by 35% between 1981 and 1991; and in Zurich by 30% between 1985 and 1990. In all these cities, public transport is provided by a combination of buses and rail-based systems. During the early 1980s, passenger journeys increased in the English metropolitan areas, stimulated partly by fares which fell by an average of 4% a year between 1982 and 1985/86. On deregulation, when fares also increased sharply, the number of passenger journeys fell markedly. In London (without deregulation) they remained broadly level. In the shire counties a slight increase between 1982 and 1983 has been followed by a sustained decline.[21]

11.36 Quicker, more reliable journeys can stimulate bus patronage. This much is clear from the 'red routes' pilot scheme introduced in London in 1991 to speed up all types of traffic. For example, following the designation of the red routes, bus journeys which previously took 37 minutes on the A1 northbound became over 8 minutes quicker and 33% more reliable. Bus patronage improved dramatically. On one route, 2,000 more tickets were sold per week and, taking into account passholders, nearly 9,000 extra passengers (an increase of 8.8%) were carried. As a result, an additional express service was introduced, further increasing by 2,000 the number of people travelling by bus on the route each week.

11.37 It is therefore crucial to speed up bus journey times so that they are close to or better than those which can be achieved by car. To do this, **bus lanes should be introduced much more widely and policed more effectively than is usually the case.** Guided systems are useful where space is at a premium. **Buses should also be given priority over other motor vehicles in urban areas, including automatic (or driver activated) priority at light-controlled junctions.** Interchanges between services should be improved so that journeys involving several stages can be readily made. This is not only a question of physical structures but also of complementary and imaginative timetabling. The urgent need is for sufficient improvement to increase patronage levels at rates comparable, when allowance has been made for differences in city structure, to those seen in the best continental examples. And, to achieve the maximum advantages, these improvements should precede car restraint measures (11.48–11.50).

11.38 It would be wrong to overlook the part which taxis could play in reducing dependence on the private car. Although less fuel efficient than buses or trains, they offer the flexibility of cars but with a lower requirement for space (because they do not occupy parking spaces during the day). The availability of reliable and affordable taxi services could help to free households from the bonds of car ownership in both urban and rural areas.

11.39 It is also desirable to provide services which bridge the gap between buses and taxis, as an alternative to the car for the disabled and others who do not find it easy to get about. Some local authorities use taxicards or tokens to subsidise the use of taxis. Dial-a-ride schemes are operated in many areas. On these, the fares charged are comparable with those on other services (despite higher operating costs). Users normally have to register with the operator and the service can then be booked in advance. There is undoubtedly scope for more dial-a-ride services to meet such needs. Similar services may also be useful in places where population density will not support frequent scheduled services (11.7).

11.40 Greater stability and regularity of services will result from implementing the recommendations we make later for a new approach to bus service provision. To meet the objectives of making motor traffic less dominant and encouraging the use of more environmentally friendly modes of transport will probably require new services where none has previously operated, for example in areas such as residential estates, and across normal traffic routes (eg to form orbital services). These services will probably require financial support for some time as people adapt to new methods of mobility.

Rail-based systems

11.41 Several continental countries have invested in modern rail-based systems which have helped to encourage a greater use of public transport. Light-rail systems are appropriate for coping with peak demands of 5,000–15,000 passengers an hour in each direction (above 20,000, heavy rail is more efficient).[22] According to the Chartered Institute of Transport, towns of above 200,000 population can create enough demand to justify a light rapid transit system, especially if there are distinct travel corridors leading to a strong central business area. In Germany there is experience of joint running of light-rail vehicles on tracks also used by conventional trains. In the right circumstances, this can offer improved services with less disruption from the construction of infrastructure.

11.42 The UK has been slow to construct modern light rapid transit systems. The Tyneside Metro, the Docklands Light Railway in London, and the Manchester and Sheffield trams remain exceptions. There have also been some extensions to the heavy rail London Underground system. The Manchester system cost about £140 million for a 31 km route. There are plans to extend the network to Oldham and Rochdale and a system of about 100 km in length is envisaged. The 29 km Sheffield system is costing £240 million. Systems of varying length are proposed for the West Midlands, Croydon, and several other cities.

11.43 Diverting traffic to light rapid transit systems can improve the urban environment by reducing local noise and atmospheric pollution and by improving safety. These advantages are strengthened if streets are closed to other traffic (except for appropriate access to frontages). Tram systems are popular with the public, and are one of the most efficient travel modes in terms of primary energy requirements. Per passenger-kilometre and with 50% occupancy, they use three-quarters of the energy used by a similarly loaded double decker bus, and only about a tenth that of a car used for urban commuting (assuming an average of 1.3 occupants). The construction of rapid transit systems can stimulate complementary improvements to town centres as well as provide greater certainty of services, thereby encouraging developers to provide new facilities. They can also lead to improvements in other means of transport. In Grenoble, for example, the perception that the new tramway had downgraded the bus network in the public mind has led to improvements to the quality of bus services, route by route.[23] Other improvements have been made to the urban environment and much of the success of the system results from the comprehensive nature of the measures which have been introduced. There are plans for extensions to the tramway to link into urban redevelopment schemes.

11.44 The House of Commons Select Committee on Transport found that new rapid transit systems attract some passengers from cars (in Newcastle 8% of Metro users had formerly travelled by car) and liberate road space. Complementary measures are, however, needed to prevent released road space from being taken up by cars as suppressed demand is released, if the maximum environmental advantages are to be realised. Segregated tracks for the trams and other limits on space available for driving and parking cars are therefore desirable. And we would like to see bus services used as feeders, to complement the trams, rather than as competitors.

11.45 Rail systems have some environmental drawbacks. The vehicles are much heavier than buses and require the diversion of utility services (gas, electricity, water). Segregated tracks can separate communities whilst tracks in the street can be a hazard to cyclists and pedestrians. Overhead equipment can be visually intrusive although this disadvantage can be minimised by careful design. The reduction in gaseous emissions per passenger-mile is to some extent offset by point source emissions from electricity generation.

11.46 The main disadvantage of rapid transit systems is the high capital cost of construction. It has been suggested that this could be reduced, perhaps by as much as half, by building single tracks with passing places on routes with fairly low service frequency, using cheaper vehicles and reducing construction time so as to produce an earlier income stream to meet what would be lower debt charges.[24] Economising at the expense of quality could be self-defeating and, in towns without the density of population necessary to sustain a light-rail system, it may be preferable to increase public transport's market share more effectively by improving the reliability and frequency of bus services. Dedicated busways, whether guided or not, could provide significant advantages in these respects.

11.47 Government grants towards the capital expenditure have been critical but because the available resources have been limited (to about £50 million a year) projects have been deferred. We consider that the long-term benefits of modernising transport systems, preserving the viability of city centres and reducing pollution would justify much higher levels of investment than hitherto. **We recommend that the government make more resources available for new light-rail systems so that they can be built within a reasonable time in those conurbations for which they are an integral part of an overall transport strategy.**

Restraint of road traffic

11.48 Investment in public transport, and cycling and pedestrian facilities will go some way towards our objective of improving the quality of life in towns and cities by reducing the dominance of motor traffic. For example, first reports suggested that the Manchester tram system was being used by more people than had been anticipated (about 12½ million instead of 11 million a year) and that over 40% of passengers have cars available which might have been used for their trip.[25] On the other hand, most evidence suggests that car traffic will not be noticeably reduced solely by measures to encourage people to walk, cycle or use public transport.

11.49 This conclusion is supported both by experience and by some modelling. For example, even the well developed and widely used public transport system of Zurich does not appear to have reduced overall road traffic levels (although they have remained stable for some 10–15 years and the city centre now presents a high-quality environment, with very little traffic noise or pollution from motor vehicles).[26] The construction of the tramway in Grenoble increased public transport's share of traffic from 18% to 20% between 1985 and 1992 but, whilst there was a fall in the proportion of journeys made by car, it was by only 1% (from 75% to 74%) and was in the context of an increase in the total number of journeys (from 831,000 to 908,000).[27] Modelling carried out for the Department of Transport (which we consider in more detail later in this chapter) also suggests that improved facilities alone would not have a large effect, at least in the short to medium term (about 10 years).

11.50 People will continue to use cars while they are perceived as cheap, convenient, more comfortable and faster than public transport. To avoid the risk that the benefits of improved public transport might be offset by increases in the total amount of travel, investment in improved public transport facilities should be closely linked to policies for restraining the use of private vehicles in urban areas. To gain acceptance for these policies of restraint, there needs to be a clear commitment to the alternatives in a package of measures. This will require a large commitment of confidence and resources in the early years.

11.51 Some of the measures already advocated (pedestrianisation, cycle and bus lanes) will take road space now occupied by other traffic. Highway space may need to be reduced further in some areas in order to reduce the dominance of traffic and enable vehicle use to be better integrated with other street uses. This is likely to be the case in primarily residential and shopping areas; in some places, traffic calming measures are already having that effect. The phasing of traffic lights could be used for a similar purpose. Areas with unavoidably high levels of through traffic may, on the other hand, benefit from a smoother traffic flow which is less polluting. Care is needed to avoid spreading traffic pollution over still wider areas and it may be helpful if some streets or wider zones were open only to cycles and neighbourhood electric vehicles. In other places, there may be no need to reduce road capacity in order to meet environmental targets. Different approaches may be needed in different parts of the same town. Local authorities should therefore set themselves environmental targets which will contribute to the achievement of those we have recommended on a national basis, and then work consistently for their attainment. Often these targets will need to be set on a regional basis and joint action will be needed by authorities in the region.

Traffic calming

11.52 Many local authorities are undertaking programmes of traffic calming in order to reduce accidents. The available techniques slow vehicles by such means as road humps, textured and coloured road surfaces, and chicanes, and the evidence is that they succeed in reducing accidents. In Oxfordshire, for instance, monitoring of traffic calming schemes that have been in place more than a year has shown a reduction of 59% in all accidents and 75% in pedestrian accidents.[28] They also have the advantage of influencing non-essential traffic away from calmed areas. Care in design, judicious tree planting and the use of high-quality materials can also greatly improve the amenity of residential, business and shopping areas.

11.53 Physical calming measures are sometimes criticised by bus operators and the emergency services for slowing down their vehicles or producing an uncomfortable ride. Further technical developments including road cushions (which can be straddled by the wide wheel-base of heavy vehicles but not by cars) may provide satisfactory solutions to this problem.

11.54 In some circumstances, badly designed traffic calming schemes can worsen the environment. The use of cheap materials and design has meant that the worst schemes are visually intrusive. Inappropriate surfaces can increase noise. If speed humps are placed too far apart vehicles accelerate and decelerate between them, and this causes noise and increases exhaust emissions. There is considerable technical experience available on how to avoid all these problems, provided that suitable design standards are used, especially with a variety of different measures in combination. Until recently much of the initiative for traffic calming came from local authorities, which produced their own guidelines[29] or relied on material produced by voluntary organisations[30], and from technical manuals produced by consultants.[31] All of these rely extensively on foreign experience, especially from Germany and the Netherlands, sometimes using direct translations of foreign guidelines.[32] The government has now provided a legal framework permitting the introduction of traffic calming features which comply with regulations and the Department of Transport has issued technical advice in Traffic Advisory Leaflets. It is also undertaking research into the effectiveness of traffic calming features. The most important needs now are to ensure that sufficient funds are available to enable state-of-the-art designs to be used and to give active encouragement to the principles of changing the 'balance of power' between vehicles and pedestrians on streets used by both.

11.55 Some continental cities have now started to rely on 30 kph (18 mph) speed limit zones in residential areas to achieve some of the benefits of traffic calming. The reasons lie partly in the practical disadvantages to the bus and emergency services referred to above and partly in the cost of installing physical measures and the time it would take to implement them in all residential areas where they might be beneficial. Lengthy delays in introducing physical measures could give rise to political difficulties in assigning priority. This seems to us to be a reason for increasing the resources devoted to traffic calming so that measures can be introduced quickly. Hanover has adopted a pragmatic approach in this respect: installing 30 kph zones quickly by reliance on regulatory controls, but subsequently introducing physical measures where traffic speeds within the area remain too high or where there is a residual accident problem. The Department of Transport is studying the effectiveness of the 20 mph zones now being established in this country. **If they are shown to be effective, we recommend they should be used more widely.**

Parking policy

11.56 As early as the 1960s, the Buchanan Report[33] emphasised the immediate importance of parking policy as a means of influencing traffic demand. This tool has been virtually the only pro-active management technique used in the UK for the past 30 years and the government has recently reaffirmed its value in encouraging the use of alternative modes and restraining particular types of car journey such as commuting.[34]

11.57 Controls on public car parking spaces (both off and on-street) have had some success in influencing the use of spaces in prime locations and, when adequate enforcement is available, have contributed to maintaining the capacity of the existing road system.[35] To take a single example of a widespread practice, the 5,000 public car parking spaces in Cambridge are currently managed through the price mechanism to encourage short-stay users (who are usually shoppers and travel off-peak) and discourage the long-stay users (who are usually commuters and travel within the peaks).[36] Limited space is thus made available to a larger number of users who contribute to the commercial viability of the city but arguably at the cost of a higher number of journeys than would otherwise be possible and of higher off-peak traffic levels.[37] This outcome will be desirable where, on balance, the overall environment is improved.

11.58 On-street parking controls have often not been deployed to maximum effect because of inadequate resources for enforcement. The recent transfer to London Boroughs of responsibility for enforcing parking restrictions is a step in the right direction. We welcome the change in principle, and look forward to its extension to the rest of the country. Enforcement should be in the hands of the body responsible for other aspects of transport policy in the same area. Because 'control of parking is one of the weapons of a co-ordinated plan'[38] that will, in most cases, be the strengthened and extended Passenger Transport Authorities and Executives we have recommended. A high level of enforcement against illegal parking on footways, cycle routes and bus lanes, will be essential in order to achieve the maximum environmental advantage from parking policies.

11.59 The effectiveness of policies on public parking will be limited by their lack of influence over a large proportion of existing parking spaces. More than half the off-street car parking space in most large towns is privately owned and used and therefore uncontrolled.[39] Cambridge, for example, has up to 40,000 private non-residential car parking spaces, up to 17,000 of them in the central area. One survey suggests that 81% of people who normally drive to work are provided with free parking there (the Midlands was reported to have the highest provision — 87%).[40]

11.60 Local planning authorities have power to specify the number of parking spaces to be provided in new developments. They exercise those powers very differently at present. The government recommends reduced requirements for locations which have good access to means of travel other than the car and stresses the desirability of minimising provision rather than maximising it.[41] This advice is welcome.

11.61 The London Planning Advisory Committee (LPAC) has published a parking strategy for London. Its concern is a useful acknowledgement that parking policy is not independent of other land use and transport policies and that both private and public (off and on-street) parking must be considered together as part of a borough-wide strategic parking plan. The solution it has selected would in effect allow different local authorities to continue to use levels of parking provision as a means of competing with each other for increased trade. In effect, an attempt by one borough to manage traffic levels through parking policies could be readily undermined by more generous standards adopted in a neighbouring borough. This is a serious flaw in the approach. The same effect is found in towns outside London with equally serious consequences.

11.62 We believe that policies on parking provision must be consistent within regions, and with national priorities, so that advances in one town are not undermined by conflicting action near by. We would expect to see the policies used in ways which improve the economics of public transport by stimulating its greater use. **We therefore recommend that the government should work closely with regional planning conferences, PTAs/PTEs and the private sector to develop and implement comprehensive parking strategies designed to restrain use of private cars.** Both the level of charges and the extent of provision will be essential elements in any strategy.

11.63 Controls on the use of public parking spaces can be introduced quite quickly and should affect traffic flows almost immediately. Less generous provision of private spaces in new commercial development will eventually have an effect but not for many years, since the change is entirely dependent on the pace and scale of redevelopment. Meanwhile, policies are needed to counter the incentive to use cars which comes from the substantial amount of existing private non-residential parking space. LPAC has recommended financial measures, urging employers to charge their staff for parking at work (a practice already adopted by a few). It has also suggested that private non-residential spaces might be taxed; that retail parking should be treated as public parking and subject to the same restrictions; and that revenue generated from parking might be used more often to improve public transport services. It goes without saying that controls of this sort will not find easy acceptance if the public sector — including the local authorities implementing the parking policies — are not themselves in the lead in implementing such policies at their own premises.

11.64 These ideas, and others, bear further consideration by the government. For example, in suitable towns access could be restricted during rush hours to permit-holders (including residents and businesses). Such a scheme (based on one operating in Milan) has been proposed for Norwich. It would also be possible to empower local authorities to levy an environmental charge on private non-residential parking spaces. This could discourage the potential users from driving, whilst the revenue could be used to improve public transport. Changes such as these are likely to be unpopular with those who see their privileges diminished and, in order to make them more acceptable, they might be phased in gradually as alternative means of access are improved. In the long term, such policies could help to reinforce changes in land use which might reduce the need for transport.

11.65 A more unconventional approach has been adopted in Southern California, where from 1 January 1993, in order to discourage car use, some companies are required by law to offer cash to their employees in lieu of subsidising their public parking charges.[42] The employees benefit from the increased choice they are offered and can take full advantage of the incentive not to drive to work. It is claimed that without parking paid for by their employers, 27% of Americans who drive to work would choose other means of travel. The government should study these 'cash-out' measures in order to see whether this approach could also be adopted in the UK. Some features of the Californian law (for example, the exemption for companies which own their parking facilities) would not necessarily be appropriate in the UK. Other variations on the 'cash-out' programme exist and could also be tried out here. The city of West Hollywood offers cash to its employees to leave their cars at home and the ARCO Transportation Company in Long Beach charges single occupancy drivers the entire parking cost but, in contrast, reimburses the cost of monthly travel passes and provides other benefits to those who car-share. In Los Angeles, a subsidised Guaranteed Ride Home programme enables car sharers to return home by taxi or, for long journeys, hire car if their normal lift is suddenly unavailable (eg because they have to work late). Car sharing arrangements appear to be more successful in California than in the UK (perhaps because of generally longer commuting distances) but we would welcome further attempts to stimulate their development. Flexible working hours, and full or part-time teleworking, could attenuate still further the traditional rush hour.

11.66 'High vehicle-occupancy' lanes, which may be used only by vehicles carrying two or more people, are well established on motorways in the Los Angeles and Washington areas and have recently been introduced in the Amsterdam area. The motives have been to encourage people to travel in ways that are more fuel efficient and/or produce fewer pollutants. Whilst these are desirable ends, the designation of high vehicle-occupancy lanes would reinforce the habit of commuting by car, risk undermining public transport and be inconsistent with the aim of developing of land use patterns which reduce the need to travel. They may have merits in some circumstances (12.36) but we do not consider that they will generally be appropriate in the UK.

Road pricing

11.67 There has been considerable effort in recent years in investigating whether road pricing offers a cost-effective means of reducing demand for congested road space. The policy would be to increase substantially the cost of using that road space, whilst leaving the cost of travel on other roads undisturbed. At present, urban road pricing schemes usually require a payment to take a vehicle into an area for a certain period of time (area licensing, eg Singapore) or the payment of a fee when a boundary is crossed (cordon pricing, eg Bergen and Oslo). Payments may be made electronically or by purchase of a permit.

11.68 Electronic systems can be implemented in two basic ways. The first identifies vehicles automatically and records centrally a congestion charge for each trip. Unless anonymity is preserved, centrally held accounts may raise fears of intrusion on personal freedom. Enforcement may not be straightforward. To be effective from an environmental point of view, the system needs to give an immediate signal to the vehicle user of the congestion and environmental costs of the journey, perhaps on a roadside sign and/or an in-vehicle unit. The second system does not record charges centrally but deducts the cost of using specific roads from a stored value medium (like a telephone card). This approach can be extended to smart cards which automatically debit the cost of trips from bank accounts or charge them to credit card accounts.[43] Again, the user needs to be made aware of the cost of the individual journey. Enforcement using existing technology would present difficulty, as the number plates of cars not carrying the required electronic tag would have to be photographed, and the user traced. There has been a failure rate of 20% in enforcing speed limits using this technique; this results from technical failures to read the number plate properly, incorrect vehicle records, and the presence of effectively untraceable (eg foreign registered) vehicles.

11.69 Possible disadvantages of urban road pricing are that at price levels which are likely to be politically acceptable, environmental degradation may remain unacceptable, especially in areas with high suppressed demand; the redistribution of traffic which would result may increase overall environmental damage; and it may lead to the expansion of urban areas with adverse consequences for urban sprawl.[44] Most thought in the UK has been given to the possibility of congestion charging in London, which is probably not typical of conditions in many other urban areas.

11.70 Urban road pricing would, however, help tackle specific congestion problems. It could help to discourage some car or commercial vehicle journeys and tilt the balance of advantage between private and public transport (in terms of price and reliability). To the extent that tolls could be set at levels which reflect the cost of environmental damage, they offer a means of charging for that damage, with the highest charges being set in areas which suffer the greatest damage. It could raise significant revenue for investing in public transport and environmental improvements.

11.71 In the longer term, the further development of variable electronic schemes may enable universal road pricing to be introduced. This could face road users with the real costs they impose on the environment, varying from place to place and time to time. It could then be a useful means of influencing traffic growth in the peripheral areas of towns and cities (and perhaps sensitive rural areas) which are less amenable to conventional measures. Meanwhile, **we recommend that decisions to introduce road pricing schemes should be made locally after evaluating the environmental effects, including those on adjoining areas.** Like other restraint measures, they should be considered in the development of regional land use and transport strategies so as to ensure the maximum degree of consistency between the policies of neighbouring areas. **The revenue should be retained by the local authority introducing the scheme and used to finance public transport or infrastructure improvements which are not environmentally damaging.**

Treatment of income

11.72 The application of traffic restraint measures in urban areas would be likely to drive business away from them unless steps were also taken to reinforce their attractiveness. This is a particular danger if the presence or absence of traffic restraint is used as a competitive measure by neighbouring towns. We argued earlier that traffic restraint was an essential component in policies to encourage greater use of public transport. The converse is also true. In this sense, improvements to public transport and to the local environment are required to improve access to, and the attractiveness of, destinations as restraint measures are introduced. Some measures (notably parking charges and road pricing) have the potential to raise large incomes for the towns introducing them. We consider it essential that income from parking and road pricing charges should accrue to the local authority introducing them and be used to benefit the area.

11.73 Priorities for expenditure will vary from place to place and should be decided by local authorities. They should include improvements to infrastructure in ways already discussed. We would expect bicycle lanes, pedestrian schemes, bus lanes and segregated bus tracks, traffic calming, and noise reduction measures (where acceptable noise levels cannot be achieved from traffic reduction alone) to have high priority. So, too, should the improvement of public transport interchanges, provision of comprehensive information systems and, in suitable locations, the construction of light-rail systems. It may be beneficial, in a limited number of circumstances, for traffic to be diverted through tunnels (eg to preserve buildings or topographical features). Provided this could be done without encouraging traffic growth, income from restraint measures could usefully contribute to the cost of tunnelling. More general environmental improvements to town centres and perhaps the employment of town centre managers might be desirable.

11.74 We envisage that some transport operations should be supported with public contributions (7.67). In order to be operated as a comprehensive network, some additional rail and bus services will need support, possibly for quite long periods. It is also desirable to improve the attractiveness and comfort of buses and trains in order to attract passengers from cars. Seat widths, which are often inadequate for more than the briefest journeys, cycle and luggage space, and leg room could all be improved. If these improvements are not judged to be commercially justified, they might be financed partly from the income derived from traffic restraint measures. One disincentive to the use of trains or coaches on interurban journeys is the inconvenience of the stages between station and home or eventual destination. Better bus and tram services to railway stations would help but more imaginative approaches are also needed. Discounted taxi fares could be made available for example. Smart cards could be developed for use in taxis as well as buses and trains, so that all stages of a multi-modal journey would be covered. A combination of taxis, subsidised for the journey to and from the station, bicycle hire at stations, and connecting bus services could form part of a strategy for encouraging more medium and long journeys to be made by rail or coach.

Effects of measures

11.75 All the measures outlined in this chapter will contribute to the objective of reducing the dominance of motor traffic, at the same time as substantially improving the quality and conditions of alternative methods of transport, contributing to the targets for traffic and for improving the environment.

11.76 Research for the Department of Transport has assessed the effectiveness of public transport improvements and other measures in solving urban transport problems.[45] The model tested several possible policy measures against a base case forecast for a large provincial city in the year 2000, which assumed the continuation of existing policies. The level at which each policy measure was applied was not optimised in any way but was purely illustrative. The results of the simulations are expressed in terms of carbon dioxide equivalent reductions compared with the base case. The first options comprised improvements to public transport such as bus priority measures, light rapid transit networks, and substantial fare reductions. For example, a 50% reduction in all public transport fares (producing an additional annual subsidy requirement of about £15–20 million) reduced carbon dioxide equivalent emissions by 3.5% compared with the base. A light-rail network of six lines (totalling 60 km in length and with a capital cost of around £250-£300 million) reduced carbon dioxide equivalent emissions by 3.1%.

11.77 A second set of options considered car restraint on its own. The specific measures involved were pedestrianisation, reductions in car parking supply, and increases in public parking charges. Even though the effect of parking charges is greatly reduced by the proportion of people who park free of charge at work or on the street (75% in the city modelled) a doubling of parking charges from a base case of £2.50 a day reduced carbon dioxide equivalent emissions by 3.6%.

11.78 Finally a combination of car restraint and improved public transport was considered. Here cordon charging, reduced parking provision and light rail construction affected emissions significantly, reducing carbon dioxide equivalent emissions by 23% compared with the base case. The results were caused by the reduction in traffic entering the central area and by a significant increase in peak central area car speeds (forecast to rise from 23.6 kph to 30.1 kph). The Department told us that about 15% of this increase was accounted for by the effects of the light-rail network and the remainder by the measures to restrain traffic.

11.79 The Department concluded that, taken separately, extensive car restraint measures are more cost-effective in reducing congestion and emissions than measures to enhance public transport. It is also important to note the assessment (supported by the model results) that restraint measures on their own, by increasing the overall generalised cost of travel to the city centre, will tend to reduce trips to the urban centre to the detriment of the local economy. We have already referred to the risk that trips will not be reduced overall but redistributed to outer areas, thereby offsetting some of the benefits of the policy. Measures to improve public transport work in the opposite direction and help to counteract these effects.

11.80 More recent work modelling several towns has shown a similar pattern of results: car restraint is generally more effective than public transport improvement in reducing carbon dioxide equivalent emissions. In this case, measures were found to have different degrees of effectiveness in each town because the underlying conditions varied from place to place. For example, a high percentage increase in parking charges is more effective in a town where parking charges are already high than in a town where they are low. Unsurprisingly, restraint measures were more effective in reducing car trips in central than in outer areas (where fuel price increases were more significant).

11.81 In all, these considerations reinforce the importance of a balanced approach comprising both direct restraint on cars and improvements to public transport. They also indicate that decisions to introduce such policies should take full account of the likely effects over a sufficiently wide area.

11.82 This evidence of the effectiveness of combinations of measures is supported by experience on the continent. In Copenhagen, a city of 1.7 million people, road building schemes were abandoned in the early 1970s, large numbers of bus priority lanes introduced and a comprehensive network of segregated cycle paths built. The result is reported to be a 10% fall in traffic since 1970 and an 80% increase in the use of bicycles since 1980. About one-third of commuters now use cars, one-third public transport, and one-third bicycles. Cycling accidents have decreased slightly, despite the increase in mileage, because of the network of cycle paths.[46]

11.83 Few cities have, as yet, attempted to consider the policy implications of attempting to meet environmental targets. One exception is the City of Newcastle upon Tyne which commissioned a study of energy use in the city, partly in order to examine the implications of specific reductions in greenhouse gas emissions for future traffic levels.[47]

11.84 The study assumes some reductions in greenhouse gas emissions from the continuation of present policies (including improvements in vehicle design and increased use of diesel engines). It calculates, however, that to achieve reductions in emissions of 20% by the year 2000 and 40% by 2010 requires not simply the cancelling of growth in car use but a 16% reduction compared with 1990 levels by 2000 and a halving by 2010. The 2000 target means a reduction of 1,100,000 vehicle-kilometres per day. On the assumptions that the load factor of Newcastle's buses is the same as the British average and that the additional journeys generated by this level of restraint could be distributed equally throughout the day, such an increase might be within the existing public transport capacity of the city. Journeys foregone, or transferred to foot, cycle or Metro would facilitate the transfer. In reality, an even distribution of journeys would be most unlikely and some additional capacity would probably be required.

11.85 The targets modelled in this study are for deeper and more rapid reductions in carbon dioxide emissions than those we have proposed. Nevertheless the approach demonstrates how radical are the measures which could be required, particularly since the area modelled covers only central Newcastle. Growth in traffic around the periphery of the city would need to be addressed and a clear policy implemented to contain pressures for higher growth in development. Traffic restraint would be inevitable. In principle, the measures required are consistent with Newcastle City Council's Unitary Development Plan which proposes restrictions on vehicular access to the city centre. Newcastle has introduced its first traffic lanes reserved for service vehicles and buses, and intends to extend them. The overall aim is to give essential service traffic, buses, cyclists and pedestrians priority over other road users in the city. When the 10,000 public car parking spaces in the controlled area are full, computerised indicators would redirect motorists to park and ride facilities. A permit scheme for essential business car traffic is being considered. Whether these measures will achieve the targets in the carbon dioxide reduction plan remains to be seen. A further phase of the energy study is to assess the effects of carrying out the proposed reductions on the nature of the city centre and the business carried on there. We expect that success of Newcastle's policies will depend critically on the extent to which the actions of neighbouring localities and central government complement what it aims to achieve.

Conclusions

11.86 We believe that our environmental targets could be met despite the growth which has already taken place in local traffic and the implications of the Department of Transport's traffic forecasts. Some local authorities have begun to improve the urban environment in ways which will encourage greater use of non-motorised modes of transport. Voluntary bodies and other public organisations have also shown imagination and determination in, for example, improving facilities for cyclists and walkers, and promoting rural rail services. There are good continental examples of how the worst effects of motor traffic can be reduced in urban areas.

11.87 The priority is to press ahead urgently with the kinds of measure described in this chapter. This means not only developing a more attractive environment for pedestrians and cyclists and a more effective public transport system but also introducing more positive means of traffic management. Both aspects are essential. The precise form of the measures is essentially a matter for local decision, with the wide and active involvement of the local community and within a framework set by central government to ensure reasonable consistency and complementarity from locality to locality. Central government must also ensure that its transport financing policy is consistent with the priorities identified in this chapter.

11.88 In order for car restraint measures to be publicly acceptable, alternatives should be in place without delay. This will call for considerable political determination, especially in the face of competing demands for public expenditure. Although the sums will not be insignificant, they are justified by their environmental benefits and are not inordinately large when set against present expenditure on roads.

11.89 Many of the measures described in this chapter can be introduced quickly and will, in five to ten years, be making a contribution to environmental targets. They are not the only ways of reducing the number and length of local journeys. In particular, journeys made in the outer suburbs and around conurbations may need to be additionally influenced by appropriate land use policies and higher motoring taxes. It may take two or three decades for this combination of measures to be fully effective. Interactions between measures can be complex and their effects will need careful study to understand their effectiveness in restraining traffic and their wider effects on the shape of towns and the countryside.

Chapter 12

LONG-DISTANCE TRANSPORT

12.1 This chapter considers long-distance transport: the road network of motorways, trunk roads, and other major roads; interurban rail and express coach services; and domestic air services. Although it concentrates on personal travel, it also takes into account the conclusions and recommendations about freight transport put forward in chapter 10. Long-distance transport cannot be considered in isolation. Most vehicles on the road network are making local trips. Interurban trips by road usually involve some use of local roads as well, if only at their beginning and end. Where coach, rail or air is the main mode for personal travel, another mode or modes usually has to be used for the first and last stages. However, long-distance transport raises specific and important environmental issues.

12.2 Longer-distance journeys represent a small proportion of total journeys made in Britain, but a substantial, and increasing, proportion of total distances travelled. In 1989/91, 6% of journeys of more than a mile extended to 25 miles or more, but accounted for 44% of distance travelled. In 1985/86 they accounted for 5% of journeys and 41% of distance travelled.[1] Table 12.1[2] shows why people make longer journeys, and the extent to which the reasons differ as the length of journey increases. The proportions of journeys made for business, social or entertainment reasons or for holidays increase with journey length, whereas commuting is more important for journeys of 25–50 miles.

12.3 The objective identified earlier (10.1) was to increase the proportions of personal travel by environmentally less damaging modes and to make the best use of existing infrastructure. Although car travel is the main mode for 80% of journeys of 25 miles or more, rail has a larger share of this market than is the case for journeys under 25 miles: in 1989/91 rail was the main mode for 11% of journeys of 25 miles or more, with coach the main mode for 4%.[3]

12.4 It was noted earlier (6.46) that rail transport has a higher overall energy efficiency than road transport. Table 12.2[4] compares the energy efficiencies of types of vehicle used for long-distance personal travel. The least energy-efficient mode is air and, because of the amount of fuel which has to be used in take-off and climbing to cruising height, aircraft use relatively more energy on internal flights than on longer flights. Second-worst is a large car. The most energy-efficient mode is express coach; it looks even more attractive in energy terms because of the relatively high occupancies achieved. There is little difference in energy efficiency between various types of train and a reasonably efficient present-day car, provided that in each case all seats are occupied. At present average occupancies, most types of train are more energy efficient than most cars.

12.5 One effect of initiating a virtuous circle by making improvements to public transport should be to increase average occupancies on trains, so giving them a more clear-cut advantage over cars in terms of energy efficiency. It also has to be borne in mind that the relative energy efficiency of different modes will be affected by technological developments. We have in particular proposed as a target a 40% improvement in the fuel efficiency of new cars by 2005 (8.59). There is less information about the scope for increasing the energy efficiency of trains, but it has been estimated that an improvement of 10–15% could be achieved by 2010, primarily through reduced weight and better aerodynamics.[5] On balance, and allowing for higher average occupancies, trains will still possess some advantage over cars in terms of energy efficiency under most circumstances. It is unlikely that the improvements in prospect in the energy efficiency of aircraft (5.30) will alter the relative positions of aircraft and trains as shown in table 12.2.

12.6 Energy efficiency provides a reasonably good basis for assessing the carbon dioxide emissions from different modes of transport, but other factors also have to be taken into account in assessing which

Table 12.1
Why people make long journeys (1989-91)

	25 to under 50 miles	50 to under 100 miles	100 miles and over
	percentages of journeys		
total leisure, including	41	54	67
social/entertainment	*27*	*30*	*34*
holiday/other	*14*	*24*	*33*
other personal business	11	9	7
commuting	25	15	5
shopping	9	3	2
business	13	17	18
education	2	2	0
all purposes	100	100	100
average number of journeys per person each year	27.6	11.7	6.1
average number of miles travelled per person each year	941	804	1,068

modes do least damage to the environment. Table 10.2 compared emissions of airborne pollutants from different modes of freight transport, and showed a clear advantage for rail over road. It is likely that rail's advantage in the case of passenger transport would not be as large (just as its advantage in energy efficiency in the case of passenger transport is smaller than that shown in table 10.2). However, rail has the important advantage in environmental terms that it can use electric power, and that electricity can be generated from a variety of sources; electrically powered cars, competitive for long-distance travel, are only a long-term possibility.

12.7 Another important environmental consideration is the extent to which different modes require new infrastructure to be built for longer-distance travel. Both new roads and new railways cause the types of environmental damage which were discussed in chapter 4. Whereas the government has a major road building programme, much of the interurban railway network is now substantially underused, and only one major new railway (the Channel Tunnel Rail Link) is at present proposed. The Department of Transport (DOT) has promoted the widening of existing motorways as less damaging than building new roads across open country, but the motorway widening programme would itself require significant amounts of land and have damaging consequences for the environment.

12.8 We referred earlier (4.36) to the differences between modes in the risk of death or serious injury. The death rate on motorways is relatively low, however, and close to that for rail. For car users the rate is about one-third the rate for cars on other main interurban roads. Air is the safest mode. Between 1983 and 1992 there were 0.2 deaths per billion passenger-kilometres in the case of air and 1.0 in the case of railways. Information on vehicle occupancy rates on motorways is limited, but estimates suggest that in 1992 and 1993 there were 1.5 car occupant deaths and about 1.0 coach passenger deaths per billion passenger-kilometres.

Table 12.2
Energy efficiency in longer-distance personal travel by mode

energy consumption in megajoules per passenger-km

	at typical occupancy		with all seats occupied
express coach	0.3	(65%)	0.2
British 125 train (diesel)	0.8	(50%)	0.4
British 225 train (electric)	1.0	(50%)	0.5
French high-speed train (electric)	1.1	(50%)	0.5
small diesel car (1.8 litre)	1.2	(35%)	0.4
small petrol car (1.1 litre)	1.4	(35%)	0.5
provincial/suburban multiple unit train (electric)	1.7	(22%)	0.4
large diesel car (2.5 litre)	1.8	(35%)	0.6
large petrol car (2.9 litre)	2.8	(35%)	1.0
air, internal flights	3.5	(65%)	2.3

12.9 In the light of the environmental considerations discussed above, our conclusions about the relative advantages of different modes of longer-distance personal travel, which are taken into account in the remainder of this chapter, are that:

provided they are designed to meet the latest emission limits and are properly maintained, express coaches are less damaging to the environment than other modes;

under most circumstances trains, which increasingly use electric power, are likely to be less damaging to the environment than cars;

aircraft are more environmentally damaging than other modes.

12.10 The next section considers the likely future demand for longer-distance travel in Britain, taking into account the effects of measures recommended in earlier chapters of this report. We then examine the scope for augmenting the capacity of the road network without building new roads or additional lanes, and for using the available capacity more effectively or in ways that would produce lower levels of emissions. The core of the chapter is a review of the scope for longer-distance travel to transfer from cars and aircraft to environmentally less damaging modes. We then consider what overall targets should be set for mode switching, the public expenditure consequences of such targets, and what the future balance should be in transport investment.

The demand for longer-distance personal travel

12.11 Alongside its convenience, the perceived cheapness of travel by car has encouraged people to live further from their work and make longer journeys for leisure purposes. Although there is a great deal of uncertainty, our estimate based on earlier analysis (7.54) is that, if fuel prices double in real terms by 2005 and are maintained at that level, road traffic in 2020–25 will be only about 80% of what it would have been in the absence of such fuel price increases. We also concluded previously (2.30) that, of the forecasts of road traffic made by DOT in 1989, the low forecast has a higher probability than the high forecast. On the basis of the low forecast, and independently of other measures, a doubling in the real price of fuel by 2005 should limit the overall increase in road traffic between 1995 and 2025 to about 30%, in other words to less than 10% a decade.[6]

12.12 DOT's published forecasts of future road traffic are made at national level and do not distinguish between local journeys and longer-distance travel or between different types of road. Government policy provides for only a very limited programme of major road improvements in urban areas (6.21) and assumes there will be only limited growth in traffic in inner urban areas.[7] Traffic on motorways has shown above average growth in the recent past: during the 1980s it grew by 7.8% a year.[8]

12.13 Concern has been expressed that slower than average growth in urban areas might be accompanied by faster than average growth in the countryside.[9] The policies recommended in chapter 11 for local journeys would have the effect of significantly limiting the overall growth in road traffic in built-up areas. We discussed earlier (6.11–6.12) the possibility that, in the absence of government intervention, congestion in existing urban areas might stimulate further transfers of activity from them. However, the positive policies for transport and land use we have recommended should be sufficient to maintain the overall attractiveness and prosperity of urban areas, while at the same time changing the relative importance of alternative modes of travel. In that case, traffic restraint in urban areas should not have the effect of further increasing traffic levels outside urban areas; but it remains the case that DOT's forecasts already imply faster than average growth on interurban roads. Unless additional measures are taken to secure the transfer of some demand to other modes, there would continue to be significant increases in traffic on interurban roads even with the increase in fuel prices we have recommended.

12.14 The effects of other policies and trends on longer-distance personal travel are more difficult to assess. The impact of land use policies integrated with transport policies in reducing the number and length of commuting journeys of more than 25 miles may not be very large. Land use policies may have more effect in making leisure opportunities more readily available closer to centres of population. Wider exploitation of telecommunications (6.63–6.66) may remove the need for some business journeys and enable more people to avoid commuting journeys by working at home.[10] However the overall impact of this form of technology on longer-distance travel within the UK is likely to be modest. In some cases it may even lead to an increase in such travel by making it easier for people to live a long way from their employer's offices or cover a wider territory. Long-distance leisure travel is unlikely to be significantly reduced by technological developments. To summarise, other trends and policies may exert a modest downward pressure on longer-distance personal travel, but it is impossible at this stage to quantify the effects.

Making the best use of the existing road network

12.15 Some stretches of the motorway network now experience peak hour or chronic congestion (6.25–6.26), as do stretches of some other primary roads. Even with a large increase in fuel prices it is clear that the demand for longer-distance personal travel will increase, as will the demand for freight transport (10.4). Without building relief roads or additional lanes there are a range of measures which could be taken to increase the effective capacity of the existing road network or to control the use made of that capacity. We look first at measures to control access to the primary route network, then at measures to improve traffic flow.

Controlling access

12.16 The case on efficiency grounds for pricing the use of roads under congested conditions was discussed earlier (6.55). Another reason for charging for the use of infrastructure is to provide income to cover the costs of its construction and maintenance: this is an established practice in the UK and elsewhere for tunnels and major bridges, and is also followed in some countries for certain roads. Charges are made for the use of motorways in France, Italy and Spain, and for some roads in Portugal and Greece. In France, tolls average 5 pence a mile for cars and in total yield £2 billion a year; in Spain tolls for cars are 10 pence a mile. The German government has been moving towards tolling, with the aim of obtaining a contribution towards the cost of the motorway system from foreign lorries in transit, as a replacement for the *vignette*; a small trial is under way of an electronic charging system for all vehicles.

12.17 In May 1993 the government published a Green Paper, *Paying for better motorways*, which set out options for charging for the use of motorways, without a definite commitment to making charges.[11] The Green Paper did not propose a particular level of charge, but estimated that a charge of 1.5 pence a mile for cars and 4.5 pence for heavy goods vehicles would yield about £700 million a year. Subsequently the government has ruled out the options of conventional toll booths and motorway access permits, emphasising a preference for electronic tolling when technology permits. It estimates this would be in about five years' time. The House of Commons Select Committee on Transport produced a report on motorway charging[12] in July 1994. The Committee's overall conclusion was that: 'The Government will have to provide a great deal more information before Parliament and the public can be persuaded that proposals ... for tolling are either desirable or workable.'

12.18 As the title of the government's consultation paper indicated, motorway tolling has been presented as a means of financing motorway building, on the ground that those who benefit from such provision should bear the cost. The revenue can be allocated in this way because motorway tolls are regarded as a charge rather than a tax; the Treasury has traditionally opposed allocation of tax revenue to specific expenditures. While not opposed in principle to motorway charging as a means of financing the road programme and of controlling congestion, the Select Committee considered that the government's position was no guarantee that additional income from charges would not be offset by a reduction in the resources made available by the Exchequer. However, we consider that, in view of the probability that the provision of additional road capacity in itself generates additional traffic (6.33), it would be undesirable on environmental grounds to use the revenue from motorway tolling to finance the building of new roads. **We recommend that, if motorway tolling is introduced, the net income, after meeting the costs of maintaining and operating the present motorway system, be used to help meet the public expenditure costs of introducing a sustainable transport policy, including the measures recommended in this report to improve public transport.**

12.19 In principle, motorway tolls could be set by reference to the environmental costs, as well as the congestion costs, imposed by a trip on a particular stretch of road at a particular time, supplementing in this respect the function of increased fuel prices as an expression of the environmental costs imposed by road traffic generally (7.41). Situations in which road vehicles impose unusually heavy environmental costs may, however, arise more often in urban areas than on interurban roads. Motorway tolls could make a more direct contribution to a sustainable transport policy if they were set at a level designed to divert traffic, or a specific category of traffic such as the heaviest lorries, to environmentally less damaging modes. At the moment both these possible bases for setting motorway tolls appear to be ruled out by EC law, which requires the rates charged for heavier goods vehicles must be 'related to the costs of constructing, operating and developing the infrastructure network concerned'.[13] Studies commissioned by DOT of the likely local effects of motorway tolls suggest that the diversion of traffic from motorways to rail or buses would in any event be small.[14]

12.20 Rather than reducing the number or length of trips, or bringing about a switch to public transport, the result of levying charges for the use of certain roads in regions with a relatively dense road network might be that road users would find alternative routes. Studies for DOT suggested that charges of 1.5 pence a mile for cars and 4.5 pence for heavy goods vehicles might cause diversion of 10% of motorway traffic onto other roads at peak times.[15] As these routes are likely to be less suitable for heavy traffic, particularly goods vehicles, there could well be an increase in accidents, in the severance of communities, and in the extent to which people are exposed to noise, vibration and pollutants. A recent large increase in the toll for the Severn Bridge (which is levied only on westbound vehicles) provides an example, in that the numbers of westbound heavy goods vehicles using rural roads on the west bank of the Severn increased significantly. Drivers chose to make a lengthy detour even if this increased the total cost of a trip: one reason given for this behaviour was that many had to pay the toll themselves and claim the amount from their employers. The Select Committee considered that: 'Until the Government can demonstrate clearly how it proposes to tackle the question of diversion, the tolling of motorways should not go ahead.' We endorse that view. **In view of the potential damage if traffic transfers to environmentally more sensitive routes, we recommend that any proposal to introduce or vary tolls for a road, bridge or tunnel be the subject of an environmental assessment before it is approved.**

12.21 An alternative to using tolls to reduce congestion would be to apply traffic management measures to limit access to the most congested sections of the primary route network. Although the main purpose of this network is to provide for through traffic, many vehicles on it, and in particular on the most congested sections of motorway, are making relatively short local journeys. Sometimes local traffic on a primary route is the result of large-scale developments of the kinds discussed earlier in this report (2.21–2.23, 9.8–9.12) or traffic tailing back onto the primary route from a nearby junction or shopping centre. The A1 on Tyneside provides an interesting example. Congestion on this road, caused by predominantly local trips to and from the MetroCentre and a large industrial estate, led DOT to promote a controversial scheme for bypassing what was itself in origin a bypass. **We recommend that, where a route designed for long-distance traffic is overloaded as a result of heavy local traffic, an assessment be made of the advantages of closing selected entry points, completely or at certain times of day, as an alternative to widening it or constructing relief roads.** This does not mean that we consider local traffic should necessarily be excluded from stretches of primary route in all circumstances. The right solution will differ from case to case. If the access problems of which congestion is a symptom are predominantly local, however, their solution is more likely to lie in policies of the kinds recommended in the previous chapter than in major road building.

Improving traffic flow

12.22 Various management measures can be taken to improve traffic flow and so increase the effective capacity of primary routes. The flow of traffic onto motorways at peak periods can be eased by reducing access ramps to a single lane or by 'ramp metering' (using a traffic light to feed vehicles onto a motorway).[16] It has been estimated that ramp metering would increase the capacity of the M25 by 3–4% and reduce by about a quarter the length of time each day during which traffic is at saturation level on the south-west sector.[17] However, extra road space would be required at junctions to hold traffic waiting to join the motorway, and the check in speed and subsequent acceleration caused by the traffic light would also lead to some increase in emissions.[18]

12.23 Another option for easing congestion on heavily used motorways by producing a more even traffic flow is to reduce the speed limit from 70 mph to 50 mph at peak periods. The Highways Agency is carrying out a pilot scheme on the M25, which has the potential to increase capacity by 5–10%.[19] Lower speed would bring direct environmental benefits through a reduction in the number and severity of accidents, improved fuel efficiency (see figure 8-I) and lower emissions of nitrogen oxides (3.11). Traffic would also be quieter, as the sound from contact between tyres and the road is closely related to speed.

12.24 The present 70 mph speed limit is not well observed: the average speed of cars on motorways is 70 mph; more than half exceed the limit, and a fifth of cars travel at over 80 mph.[20] Effective enforcement of the 70 mph limit on dual carriageway roads and the 60 mph limit on single carriageway roads would reduce casualties and would also lower carbon dioxide emissions from road vehicles by about 3%.[21]

12.25 More effective methods of enforcement have become available. EC law now requires the fitting of speed limiters to prevent new heavy goods vehicles over 7.5 tonnes from exceeding 60 mph and coaches from exceeding 70 mph (65 mph from 1996).[22] New requirements will apply to HGVs over 12 tonnes from 1996 (10.46). No country has yet required speed limiters on cars; any such requirement would have to be an EC-wide measure, or else be so applied that it could not be construed as an anti-competitive measure. In view of the availability of other methods for enforcing speed limits, we do not believe the potential benefits of fitting speed limiters to cars would at present justify the cost. In one recently introduced method the number plate of any vehicle exceeding the speed limit is photographed. This is mainly intended to deter; signs warn of cameras ahead. Initial results are encouraging, with a drop in accidents and speeds down by 10 mph. Further trials are planned in various parts of the country. In another system the registration numbers of cars exceeding the speed limit are displayed on roadside boards as a warning to the drivers. **We recommend that increased effort be devoted to enforcing speed limits, making full use of new technology.**

12.26 It can be assumed that any reductions in speed limits to improve traffic flow would be applied selectively, and they are therefore unlikely to have a significant effect on fuel consumption and emissions. If the speed limit was reduced to 55 mph for all types of road (the limit introduced in the USA in the 1970s following the first oil shock), and if it was effectively enforced, it is estimated that a further reduction of about 3% in carbon dioxide emissions could be obtained.[23] **We recommend that the case for reducing the general speed limit be considered after 2000 in the light of progress towards our targets for reducing carbon dioxide emissions from surface transport.**

12.27 Large-scale programmes are being carried out in the EC, the USA and Japan to develop 'intelligent vehicle highway systems' which apply electronic technology to the management of road vehicles. Such systems take various forms; one is the computer-linking of traffic signals, already used on a large scale in London. Another is the use of variable message signs to give drivers warning of congestion ahead: widely used in Japan, these have only recently been introduced on a pilot scale in the UK. It has been estimated that an automatic congestion detection and warning system on the M25 could increase traffic flow by 4.5% and halve the number of flow disturbances, as well as reducing accidents by almost a quarter.[24]

12.28 Research in progress on the application of electronics involves 'intelligent' cars using dynamic navigation. A system is already available which keeps a car a specified distance from the vehicle in front. Completely automated guidance systems would improve safety, and reduce fuel consumption and pollution by ensuring that vehicles travel at a uniform safe speed. They would also increase road capacity perhaps threefold by permitting vehicles to travel closer together. Such systems could also select the least congested route for a journey. They are some years away, and both the infrastructure and the additional equipment required in each car are likely to be expensive. Moreover there is the problem of mixing intelligent and ordinary vehicles on the same road during a probably lengthy transition period. These problems are being addressed in the research programmes. There is a legislative requirement on the US government to demonstrate automatic control of a convoy of vehicles ('platooning') by 1997. **We recommend that the government investigate the potential of automated guidance systems for increasing the capacity of motorways; and that the introduction of such systems should depend on an assessment of their environmental impact, including their impact on the total volume of road traffic.**

12.29 Any measures which increase the capacity of the road network may lead to additional traffic. We emphasise that, in supporting certain measures which could increase capacity in the primary route network, our aim is not to encourage an overall increase in road traffic, but to provide some flexibility to resolve local shortfalls in capacity without the need to take land for the construction of additional lanes or relief roads, and at the same time to obtain whatever specific benefits these measures would bring in terms of safety, noise and emissions. We regard this as entirely consistent with a sustainable transport policy.

Scope for mode switching

12.30 We now review the scope for increasing the proportions of longer-distance personal travel undertaken by rail and express coach. The policies recommended towards private road use will have an important effect. Some journeys which are no longer made by car as a result of the increase in fuel prices will be made by public transport instead. Studies of the effect of past changes in the cost of car travel have produced a wide range of findings. In three studies, the estimated increase in travel by public transport (bus and rail) in response to a 1% increase in petrol prices ranged from 0.08% to 0.8%, with a mean of 0.34%.[25] Insofar as it is valid to extrapolate from the effects of relatively small increases in fuel prices, these findings imply that the increase in public transport traffic following a doubling of fuel prices would lie in a range from 8% to 80%. We conclude that, in view of the extent of the uncertainty, it would not be prudent to rely solely on the recommended increase in fuel prices to bring about a substantial switch to public transport.

12.31 Switching from car to public transport for longer journeys will also be encouraged by improvements in public transport services and reductions in public transport fares. The importance of these factors is discussed below. They will have the side-effects of increasing the length of some journeys by public transport and stimulating some new journeys that would otherwise not have been made. This extra travel will bring personal benefits that must be set against the additional environmental costs. Nevertheless, it will be important to try to ensure that measures to make public transport more attractive do not become counter-productive by themselves encouraging transport-intensive lifestyles, for example, by stimulating long-distance commuting by rail.

12.32 Other important factors which will facilitate mode switching are land use planning policies and the quality of transport interchanges (11.37); longer journeys by public transport will almost always involve use of more than one mode, or at least changes between services within a mode. As with local journeys, land use planning policies are likely to have a more immediate impact on the choice of mode for travel than on the overall demand for travel (9.28–9.33).

Long-distance coach travel

12.33 Although only 4% of longer-distance journeys are by coach there has been very rapid growth: between 1975/76 and 1989/91 the average number of journeys made by coach each year doubled and there was a quadrupling of passenger-kilometres travelled.[26] This reflects the deregulation of coach services in 1980 and the completion of the national motorway network. The privatised company National Express remains the dominant operator in England and Wales. Commuter coach services into London have expanded particularly rapidly, although they still have only a small share of the London commuter market.

12.34 Coach fares are lower than rail fares, especially at peak periods, and for an individual traveller may be competitive with the cost of fuel for travel by car. The evidence is that demand for coach travel is quite responsive to price: a review found that for a 1% change in fare levels inter-city bus travel showed a variation of between 0.32% and 0.69%.[27] Coaches are reasonably comfortable and the possibility is being considered of premium services providing a higher standard of comfort. On busy routes like that between London and Oxford (see box 12A) coaches can provide an attractive and convenient alternative to car or rail. On other routes services are infrequent and travel by coach would not be an option considered by most people at present. Coach stations are often inconveniently located, poorly designed and unattractive. It is difficult to obtain information about services and it may not be easy to book a seat.

12.35 Coaches can adjust flexibly to changes in demand and changes of route. As well as having an inherently higher energy efficiency than other modes of longer-distance travel, they also gain in energy terms by having a high average occupancy. The environmental acceptability of diesel engines has been discussed in chapter 8, in the light of the concern the Commission expressed in its Fifteenth Report about emissions from heavy diesel vehicles, including buses. In contrast to many buses, express coaches are usually relatively new and well maintained. Provided controls over emissions are tightened in the way we have recommended, expansion of express coach services will be advantageous in environmental terms because of their high energy efficiency. There has recently been public concern about accidents involving coaches (not usually scheduled services) but the death rate for coach passengers is similar to that for car or rail (12.8) and the fitting and wearing of seat belts would reduce casualties.

12.36 Expansion of coach travel is likely to come about through market forces. That perhaps provides the best guarantee that innovations will be introduced, such as new routes and premium services, to appeal to a wider range of potential passengers, but government should take certain measures to facilitate expansion. **We recommend that transport and land use planning recognise the role of express coach services, and provide full facilities for them in traffic management schemes and at transport interchanges.** Coach services would benefit from the wider provision of bus lanes in urban areas which

BOX 12A **OXFORD–LONDON COACH SERVICE**

There are two direct coach services between Oxford and London Victoria: Oxford CityLink 190 and the Oxford Tube. As each service runs every 20 minutes during the day, there is a coach every 10 minutes. At peak periods the Oxford Tube service runs every 10 minutes, giving a combined service of 9 coaches an hour. During the night there is a coach at least every hour.

One company advertises a journey time of 90 minutes, the other approximately 100 minutes. Both warn that the journey may take longer when heavy traffic causes congestion.

The adult fare, for a distance of 57 miles (92 kilometres), is £5.20 single, £5.50 or £6 for a 24-hour return and £7.50 for a period return. For students, children and senior citizens the corresponding fares are £4.70, £5 and £5.50. One company offers 6 return trips for £27, the other offers through tickets from local bus stops.

Coaches gain in convenience by starting in Oxford city centre, making six other stops in Oxford and making at least one other stop in central London.

A number of other coach services run between Oxford and the London area, making intermediate stops or serving airports.

For comparison, fast trains run every hour from Oxford station (half a mile from the city centre) to London Paddington and take about an hour. There are some additional fast trains; and during most of the day a stopping train every hour which takes almost 90 minutes. The standard adult fares are £13.80 single, £26.40 day return and £11.20 cheap day return (not available for arrival in London before 10 am Mondays to Fridays).

Information for April 1994

we have recommended (11.37); and would benefit from the introduction of high vehicle-occupancy lanes (11.66) on interurban roads, which could improve reliability. **We recommend that, in considering whether high vehicle-occupancy lanes should be designated on interurban roads, highway authorities should place considerable weight on the potential benefits to express coach services.**

Transfer from car to rail

12.37 Although rail travel increased in a number of European countries (including Britain) during the 1980s, it was only in the Netherlands that it grew rapidly enough to maintain its share of personal travel, which is broadly similar in all the main western European countries (table 6.3). Rail is more competitive with cars over longer distances because it can achieve higher speeds. Over shorter distances any advantage in speed is likely to be cancelled out by the time spent travelling to and from the station at either end. Over distances of up to 100 miles coach journey times are often competitive with rail. The French and Spanish high-speed trains discussed below have been successful in attracting passengers who would otherwise have travelled by car, but this has occurred over distances which are long by British standards (broadly equivalent to London-Edinburgh/Glasgow).

12.38 Rail has complex pricing structures designed to boost occupancy at off-peak times, but within the constraints imposed by the need to cover the whole of the direct costs, including any profit margin (7.23). There is evidence that rail travel, at least in some segments of the market, is very responsive to price changes. A review suggested that for a 1% change in the price of interurban rail travel the variation in traffic was between 0.32% and 1.20%.[28] Another review found similar values.[29] While part of the effect of changes in the price of interurban rail travel is to alter the proportions of traffic carried by rail and by car, part will be additional or forgone journeys. There will also be a change in the respective shares of rail and express coach.

12.39 Other factors are important in increasing the relative attractiveness of rail. Potential passengers are influenced by the frequency and speed of services. The perceived relative reliability of car and rail travel and the overall journey times will also be important considerations. Rail operators now place much more emphasis on punctuality and their reputation should improve in that respect. The liability of rail to disruption by industrial action may be a factor, but may have less influence on potential passengers than on freight shippers. On the other hand the risk of delays through congestion at busy times is a factor favouring rail over both car and long-distance coach. A further advantage of rail travel is the possibility of doing other things, such as working or eating, during the journey. Modern rolling stock has a high standard of comfort; many people find coach travel less comfortable.

12.40 As with local journeys, convenience is an important factor in long-distance rail travel, in terms of the essential attractiveness of the service and in its effects on overall journey duration. As a high proportion of rail journeys involves the use of another form of transport, improvements in connections with other modes may be as important a factor as reductions in the time taken on the rail portion of the journey. Aspects of convenience include the provision of secure and cheap car parks at stations with sufficient capacity, clear timetables and co-ordinated services, good links with other public transport services, and booking arrangements. **We recommend that the rail regulator require rail operators to make comprehensive information easily available at stations and on longer-distance trains about public transport connections at destination stations.** Connecting systems of types described earlier (11.74) could play a valuable role.

12.41 The leasing of some railway stations to private sector companies may lead to new development opportunities to attract passengers, by providing shops and other facilities and by attracting custom away from alternative facilities less accessible by public transport. This is to be welcomed where the result is either to reduce the total amount of travel or to attract it to environmentally less damaging modes.

Transfer from air to rail

12.42 Air travel accounts for only about 1% of domestic travel, but has been expanding rapidly (2.3). Over three-fifths of visits abroad by UK residents, which have increased even more rapidly (figure 5-I), are to EC countries[30], and a substantial proportion of them to cities within 600 km of London. The completion of the Channel Tunnel makes it possible for rail to compete for this latter segment of the market, which is broadly comparable in size to domestic air travel.

12.43 A major factor in competition between air and rail is the speed of trains. Until the new rail link to London is opened at about the turn of the century, rail services through the Channel Tunnel will travel at less than 150 kph on the British side, although they will travel at up to 290 kph on the continent. There are expected to be 3,000 km of high-speed rail line in continental Europe by 1996. Over the distance of about 500 km between Paris and Lyon the TGV has now been operating for 10 years; it now carries 92% of travellers, the number of passengers carried by air has halved and car travel between the cities has noticeably diminished. Because of the growth of traffic on the TGV, the total numbers travelling between these cities have, however, increased. Similar effects have been noted elsewhere. The introduction of the AVE high-speed line increased the share of passengers travelling by rail for some 600 km between Madrid and Seville from 20% to 45% in one year. Of the traffic attracted to the AVE services, 32% diverted from air, 25% from car, and 14% from other rail services; 26% was new traffic. Not all this increase in traffic can be regarded as generated by the new service; some will have diverted from other destinations, and the net increase in traffic on all routes will be smaller.

12.44 Eurotunnel's targets are to capture 60% of present London-Paris air traffic for Channel Tunnel rail services by 1996 and 57% of present London-Brussels traffic.[31] Plans by the European Commission for a Trans-European high-speed rail network include expansion to 7,400 km by 2000 and an eventual network of 20,000 km. They now include Eastern Europe. A further 15,000 km of existing lines could be upgraded for high-speed use.

12.45 For domestic flights, UK air fares are between first class and standard rail fares. Within Europe, it can be expected that air fares will be reduced as European air travel is progressively deregulated, and in response to faster rail services and new services through the Channel Tunnel. At present, much of the market is business travel for which differences in cost are a relatively unimportant consideration. Journey times are much influenced by the time taken to pass through terminals at each end; for that reason rail may turn out to be more competitive for international journeys than for domestic journeys over a similar distance. Because rail stations are generally centrally located and airports are peripheral, rail has an advantage for journeys to Brussels, Amsterdam and Paris, and on any domestic journey to a city centre. The convenience of inter-modal links is also an important factor.

12.46 The transfer of long-distance traffic from air to rail would reduce emissions of pollutants and the disturbance caused by aircraft noise (although there would be little effect on numbers of night flights). Against that a number of people would suffer an increase in train noise. It is unlikely that the resulting reduction in aircraft movements would have a decisive effect on the justification for additional terminals, runways or surface links proposed or under study at airports in south-east England (5.7).

12.47 Although trains are much more energy efficient than aircraft (table 12.2), raising train speeds carries an energy penalty. For a given design of train covering a given distance, the energy required to overcome air resistance increases as the square of the velocity. Whereas at low speed energy requirements are determined primarily by the weight of a train and rolling resistance on the track, overcoming air resistance ('drag') uses about two-thirds of the energy required at 100 kph and about 90% at 200 kph. For diesel multiple unit trains on the London to Bristol line an increase in the maximum speed from 185 kph to 200 kph reduces the journey time by 1.3% but increases energy consumption by 7%.[32] Noise also increases with speed, and has been a very controversial issue in the case of the Japanese bullet trains. These adverse environmental effects can be mitigated to some extent by improved design and, in the case of energy efficiency, by higher occupancy as a result of a more attractive service. The French TGV has been able to achieve a typical occupancy of 70% in practice, against only 40% for UK InterCity services. This converts what would be an inferior energy efficiency at full occupancy into an energy efficiency which is said to be up to 40% better than for InterCity services.[33]

12.48 Further generations of high-speed trains now being planned in some countries to travel at speeds up to 400 kph would have an even higher energy requirement. They would be levitated magnetically to eliminate rolling resistance from wheels on track, but the energy requirement would in any case be related almost entirely to air resistance, as explained above. Noise would be a serious problem as would landtake for new track. To overcome these last two problems, Japanese engineers are exploring the possibility of constructing a 400 km line entirely in tunnel. The public appeal of such a service is not clear.

12.49 While very high-speed trains may have environmental advantages over aircraft, they might cause more damage to the environment than fuel-efficient cars, especially if a new track has to be constructed. As well as passengers who would otherwise have travelled by air, they would carry large numbers of people who would otherwise have travelled by slower, less damaging, rail services. It is therefore possible that their net impact on the environment will be unfavourable. **In view of this we recommend that no proposal be taken forward in the UK for trains running at more than 300 kph unless a comprehensive assessment has shown that the environmental benefits from transferred air traffic will outweigh the environmental costs of landtake, construction work, noise and the additional energy required to propel trains at this speed.**

12.50 Nevertheless we believe that government policy should be based on transferring as much traffic as possible from air to rail. This will require fast rail links not only between cities, but also directly to airports to connect with intercontinental flights. **We recommend that policy on air services should be based on discouraging air travel for domestic and near-European journeys for which rail is competitive, and that the government should support the upgrading of rail links to the main international airports in order to avoid the need for development of air feeder services from regional airports.**

Targets for mode switching

12.51 The general objective of increasing the proportion of personal travel by environmentally less damaging modes and making the best use of existing infrastructure, identified at the beginning of this part of the report, needs to be supplemented by a specific target. An overall target for increasing the role of public transport has to take into account the conclusions and recommendations about local journeys contained in chapter 11, as well as the conclusions in this chapter about longer-distance travel.

12.52 There is substantial spare capacity for personal travel on the rail network at present. Typical occupancy rates (table 12.2) show that, in theory, the number of passengers carried could be doubled on InterCity services and increased fourfold on provincial services, without running any additional trains. But part of the spare capacity is available at unpopular times, and so is not usable in practice with the present patterns of demand. Increases in average occupancy should be sought, but low occupancy at certain times is an inherent feature of any regular timetabled service.

12.53 In practice a substantial increase in rail travel would require a significantly increased service of trains. These would be difficult or impossible to accommodate at peak periods on parts of the network as it exists at present. However, with a higher level of investment, considerable increases in capacity could be possible without the need to build new rail lines. These increases would be obtained by lengthening trains and platforms, resignalling to reduce the interval between trains, and rearranging track layouts to remove bottlenecks. Investment of this nature would allow much more intensive use to be made of the existing infrastructure. Capacity would also be available to accommodate the increased use of rail for freight, to meet the target proposed previously (10.63), because freight trains would not generally need to run at the peak periods for passenger traffic. Increased use of road-based public transport would also allow better use to be made of the existing infrastructure because buses and coaches take up less road space than private vehicles carrying an equivalent number of people.

12.54 We propose the following target:

> **To increase the proportion of passenger-kilometres carried by public transport from 12% in 1993 to 20% by 2005 and 30% by 2020.**

As present use of public transport is divided fairly evenly between bus and rail, both modes will have a major part to play in achieving this target. An important contribution will be made by new light-rail systems in cities. Our expectation is that use of express coaches will continue to increase substantially through market forces. An essential element will be a large increase in longer-distance journeys by rail. We focus on this last element here because government support and action will certainly be needed in order to bring about a transfer of journeys from car to rail on the required scale. Transfer of journeys from air to rail will have significant environmental advantages but will not contribute to this specific target.

12.55 Rail operators have not up to now had a direct incentive to increase the number of passengers carried. The obligation on British Rail (BR) to achieve a specified return on investment has placed it under pressure to increase or maintain fares where that would maximise revenue, rather than to set fares to increase the number of passengers carried. This pressure applied even if it could be shown that the transfer of such passengers to rail from other modes would yield net benefits when the social benefits of the transfer were taken into account. The environmental benefits of such a transfer were not taken into account at all, at least in quantified form. In the context of environmental objectives, it is desirable that future financial arrangements should give rail operators a clear incentive to maximise the number of passengers carried. We have already recommended (7.69) that, until it is superseded by measures for achieving a major increase in the role of public transport, the present net level of public contribution should be at least maintained in real terms.

12.56 An immediate effect of the restructuring of the railway industry in April 1994 (13.48–13.56) has been to increase considerably the costs which rail operators have to meet. Railtrack, the public sector company which now owns the infrastructure, is charging the operating companies and passenger transport authorities at levels which reflect the valuation placed on its assets. For 1994/95, these increased costs are to be met by increased contributions from central government, but there is no commitment yet to similar compensation in future years. At the moment the increased contributions represent an internal transaction within the public sector. That will no longer be the case if Railtrack is privatised.

12.57 In the absence of public contributions to offset increased Railtrack charges after 1994/95, operating companies would seek to increase their income through growth in the volume of their business and to reduce their other costs through greater efficiency. As the scope for doing either of those things may be limited in the short term, they would also be likely to try to pass on their increased costs by increasing fares. In our view one key purpose for a public contribution to the costs of rail transport is to make it possible for operators to levy charges which are a closer reflection of full additional costs, rather than average costs (7.70). That would point towards a reducing level of fares in real terms. Under the new financial arrangements, however, the prospects for keeping rail fares at present levels, let alone reducing them, will be dependent on an increase in the proportion of rail costs covered by public contributions from central or local government.

12.58 It is important that any public contribution made to the costs of public transport on environmental grounds should be related to specific benefits, especially those obtained by transferring traffic to rail from cars and lorries. In the past, the effect of the criteria used in appraising rail schemes has often been that benefits of schemes for users and non-users have been taken into account only to the extent that they are captured by the fares users pay. More recently, DOT has introduced more flexibility in taking account of benefits to non-users, in particular from reduced traffic on roads. **We recommend that the criteria for the public contribution to railway costs be the environmental benefits of attracting traffic to rail from road, and whatever social criteria are considered relevant; and that existing forms of subsidy be reviewed in the light of these criteria and consolidated.**

12.59 An argument sometimes made against public contributions to the cost of public transport is that, because only a small proportion of journeys is made by public transport, only a small proportion of the population benefits. Our proposed targets, however, are for a very substantial increase in the role of public transport and in the proportions of freight transported by rail and water. These entail a fundamental change in attitudes towards the public transport system and in the extent of government commitment to it. Piecemeal improvements, not made in the context of overall targets, might not be cost-effective. With such a change in attitudes and commitment on the other hand a virtuous circle can be initiated which will lead to the creation of a public transport system that is not only more intensively used, with higher standards of service, but also much more viable and attractive.

12.60 Large and important benefits will flow from that, which can be summarised as follows:

a. improved access within towns and cities, which could not be achieved by any other means;

b. substantial environmental benefits outside towns and cities because more effective use would be made of the existing infrastructure, rather than building new infrastructure;

c. greater freedom and mobility for that half of the population which does not have access to a car.

We consider that these benefits, and the other benefits sought in our proposed targets, will sufficiently justify the expenditure required to achieve those targets.

The balance of future investment

12.61 Transport investment in recent years was analysed in table 6.1. Investment in road infrastructure has been the dominant force (excluding the purchase of road vehicles), and in 1985/86 was more than three times the amount invested in rail infrastructure. Rail investment has grown in both amount and relative importance in recent years, but in 1991/92 and 1992/93 investment in road infrastructure was still almost twice investment in rail infrastructure. Moreover, as table 6.1 shows, much of the growth in rail investment came from private financing of the Channel Tunnel link with France, which is now completed, will not benefit domestic travellers, and will to a large extent be used to convey road vehicles. Excluding the Channel Tunnel, investment in rail infrastructure was equivalent to just over a third of investment in road infrastructure in 1991/92 and just under two-fifths in 1992/93.

12.62 The UK already has a high-quality interurban road network. We concluded earlier in this chapter that overall demand for long-distance travel will be materially reduced by higher fuel prices, and that there are ways of increasing the capacity of the existing road network and using that capacity more effectively which do not involve the construction of additional lanes or relief roads. It is our view that expenditure on the trunk road programme can be, and should be, considerably reduced. A large part of expenditure on motorways and other trunk roads is taken up with maintenance of the existing network, and that must obviously continue. Bypass schemes which will bring environmental benefits to particular towns or villages should also proceed, subject to the findings of the further studies and process of review we recommended earlier (4.34). We discuss in the next chapter whether there should continue to be a central government road programme in the traditional sense. **If there continues to be a central government road programme, we recommend that, except where the construction or widening of a motorway or other trunk road represents the best practicable environmental option for meeting access needs, expenditure should be confined to maintaining existing roads and constructing new bypasses for local traffic.**

12.63 It is likely that a reappraisal of proposed road schemes will in any case be necessary if the report of the Standing Advisory Committee on Trunk Road Assessment confirms that the provision of additional road capacity generates additional traffic. The schemes in the present road programme have been assessed on the assumption that there will be no difference in the overall volume of traffic with or without the scheme. As a result, those environmental effects which vary with the volume of traffic will have been underestimated and those calculations which depend on the speed of traffic will have been biased.

Moreover, existing assessments have been based on the 1989 National Road Traffic Forecasts. The methodology of those forecasts is questionable, and is now being reviewed (6.31–6.33). The recommendations in this report are designed to lead to a different kind of transport policy, which will be environmentally sustainable, and in which road traffic would not show further growth on such a massive scale. **We recommend that road schemes for which contracts have not yet been let should be reassessed to take into account changed policies and methodology, and further progress on them should be suspended until that reassessment has been completed.**

12.64 In the case of bypass schemes, it is important to introduce further safeguards to ensure that overall environmental benefits are in the event achieved and that this is done in the most cost-effective way. We have already drawn attention to the importance of limiting new development alongside bypasses, which has been a major factor in encouraging car-dependent lifestyles. In addition, **we recommend that, where the major purpose of a proposed bypass is environmental improvement:**

- **i. the design standards should provide only for the traffic flows which will be diverted from the town or village bypassed, not for further large growth of traffic;**
- **ii. the design, funding and implementation of comprehensive traffic measures in the area bypassed should proceed on the same timetable, and be given the same priority, as the bypass itself.**

12.65 Although we cannot anticipate the outcome of the processes of review recommended above in relation to individual schemes, we envisage that, in conjunction with the other recommendations in this report, they will make possible a very substantial reduction in the present programme for road building and widening. This will bring very considerable benefits in environmental terms and free resources for investment in environmentally less damaging forms of transport. In the light of this, **we recommend that planned expenditure on motorways and other trunk roads should be reduced to about half its present level.**

12.66 The investment programmes required for increasing public transport's share of local journeys will need to be drawn up at local level, and we discuss in the next chapter the arrangements required for that purpose. Investment in rail has been restricted for some years by the financial regime applying to BR and London Regional Transport (reinforced by restrictive borrowing controls) and by the unfavourable method of appraisal applied to proposals for investment. In the case of the former BR system we hope that the result of restructuring might be to make finance more readily available in future. Some of the investment to permit increased levels of services would be required from the passenger railway operating companies, the rolling stock leasing companies or the freight operating companies, but much of the expenditure would fall initially to Railtrack; Railtrack would seek to achieve a satisfactory return on capital expenditure through its charges to operators. It may well be necessary for the government to provide long-term assurances about the future levels of public contributions to individual railway operating companies, so that they will be in a position to enter into the long-term contracts Railtrack and the rolling stock leasing companies will require before they will be prepared to undertake major new investment. In order to provide incentives for efficiency (7.71), the amounts of public contribution would be related to identified environmental or social benefits and specified in advance in franchise agreements, and where appropriate would be taken into account by bidders for franchises.

12.67 Railtrack's planned programme of capital expenditure is about £0.8 billion a year. This is required in order to maintain the present capacity of the network, although replacement will inevitably involve some element of upgrading. As a result of under-investment in the past, there is a backlog of further work which needs to be carried out in the near future. The cost of this might amount to over £1 billion, spread over four to five years. Investment to increase the capacity of the network would be in addition to these figures. Taking into account expenditure by the rolling stock leasing companies as well as expenditure by Railtrack, the total amount involved might be £5 billion spread over a ten-year period.

12.68 We believe the need is for a major injection of new investment over a specified period in order to create a much expanded and more viable public transport system, and that this is an essential ingredient in a sustainable transport policy. **We recommend a substantially increased programme of investment in public transport over a ten-year period.** The fundamental change in attitude and commitment towards public transport which we are advocating, and the redirection of investment away from the traditional form of road programme carried out by central government, will also require organisational changes, which we now go on to consider.

Chapter 13

INSTITUTIONAL DIMENSION OF TRANSPORT

13.1 In 1990, the House of Commons Transport Committee concluded that the government's approach to transport policy was not 'an adequate basis for promoting a strategic transport policy that is consistent with the requirements of sustainable development.'[1] Much of the evidence the Royal Commission received echoed this view. A disregard of the environmental consequences of growth in motor traffic, a preoccupation with road-building at the expense of other modes, and a failure to appreciate the links between land use planning and transport were felt by many to characterise recent policies.

13.2 This chapter discusses the role of some of the key public bodies which determine those policies and provide the formal institutional framework within which they are developed. It also considers the organisation of the main transport services, insofar as this may have implications for the quality of the environment.

13.3 The functions and role of institutions are important in that they may influence the content of policy and bias the outcome of policy making. Furthermore, it may prove difficult to deliver a policy, however well formulated, in the absence of an effectively aligned institutional framework. Nevertheless, we start with the general assumption that institutional restructuring by itself will not guarantee an environmentally sound transport policy. In the absence of a clearly articulated and consistent policy, which is the responsibility of government rather than the machinery of government, little will be achieved.

13.4 This chapter first outlines how we consider the existing institutional framework for transport policy should be judged and sets out the objectives which we think should guide it. It then highlights features which we regard as unsatisfactory and recommends specific measures designed to help achieve a transport policy which takes better account of environmental imperatives.

Essential criteria

13.5 We have attempted to assess the present institutional framework for UK transport policy against three main characteristics. These are, first, the ability to take account of wider and long term issues; second, consistency of policy among all the players (local authorities, national institutions, supranational and international bodies — this we term vertical consistency) and across modes and sectors at any institutional level (horizontal consistency), and third, the ability to generate and implement policies which are consistent with the principles of sustainable development.

Strategic view

13.6 It has become increasingly obvious in recent years that transport affects and is affected by many, disparate issues which need to be taken into account by the institutions responsible for developing policy. Among these issues, environmental priorities are of major importance and transport objectives should be pursued in ways which are consistent with them. To do this means having a vision of the kind of society which we want to see in 20 or 30 years time. As we have argued earlier in this report, this might be best expressed in terms of objectives and achieved by working towards specific environmental targets. Attempting to fulfil these objectives would help to stimulate creative solutions to transport problems. And progress towards meeting the targets would be a measure of the success of policies.

13.7 A strategic approach to transport policy would include systematic assessment of the implications of policy and investment decisions for the global, regional and local environment. The framework for policy development and decision-making processes should encourage local and national government to take proper account of the relative importance of individual choice and society's present and future needs. There should be no methodological bias which favours one transport mode at the expense of another and neither should there be an automatic assumption that the prime mode is motor travel by road. Modes such as walking and cycling should not be neglected, for example in transport models. At a personal level, policies should combine to lead individuals to make their transport choices in ways which are also in society's wider interests. The institutional framework should enable the Best Practicable Environmental Option to be pursued when policies are decided.

13.8 Although competition may help to stimulate innovation and efficiency, the choices to be made cannot be left entirely to market forces to determine because of the external costs unavoidably generated by transport, and because of the benefits lost if networks fragment under unrestrained competition. The choices are also too complex to be decided by central government acting alone. A more fruitful model would be for decisions to be taken, openly and following full consultation, as close to the point of use as possible but within a framework, set by central government, which enables proper account to be taken of the full complexity of transport issues. In this model, local government would have an important function in enabling effective solutions to be found to transport problems.

Consistency

13.9 This approach will not be effective unless policies are consistent, both vertically and horizontally. Unless the vertical relationships between European Community, national and local policies are complementary, contradictions will develop, and waste and inefficiency result. For example, EC policies to promote combined transport are unlikely to succeed if national policies continue to increase the economic and practical advantages of road transport.

13.10 Horizontal consistency is also necessary. For example, if motor traffic is to be less dominant in towns, the necessary policies (such as parking restraint, pedestrianisation) must not lead to more dispersed development, generating greater reliance on private transport and increasing overall pollution levels. Consistency of purpose between transport and land use planners and between neighbouring areas is necessary to avert this outcome.

Sustainable development

13.11 A crucial test of institutional arrangements is whether they facilitate action which is consistent with the principles of sustainable development or hinder it. We have already outlined ways in which present trends conflict with the concept of a sustainable transport policy and much of this report builds on that analysis to recommend new ways of resolving the transport dilemma.

13.12 Taking these factors into account, we believe that it is an important objective:

> **TO ENSURE THAT AN EFFECTIVE TRANSPORT POLICY AT ALL LEVELS OF GOVERNMENT IS INTEGRATED WITH LAND USE POLICY AND GIVES PRIORITY TO MINIMISING THE NEED FOR TRANSPORT AND INCREASING THE PROPORTIONS OF TRIPS MADE BY ENVIRONMENTALLY LESS DAMAGING MODES.**

We now consider how well the major public bodies involved in developing transport policies are attuned to deliver this objective.

The policy framework

Local government

13.13 The planning and maintenance of most of the country's road network, subsidies for uncommercial local passenger services, parking policies and traffic management schemes are the responsibility of local authorities. The present arrangements are in some ways confusing: responsibility for traffic regulation and new infrastructure provision is muddled in the public mind because, even in large cities, it is a county (or, in Scotland, regional) rather than city function. Local government boundaries rarely coincide with predominant traffic patterns. There may be a tension between local sentiment and the most efficient planning and delivery of services (which may demand quite sizeable authorities). If transport planning is to be effective, it generally needs to cover reasonably wide areas. Districts will not usually be big enough and even counties may not be optimal. To take a single example, whilst the overall traffic and environmental effects of park and ride schemes may be beneficial, the balance of advantage may differ between the city centre (which benefits from a reduction in traffic) and the surrounding districts (which may derive little advantage from car parks required in their areas). A body is needed to fill the strategic gap and take an overview of transport problems and opportunities which transcend local authority boundaries.

13.14 Seen in this light, the structural changes to local authorities which the government envisages have significant implications for the development and implementation of a sustainable transport policy. The government's intentions for the future shape of Scottish and Welsh local authorities are already clear. In Scotland, 28 single tier councils are planned to replace the existing nine regional councils and 53 district councils, while the three island councils will continue with boundaries unaffected by the reform. In rural Scotland the unitary authorities will be regional in scale but elsewhere strategic planning will depend on voluntary arrangements. In Wales, 22 unitary authorities are to be created and here, too, strategic planning will depend on voluntary arrangements. In both cases the government will have reserve powers for use should the need arise.

13.15 The likely outcome for England is less clear. The government's guidance to the Local Government Commission which is reviewing English local authority structure acknowledged some of the issues outlined above, pointing to the need of local authorities to take into account and balance the interests of the wider travelling public with those of the communities they serve directly. It concluded: 'An authority which is large enough to deliver most local services effectively may not be large enough to take a sufficiently broad view of its transport responsibilities. The option of joint arrangements for some or all transport functions may need to be carefully considered.'[2]

13.16 It seems unlikely, however, that the pattern emerging from the English review, of unitary authorities interspersed in some areas with two tier authorities, will remove confusion or, more important, achieve a balanced approach to transport provision or ensure sufficient integration of strategic land use and transport planning.

13.17 One likely result of these structural changes is for central government's Highways Agency to take on some of the road planning, construction and maintenance responsibilities now exercised by local government. Later in this chapter we criticise procedures for deciding central government road schemes, and recommend substantial changes in order to strengthen the links between transport and land use planning. Given this, we would find any increased transfer of highway responsibility from local government to the Highways Agency unsatisfactory, unless the types of modifications we propose for trunk road schemes are implemented.

13.18 In any event, a body will still be needed to take a strategic overview of transport problems and solutions and to have them implemented by the providers of transport and transport infrastructure, whether public or private. It will need to work closely with land use planners at the strategic and regional levels and with the government's integrated regional offices but would not necessarily have to be responsible for land use planning matters. Various models suggest themselves.

13.19 A radical approach, consistent with the establishment of the Committee of the Regions under the Maastricht Treaty, would be an elected regional tier of government with transportation, strategic planning and other functions which need to be exercised over a wide geographical area, together with single tier authorities at grass roots level. Whatever its merits, the implications of this course go too far beyond the boundaries of our study for us to be able to comment on them.

13.20 The government's solution is to rely on voluntary co-operation between local authorities. We received considerable evidence, however, from those working in various statutory and non-statutory bodies that they can operate much more effectively when they have a statutory basis, even where their functions are purely advisory. The South East Regional Planning Conference (SERPLAN) told us that it suffered in comparison with, for example, the London Planning Advisory Committee (LPAC) because unlike LPAC it is not a creature of statute. Nor is the evidence provided by existing voluntary arrangements encouraging. The unitary metropolitan districts in England have found difficulty in making joint arrangements work since the abolition of their counties in 1986.[3] In voluntary joint committees, it is precisely the strategic overview which risks being lost when the economic self-interest of an individual council is under threat. In London, the government has had to involve itself directly in traffic problems, with the designation of a minister in the Department of Transport and the establishment of a Traffic Director for London. There is also a Cabinet sub-committee which deals directly with more general London questions.

13.21 It is obviously feasible for central government to rely on reserve powers to cope with the possibility of a breakdown of working relationships between authorities. But we consider that an administrative structure which depends on such arrangements is misconceived. The Secretary of State is unlikely to use reserve powers allowing him to establish cross-boundary strategies except in very extreme circumstances. For him to act otherwise would be an undesirable centralisation of decision-making, inappropriate for resolving difficulties at a local or regional level. The alternative, of powers which allow the Secretary of State to require local authorities to discharge jointly certain functions, is hardly more attractive. The administrative arrangements for developing local transport policies should be those which encourage vigorous and purposive searches for sustainable solutions. What seems likely to emerge from the present local government reviews will be arrangements which rely on fudging and compromise.

13.22 We consider that statutory joint boards, operating within clearly defined powers, might overcome these difficulties. The authorities could be required by law to devise a framework within which a co-ordinated network of public transport services would be provided; to develop ways of increasing patronage; to set quality targets for services; to take fully into account the actual and planned land use policies of the constituent councils; and, where appropriate, to enter into agreement with the rail franchising director to secure rail services.

13.23 It is unnecessary to create wholly new statutory authorities since a model already exists in the Passenger Transport Authorities (PTAs) and their Executives (PTEs). These bodies were originally established under the Transport Act 1968 'to secure or promote the provision of a properly integrated and efficient system of public passenger transport to meet the needs of [the] area with due regard to the town planning and traffic and parking policies of the councils of constituent areas and to economy and safety of operation'.[4] Their original composition and powers have been changed. The members of the PTAs are now drawn from the district councils for the area they cover (from the Regional Council in Strathclyde). PTAs have lost their formal powers over bus services which can be provided commercially but they can subsidise non-commercial services (after competitive tender) to secure specific fare and service levels. PTAs and PTEs have been established in seven conurbations.

13.24 **We recommend that additional PTAs and PTEs should be established, including one for London, to ensure the provision of a co-ordinated and efficient system of public transport throughout their areas.** Some would be responsible for predominantly rural areas and would have different preoccupations in some respects from those covering major conurbations. There should be a PTA for London which should cover broadly the area within the M25. All would need powers such as those outlined in the two previous paragraphs together with powers to regulate on-street and other public parking. These authorities would need to work closely with district councils, which would remain responsible for the important non-strategic elements of transport planning and provision in their areas. The composition of the PTAs, largely drawn from local authorities, would strengthen the links with district councils, although membership might be drawn wider, possibly including, say, the franchising director for railways. To mark their new role, it may be advisable to rename these bodies — perhaps as Local Transport Authorities — but for convenience, we continue to refer to them as PTAs and PTEs.

The planning of transport infrastructure

13.25 Strategic planning authorities (the counties and metropolitan districts) are increasingly removing the barriers between their highways and land use planning departments. About a quarter of English county councils (but a smaller proportion of London boroughs (1/8) and metropolitan districts (1/7)) appear to have such combined departments. The absence of organisational integration does not, of itself, mean that policies cannot be co-ordinated. Surrey County Council, with separate highways and planning departments, produced a transport plan designed to move forward the strategy set out in the structure plan and to feed into structure plan reviews and into local plans. Whether or not highways and planning departments are combined, effective mechanisms are needed within local authorities to ensure that the reciprocal influences of transport infrastructure and other forms of development are fully evaluated as plans are drawn up. The public should also have an opportunity to probe these issues, as is the case when the development plans are open to objections.

13.26 Institutional change could strengthen the land use planning process at regional level and better integrate it with the planning of transport infrastructure and services. The regional planning conferences established for the English regions have an important function in developing policies and providing advice for the Secretary of State to consider in preparing regional planning guidance. SERPLAN, as the longest established of these bodies, has played a valuable role in bringing together the disparate interests of local authorities in the South-East and in advising the Secretary of State. Planning guidance is now being prepared or has already been produced for all the English regions. We consider that the informal basis on which SERPLAN and the other planning conferences are established could be improved. The aim would be to ensure reasonable consistency of scope and approach, and give the conferences a certain measure of independence from their constituent local authorities.

13.27 **We recommend that the existing regional planning conferences should be strengthened and put on a statutory basis for each economic planning region.** With a remit covering both land use and transport planning, these conferences should be the place in which the need for major road schemes affecting more than one authority is decided. They should be given a duty to consider and advise the local authorities in their area on matters of common interest, including land use, and, in conjunction with PTAs, transportation, parking and highways; to inform the Secretary of State of those authorities' views; and to liaise with neighbouring authorities. In Scotland the Regional Councils are at present the strategic planning authorities. We believe that planning conferences will be needed there, too. Given the significance of the policy issues with which they will be concerned, it is important that the regional conferences operate as openly as possible.

Air pollution control

13.28 The contribution of transport to the episodes of poor air quality experienced in some areas was discussed in chapter 3. Better integration of transport and land use planning would be one part of a broader framework designed to achieve long-term improvements in local air quality. In this context we welcome the setting up of an Air Quality Network and Unit by London local authorities to co-ordinate monitoring and provide prompt information to doctors and environmental health officers. Local authorities in other areas have also formed pollution groupings to share information and expertise. **We recommend that the government explore with the Institution of Environmental Health Officers, other professional bodies and local authorities how such forms of collaboration can be built on and encouraged in other conurbations.**

13.29 Measures recommended earlier in this report will help in the attainment of long-term targets for air quality but procedures for dealing with pollution alerts will still be needed. Such procedures have existed in some countries for several years and are being introduced elsewhere. The EC Ozone Directive[5] requires the public to be informed and guidance issued if a threshold is exceeded. The Department of the Environment issues daily Air Quality Bulletins giving UK concentrations of nitrogen dioxide, sulphur dioxide and ozone over the previous 24 hours and air quality forecasts for the next 24 hours, using the bands shown in table 13.1.[6] Some local authorities publish information based on their own monitoring data; and sometimes insert a further band between 'good' and 'poor'. DOE intends to publish a consultation paper on standardisation of air quality bands. The DOE bulletins give advice to members of the public with respiratory problems on action they should take if air pollution levels enter the 'poor' or 'very poor' bands. Only in the unlikely event of nitrogen dioxide levels exceeding 1,128 $\mu g/m^3$ does the Advisory Group on the Medical Aspects of Air Pollution Episodes consider a more general health warning would be necessary.[7]

13.30 As the government acknowledged in its recent discussion paper[8], a joint approach between central and local government is needed in order to achieve the necessary improvements in air quality. **We recommend that local authorities be given new duties to assess ambient air quality, the sources of air pollution in their area and the risk of pollutant concentrations exceeding threshold levels to be set by the government. They should also have the duty of drawing up and implementing an air quality management plan if they consider this necessary in order to prevent any of the thresholds from being breached.** Emissions inventories and air quality models are likely to be essential elements

Table 13.1
Air quality: categories used in DOE bulletins

in ppb (μg/m³)

	very good	good	poor	very poor
nitrogen dioxide	<50 (<96)	50-99 (96-190)	100-299 (192-575)	≥300 (≥577)
sulphur dioxide	<60 (<158)	60-124 (158-326)	125-399 (329-1050)	≥400 (≥1053)
ozone	<50 (<100)	50-89 (100-178)	90-179 (180-358)	≥180 (≥360)

Conversion factors used as in table 3.4.

in local authorities' assessments. The plan should go on to specify how the long-term thresholds are to be achieved. This will be done partly by means of emission limits applied under existing legislation but local authorities should also be required to take air quality explicitly into account in preparing and reviewing their traffic management plans and their structure, local and unitary development plans. The government should consider whether local authorities require further powers to ensure both that air quality does not deteriorate to dangerous levels and that the long-term thresholds can be met. The plan should also specify what action should be taken if acute episodes of poor air quality are forecast. Regular reassessment will be necessary to monitor the effectiveness of the management plan (or the need to draw one up in areas not originally considered to suffer from poor quality air). Because of air movements, the areas worst affected may not be those which contribute most to the pollution. For this reason, the plan will need to be drawn up, and action undertaken, jointly with neighbouring authorities. The government should establish effective means of ensuring the requisite co-operation, perhaps including an arbitration procedure.

13.31 For operational purposes DOE has suggested that three thresholds (alarm, long-term acceptability and background) might be appropriate. The Department defines the 'alarm' threshold as that above which there is a serious risk of widespread health effects in a significant proportion of the population (not necessarily particularly sensitive groups). We consider that action should be put in hand to reduce the worst effects of acute episodes before this level is reached and a greater degree of disaggregation would therefore be preferable. **We recommend that, for the major pollutants including carbon monoxide, nitrogen dioxide and PM10, the government should set thresholds at levels which enable action to be taken to prevent air quality from deteriorating to a dangerous level.** In each case, the threshold would come between the 'alarm' and 'long term acceptability' thresholds and would serve as the trigger for local authority action. This action will depend on the nature of the threat but might include warnings to susceptible people to stay indoors and advice to motorists not to use their cars. In some cases ways of reducing car use might be agreed in advance with major employers. We expressed in chapter 3 our views on the long-term targets which should be pursued. The strategy required to achieve the targets will include the setting of thresholds for each pollutant based on human health criteria or critical levels for sensitive ecosystems. The government should set health-based targets which would apply to the whole country. It should also set critical levels for different types of local environment. Targets (which may differ from place to place and should be subject to government approval) could then be proposed by local authorities to take account of special local circumstances, or to enable different areas to reach long-term targets over different timescales in the most cost-effective way.

The financing of local transport

13.32 The main vehicle for strategic planning at a local scale is the structure plan (or part I of the unitary development plan (UDP)) and mechanisms exist to ensure consistency with regional and national policies. In principle, the transport elements of the strategy should feed through into authorities' Transport Policies and Programme submissions (TPPs). In the past, however, TPPs have been little more than the mechanism by which local authorities have bid for resources for their planned road schemes. There is a gulf between the long lead time (and the expertise) needed to draft and update structure plans and the annual cycle of TPPs; and it has been suggested that a lack of resources or direction has reinforced a tendency in some authorities to resubmit a minimally updated TPP each year instead of adopting a more strategic, better integrated approach.[9] Far from being a means of supporting innovative policies to control the adverse effects of traffic and promote less harmful means of transport, TPPs have been criticised for diverting investment into road building (not least because it has appeared easier to comply with the investment criteria for roads than for public transport). Many local authorities have been content to concentrate resources on road improvements but it has also been claimed that an increasing trend towards ring-fencing of resources has limited authorities' freedom to determine their own expenditure priorities.[10] The Department of Transport's explanation is that this results largely from a transfer of resources from Rate Support Grant (which authorities could spend as they chose) to Transport Supplementary Grant (TSG), brought about to ensure that the resources were spent on the maintenance of principal roads and the strengthening of bridges.

13.33 Much of the evidence received by the Royal Commission heavily criticised the financing of local transport infrastructure for its pro-road bias. When the present system of transport grants was introduced twenty years ago, the government recognised that: 'There will be common ground between structure plans, local plans and TPPs, which should be compatible with one another. The preparation of TPPs and of the transport proposals of development plans should therefore as far as possible be integrated.'[11] Accordingly, TSG was originally paid towards capital and current spending on transport. From April 1985, however, partly in order to cut subsidies to bus services, it was restricted to capital expenditure on roads and traffic regulation, with major schemes subject to standard Department of Transport cost/benefit procedures (COBA). This placed a disproportionate emphasis on road-based solutions to transport problems, distorting the original rationale for TPPs.

13.34 During the course of our inquiry the government introduced significant changes to TPP procedures. The package approach, introduced for the 1994/95 financial settlement and extended for the 1995/96 settlement, offers greater flexibility, within the constraints of existing legislation, for local authorities to switch resources between different forms of transport. Bids for 1994/95 could cover both road and public transport investment proposals and had to be supported by 'a comprehensive transport strategy'.[12]

13.35 According to a survey of highway authorities carried out after the 1994/95 settlement, the great majority supported the package approach.[13] A quarter of the authorities which submitted package bids said that a more comprehensive approach in planning had been promoted. The measure most frequently included in the bids was bus priority, but traffic calming, cycle schemes and park and ride were also popular. The three most popular underlying objectives were improving the urban environment, reducing congestion and reducing traffic accidents. Better co-ordination of policy measures and the ability to shift expenditure away from roads, appear to have tilted the balance towards environmentally friendly modes together with increased emphasis on traffic management.

13.36 Some difficulties emerged during the course of the last settlement. The most common problem was in developing assessment methods by which to evaluate package bids. Many of the techniques which the package approach is designed to stimulate (such as traffic calming, bus priority, pedestrianisation) would score badly in the customary COBA assessment (which attaches a high value to measures to speed up traffic flows). In addition, some rural authorities feared a loss of funding to urban areas but some London boroughs also doubted whether the approach could be applied successfully to London because of their lack of responsibility for London Regional Transport. There appear to have been some instances in which authorities failed for political or other reasons to co-operate with each other in developing a strategic approach. This is inevitable when authorities see themselves as to some extent in competition with each other. We consider that the extended PTAs we have recommended will lead to closer co-operation between the parties in submitting bids. (It follows that PTAs should continue to be able to make bids jointly with local authorities.)

13.37 The main elements of the package approach have been continued for the 1995/96 settlement. The Secretary of State's objectives for local authority expenditure on transport now include more emphasis on demand strategies in urban areas, encouragement of a shift from private to public transport, and an emphasis on better provision for environmentally friendly forms of transport, as well as the more traditional emphasis on road construction.[14] There have been other changes designed to increase authorities' flexibility in using the resources on transport infrastructure of their choice. The circular explaining them also stated: 'In view of the increasing emphasis the government places on the need for alignment of transport plans with land use strategies all authorities will be expected to explain in their TPPs how their proposals fit into, and flow from, the local structure and development plans.' The government has produced much fuller guidance on how authorities should submit package bids and has sought to answer some of the fears which had been expressed. Whilst accepting that authorities will wish to look at the immediate problems of urban congestion, it also encourages them to draw up a 'positive, even a visionary, statement of objectives' and to consider how the area in question should function in 5, 10 or 20 years time.[15] There are no restrictions on the types of area for which package bids would be considered. These might include an urban area and its hinterland, or several related urban areas as well as the more obvious single town or conurbation. Rural areas are not excluded.

13.38 Three points are of particular importance. First, the government emphasises the need for transport objectives to be consistent with up to date development plans. Submissions should explain how bids fit in with both existing and proposed development plans and the guidance contained in Planning Policy Guidance Note 13. Second, authorities are advised not to consider schemes in isolation but to present different kinds of measure as part of one strategy. Third, an attempt has been made to develop a new form of appraisal with which to evaluate packages as a whole (though traditional economic and financial appraisal is still required for schemes costing £2 million or more).

13.39 We welcome the major step forward that has been taken. **We recommend that the government monitor the effectiveness of the new package approach and publish the results.** If the approach is embraced enthusiastically and allowed sufficient continuity, it may even encourage new thinking about the more intractable problem of traffic growth outside urban areas. This is a problem on which guidance is lacking and to which the concept of the package bid could usefully be extended.

13.40 **We recommend that existing legal restrictions on the use of various types of grant be modified to improve the flexibility and effectiveness of the package approach.** In particular, the continuing restriction of Transport Supplementary Grant to capital expenditure on highways and the regulation of traffic should be removed so as to increase local freedom to use resources to achieve environmental objectives. As highway schemes were completed this would free resources to be used in financing all parts of a package and should allow a higher proportion of bids to be met. Ministers are of course able to adjust the balance of resources between different grant schemes but perhaps the simplest way of ensuring flexibility would be to distribute the resources by means of credit approvals. A fresh approach of this kind should also reduce the amount of detailed central government scrutiny of local plans. As we have argued earlier, some funding should be switched from the national trunk road programme so that far greater resources can be made available for local transport schemes aimed at reducing dependence on cars and extending transport options. The effectiveness of the new method of appraising package bids remains to be seen. We consider that the Department of Transport should publish a commentary at the end of each TPP round explaining how the criteria it has adopted for judging bids have been used, so as to give the lie to any suggestion that schemes have been judged against the COBA criteria which have been traditional. When the methodologies recommended earlier (9.53) for appraising the reciprocal influences of transportation and land use policies have been developed, they too will have a role in assessing how well these factors have been co-ordinated in bids.

Bus deregulation

13.41 The absence of local authority control over London Regional Transport, referred to above, illustrates a more general weakness in authorities' ability to implement strategic transport packages. The

Transport Act 1985 removed the powers of the PTEs and local authorities outside PTA areas to provide bus services and integrate fares and services. It is not therefore possible for them to ensure the most cost-effective implementation of the measures which they adopt in pursuit of land use and transportation strategies.

13.42 Bus deregulation is based on the theory that competition is normally the best way of ensuring that services meet the needs of customers at an economically justifiable price and that deregulated companies will improve efficiency and seek to innovate in order to meet the needs of their customers. The dominant competing mode, car travel, is not, however, required to meet the social and environmental costs which it imposes. Nor is there a market mechanism for taking account of any additional adverse environmental effects of deregulated bus services as compared with regulated services.

13.43 The effects of deregulation have been well documented.[16] Fares increased sharply (by 39% in real terms between 1985/86 and 1991/92 in metropolitan areas where they had previously been cut, and 13% on average in Great Britain), many services were reduced, networks were restructured and became less stable. The general impression was of declining services. The immediate sharp drop in bus patronage is not therefore surprising. There was also a deterioration in the extent of bus-rail integration (markedly in Tyne and Wear) but there has subsequently been something of a move back towards integration in PTE areas. Competition has not generally been widespread or sustained, although there are exceptions. Where it has taken place, it has usually been on popular routes where more vehicles have not necessarily led to a better service for passengers but may have increased urban congestion and pollution. In some areas operators have moved out of bus stations, using residential roads as substitutes, with detrimental effects on amenity. In historic centres, tourist buses can create considerable nuisance. In addition, a lack of investment has kept highly polluting old buses in use for longer than might otherwise have been the case. On the other hand, deregulation has encouraged the more widespread use of midi-buses which can be deployed more flexibly than the traditional larger buses (eg in suburban areas) and may be less energy intensive when passenger numbers are low.

13.44 Rural areas have also shown mixed results. One study of four counties shows a trend towards more services in major rural settlements but a decline in more sparsely populated areas. The study suggests that the changes result from local authorities' ability to target subsidies to areas with greatest needs because individual services are now subject to contracts which provide better estimates of their costs than the previous block grant system. Subsidies for rural bus networks fell in all four counties (by between 11.8% and 32% in real terms).[17]

13.45 The House of Commons Select Committee on Transport summed up the effects of deregulation by saying: 'while bus mileage has increased, so have fares in real terms, and while operating costs are down, so is the level of patronage. Similarly, whereas in places like Oxford and Inverness, passengers have experienced genuine improvements in the quality of services, in larger cities such as Sheffield and Manchester, deregulation has led to considerable problems of congestion and unreliability.'[18] The Committee was sceptical of the government's assertion that the decline in ridership 'after allowing for the critical steep fall, is in line with long term trends relating to expanding car ownership' noting that it 'sits uneasily with experience in London and other European cities where demand for bus services has remained constant or even increased in the face of a broadly similar growth in car ownership'.[19] Finally, it was concerned about the consequences for bus patronage of the element of instability which is an inevitable consequence of deregulation. It found very little evidence to support the government's proposition that this was simply the corollary of allowing the market to detect and respond rapidly to changing patterns of demand, saying 'stability matters, since people's decisions about where to live, work and shop, and how to travel, depend on confidence that the bus service they enjoy now will still be there in six months' time.'[20]

13.46 We share the concerns of the Select Committee, especially about the polluting effects of large numbers of buses in congested streets, the longer vehicle life which tends to defer the purchase of less polluting new vehicles, and the increased incentive for potential passengers to use cars instead of buses because of the instability of services. The reduction in operating costs may have been bought at too high an environmental price. In London, private sector involvement and reduced operating costs have been achieved by route tendering and this has not produced the environmental harm of deregulation.

13.47 We make recommendations later about operating practices which will go some way to reverse the decline in bus usage. The institutional arrangements are also crucial. On the evidence we saw, **we recommend that the government reform the legal framework for bus services and base it on the franchising of routes by PTAs/PTEs to operators who meet high standards.** This change would have much in common with the approach being used for rail privatisation. It could be reinforced by a form of 'quality contract' between local authorities and operators so that priority in road use (for example through bus lanes and other traffic engineering measures) is provided in exchange for stated standards of operation and vehicles. Franchises should last long enough to give operators a sufficient financial return to be able to invest in modern buses on a continuing basis and adopt technical improvements to meet passengers' needs (such as low-floored buses). Passengers also need services which are stable for reasonably long periods and on which they can plan their daily business. Franchising would provide this stability. It could also ensure an end to congestion on popular routes and to the excessive concentrations of bus traffic, especially on residential streets, which were singled out as an unwelcome side effect of deregulation and which the Traffic Commissioner's present powers appear less than adequate to prevent. In all, this arrangement would preserve the advantages of private sector operation whilst operating within a regime designed to answer social and environmental concerns.

Rail privatisation

13.48 A key element of rail privatisation has been the separation of responsibility for land, track, stations, signalling and other rail infrastructure from the operation of trains. Railtrack is responsible for infrastructure, including all rail stations and operational land. Access to these facilities by train operators is governed by a commercial contractual regime.

13.49 The remainder of British Rail (BR) has been split into operating units which will become subsidiary companies and then be sold to the private sector, in the case of freight and support services or, in the case of passenger services, operated by BR until they can be franchised to private operators. BR's passenger rolling stock has been vested in three leasing companies which will lease it back to train operators. The railway network will be open to other operators for the provision of both freight and passenger services in competition with the existing operators.

13.50 The Railways Act 1993 also established two government appointments to oversee the restructured railway. The first, the regulator, will exercise control over all rail bodies including Railtrack and will licence all railway operators and arbitrate in disputes between Railtrack and service providers over access to the network. The Regulator will be jointly responsible (with the office of Fair Trading and the Monopolies and Mergers Commission) for the application of competition legislation to the rail industry.

13.51 The second appointment is the franchising director, who will control public subsidy for the operation of passenger services, securing services from BR and subsequently the franchisees. He will administer the process of tendering for rail franchises, specify minimum service levels and control fares where the railway 'exercises significant market power.'[21] In the metropolitan areas designated as Passenger Transport Areas, the PTAs and PTEs will act jointly with the franchising director.

13.52 It is too early to say with certainty what effects rail privatisation will have on services and on the environment. Our main concern is that railways should become more attractive for the carriage of both passengers and freight so as to exploit to the full their environmental advantages. These advantages must be given due weight in the legal and administrative arrangements under which the railways will operate after privatisation.

13.53 The benefits which are to be expected from privatisation include a greater sensitivity to the wishes of travellers and consignors of freight, more imaginative and innovative ways of increasing market share (including better collaboration with other modes) and, freed from public sector constraints, greater freedom to invest in order to realise the potential of the railways. These benefits are unlikely to materialise if private sector operators concentrate on lucrative routes to the neglect of the rest, if services operated for social reasons (eg some commuter services and the rural railways) are inadequately financed, if would-be travellers are discouraged by high fares, and if competition between operators disadvantages passengers and freight (eg by limiting the scope of tickets).

13.54 The changes in the rail charging regime have led to increased costs for PTEs: previously, BR had charged PTEs on a marginal basis for the services they specified. Restructuring has meant that the charges for infrastructure and rolling stock reflect the revaluation of the relevant assets and a commercial rate of return on those assets. The government will cover the increase in the financial years 1994/95 and 1995/96 by means of a direct payment — Metropolitan Rail Grant — and the PTAs have in turn agreed to revised agreements with BR.[22] We regard it as imperative that the future of these suburban services be secured. As we have argued earlier, public contributions to public transport are an acceptable means of obtaining environmental benefits and are an appropriate way of securing good quality suburban rail services.

13.55 As in the case of buses frequency, regularity and reliability of rail services are important factors in retaining existing users and attracting new ones. The arrangements for franchising rail services should therefore be operated in such a way as to maintain and improve these qualities. We believe that there is considerable potential to increase rail business, both passenger and freight, by providing more cross-country services, better connections between services and especially by improving the links between rail services and the hinterland they serve. In some case that will involve new or improved rail services on feeder routes. Mostly, however, other types of transport will be used to make the journey stage to or from the rail station and rail services need to be integrated with them. Some possibilities have been outlined earlier in this report. The number of bodies potentially involved in meeting these needs could be large (several rail operators, Railtrack or a private company as station operator, bus companies, taxi firms, cycle and car hire companies). Some means of co-ordination will be required if facilities are to be comprehensive and easy to use and we consider that this is a task which should be carried out by the PTEs (which will also continue to provide financial support for non-commercial rail and bus services).

13.56 The government should ensure that the fears many people have voiced about the outcome of privatisation do not materialise, so that rail can play a greater role in meeting the travel needs of the nation in a way which is compatible with environmental concerns. The crucial test is that the total public transport system should be as nearly seamless as possible for travellers. It should aim to be as attractive to use as the car, and easy to use in combination with a car. That does not demand centralised ownership or direction. But it does demand more willingness to provide the services which travellers want and in some cases a willingness for operators to co-operate, especially in the provision of complementary services. **We recommend the government ensure that the structure and operation of the privatised rail industry lead to a much greater use of rail.**

Central government: organisational issues

13.57 Our recommendations put a heavy responsibility on local authorities to reconcile conflicting pressures (environmental, social, economic, commercial) and enable service providers to create a transport network which serves society's interests. This is a reconciliation which only a government body can carry out and it is appropriate for local government because it is closer to people's needs than is central government. Local authorities will be able to respond only if they operate within the right framework. It is the responsibility of central government to create that framework. Although there have been encouraging developments recently, including the issue of PPG 13 jointly by DOE and DOT and the introduction of the package approach to local transport financing, the picture which confronts us after this part of our review is of an often inadequate framework for linking land use planning and transport planning; a weakening of the strategic level of planning; scope for inconsistent decisions by neighbouring authorities; and changes of policy by government which, until recently, have stimulated local authority road building and made an integrated public transport system less likely. Our recommendations are designed to remedy the weaknesses which lie within local authorities but changes are also required in the way central government operates.

13.58 The doctrine of collective responsibility means that once policies are agreed, they are those of government as a whole. Individual departments have their own priorities and the process of policy development normally requires a certain amount of give and take if consensus is to develop. In the Environment White Paper published in 1990, the government acknowledged that the policies of all government departments affect the environment and proposed to strengthen arrangements for interdepartmental co-ordination in order to integrate environmental concerns more effectively into all policy areas.

13.59 The proposals were, first to retain the standing committee of Cabinet Ministers which had drawn up the White Paper to ensure that environmental issues were tackled in a co-ordinated and effective way. Second, a nominated Minister in each department would be responsible for considering the environmental implications of all his department's policies and spending programmes. Third, departments were encouraged to set out in their Annual Reports how they were following up the White Paper and to describe any future environmental initiatives.

13.60 Although the Ministerial Committee had met only three times by May 1994, the procedure seems to have raised the consciousness of departments in respect of some issues (such as energy efficiency and the use of CFCs). The present emphasis appears narrower than originally intended, however, and seems to be largely on good housekeeping in the government estate. This is important because of the size of the government estate (over 3,000 properties, occupying some 72 million square feet) and because government departments are major employers and users of transport. Their own practices (on, for example, relocation or homeworking) should, therefore, be assessed for their transport implications so that adverse environmental effects can be minimised. **We recommend that the committee of Green Ministers publish periodical reports on the results of its scrutiny of the environmental implications of government practices.**

13.61 Many government policies affect the environment indirectly through their implications for transport. In some cases the link is fairly obvious (for example, company car allowances); in others the full effects may not be immediately apparent (for example, industrial regeneration). **We recommend a more consistent scrutiny of the transport implications of all government departments' legal and policy responsibilities so that these can be taken fully into account and their adverse effects minimised.**

13.62 Environmental consideration of the Department of Transport's policies seems to have been patchy. In its comments on the government's consultation document on motorway charging, the National Society for Clean Air (NSCA) said: 'The discussion paper offers no evidence of alternative policy evaluation, as recommended in official guidance on policy appraisal and the environment, to see whether (for instance) objectives might be better met by investing in improving public transport provision rather than building more roads. Again, this is because clear objectives have not been defined.'[23] The document did contain a brief discussion of the effects of tolling on the environment and on other modes but not the wider consideration sought by the NSCA.

13.63 This illustration appears symptomatic of a more general failure. We were surprised that, according to its supplementary evidence to the Royal Commission, the Department of Transport's method of applying the government's own guidance on the environmental appraisal of policies[24] was merely to draw it to the attention of all the relevant parts of DOT. The Department added: 'And "Questions of procedure for Ministers" requires papers for Cabinet committees to cover, where appropriate, any significant costs or benefits for the environment.' It is difficult to identify what effects this attention may have had. The Department also told us that it was not possible to make any precise estimate or forecast of the environmental impact from rail privatisation because this could be assessed only in the context of the share of the market that rail would have. It has said that it intends to examine the scope for developing an environmental appraisal of the road programme but progress is not yet evident. These responses appear to us to indicate deficiencies in the systematic evaluation of the environmental consequences of transport policy decisions and insufficient political commitment to using those evaluations to improve decision-making.

13.64 In line with White Paper commitments, the Department's 1992 Annual Report contained a description of 'Future developments: beyond 1994–95' which included a short section on the environment. The section was not repeated in 1993 but the 1994 Report contains a section describing present environmental action. Despite this, the overall impression is that success in integrating environmental concerns with transport policy has been limited and that the network of 'green' ministers could play a more pro-active part in this task.

Central government: Department of Transport

13.65 The Department of Transport was frequently criticised in evidence as being roads dominated, lacking any real appreciation of the links between land use planning and transport, and failing to provide a rounded framework of transport policy. Although there have been some signs of change during the progress of our inquiry, many of these criticisms are justified, and we have concluded that, since it again became a separate department in 1976, DOT has failed to provide this country with an effective and environmentally sound transport policy.

13.66 On 1 April 1994 the government established the Highways Agency. It is an agency of the Department, headed by a chief executive who is directly responsible to the Secretary of State within the terms of a framework document. The Agency and its present 2,400 staff are charged with carrying out the executive functions of the Department's former Highways, Safety and Traffic Command in relation to the management and maintenance of the trunk road network, and the delivery of the Secretary of State's programme of trunk road improvement schemes. The Agency's overall aim is 'to secure the delivery of an efficient, reliable, safe and environmentally acceptable trunk road network.'[25] Five objectives have been set which contribute to this aim and are backed by performance indicators. One of the objectives is 'to give full weight to both the environmental and economic costs and benefits associated with the use, construction and maintenance of trunk roads and to strike a balance accordingly.' The Agency has also been set the task of developing a system of measuring the performance of the network, including levels of congestion and journey times by 1 April 1995.

13.67 The Secretary of State remains responsible for the scale of the trunk road network; the content and priorities of the new construction programme; decisions on orders following public inquiry (jointly with the Secretary of State for the Environment); the methodology for the traffic and economic appraisal of trunk road schemes; and policy for road charging and for private finance for roads.

13.68 In brief, the Agency's task is to carry out the government's plans for new construction and maintenance as quickly and efficiently as possible and it has been set specific targets to measure its progress in so doing. It remains to be seen whether the concentration of road-building activity within the Agency will help the Department as a whole to develop a less road-dominated ethos.

Land use and transport planning

13.69 The practical implications of the interactions between land use and transport planning were discussed in chapter 9. It is important that institutional arrangements do not lead these implications to be overlooked or ignored in policy making. A major reason for creating the Department of the Environment in 1970 (out of the Ministries of Transport, Public Building and Works, and Housing and Local Government) was to bring together land use and transport planning issues.[26] In that case, the link was imperfectly established. A formal connection between the civil servants responsible for roads and land use was made only at very senior level and did not survive for long. Secondly, the Road Construction Units were allowed to remain separate from the new integrated Regional Offices set up under Regional Directors.[27] It is difficult to integrate departments with markedly different cultures, and in retrospect, it may have been unfortunate that when the DOE was set up there was little pressure to bring the highways side under wider control.[28] Moreover, the Department was simply too big to be easily integrated or to be managed effectively and a new Department of Transport was created in 1976. The rationale for this solution was that land use planning was more closely linked to local government than to transport and that in the new separation of functions it was more important to preserve the former link than the latter. Since full integration had not been secured, the particular form which this separation took was probably inevitable.

13.70 These issues are not all peculiar to the 1970s. Today, the Scottish Office, a unified department, makes no organisational link between roads and planning except at the level of the Secretary of State. Roads are dealt with by the Scottish Office Industry Department, whilst land use planning is in the Scottish Office Environment Department (alongside local government responsibilities).

13.71 One potentially beneficial result of the creation of the Department of the Environment was that, over time, some administrators experienced both highways and planning jobs. Even after the split the free movement of administrative staff between the two departments continued. In 1989, however, 'common citizenship' ended and movement between DOE and DOT became as exceptional as between other departments of state. It should be noted that professional highway engineers and planners did not generally share in this freedom of movement.

Departmental change

13.72 We considered carefully whether the Department of Transport should again be merged with the Department of the Environment. There are many arguments which favour such a change (notably DOE's responsibility for land use planning policy, relationships with local government, and environmental protection), but we were not convinced that such a restructuring would of itself bring about the sorts of changes we seek.

13.73 This should not imply that existing arrangements can continue without further significant change. We identify important changes of principle to its internal structure which we feel are essential if DOT is to be able to formulate and deliver the types of transport policy we seek. We also make recommendations which will strengthen further the links between DOE's land use interests and highway development undertaken as a result of transport policies. Without both the changes in principle and the full integration of highway development with land use procedure which we recommend, we doubt whether the Department will be capable of delivering the form of transport policy we propose.

13.74 The establishment of the Highways Agency has removed a large block of work from the central core of DOT and has led to some reorganisation of the Department's remaining activities. One eventual possibility is that the Department may in due course be wound up, rather in the manner of the Department of Energy, with some responsibilities (such as emissions control) being transferred to DOE and others (such as other aspects of regulatory responsibility and vehicle standards) going to the Department of Trade and Industry. There is a very real danger that in such a scenario the opportunity to develop a coherent framework for a longer term transport policy at central government level will be lost at precisely the time when it has become most necessary.

13.75 In the meantime, the creation of the Agency and the subsequent restructuring of Central DOT in the summer of 1994 have provided an opportunity to create a department truly concerned with transport policy. The restructuring appears mainly intended to streamline the Department's functioning with the aim of improving its efficiency by 20%.[29] Hitherto, it has been organised along modal lines, with separate Divisions responsible for highways, road and vehicle safety, aviation, railways, shipping and so forth. The new structure contains three broad groups concerned with transport infrastructure, transport operations, and other matters (mainly establishments, finance and central services) respectively. The new organisation is an improvement on what went before it, particularly in the way it deals with urban and local transport, but it still appears to contain more separation between modes at working levels than is desirable. We should like to see a much stronger commitment to developing links between interurban transport modes, as well as air and waterborne transport. There will have to be a good deal of effort to realise the full potential of the restructuring if the Department is to grapple more successfully with the difficult problems which are facing society.

13.76 **We recommend that responsibilities within the Department of Transport be restructured, from Ministerial level downwards, to reflect the fundamentally different approach which a sustainable transport policy will involve.** It will require divisions with the specific remit of bringing about the transfers of traffic to less damaging modes which are required in order to achieve environmental objectives and targets. They would fulfil this remit by, for example, the creation of appropriate regulatory frameworks; the promotion of investment in public transport; and the spreading of information. These divisions would cut across modal demarcations and, except where improvements have recently been made, supersede present Divisions. As illustrations, responsibility for urban transport might be extended to include railways and waterways; and responsibility for inter-urban transport would cover road, rail, air and coastal shipping. Consequential changes in the rules for the financing of investment would be essential. The scope, grading, career structure, staff resources, and selection of personnel should make it clear that the Divisions are of central importance to the entire mission of the Department, not 'Cinderella' divisions with marginalised authority. We welcome the changes in Ministerial responsibility which have been announced and which we hope will cement the changes in ethos and practice which we have recommended.

Road building and maintenance

13.77 The Department of Transport's executive responsibility for trunk road building and maintenance has its origins in the establishment in 1910 of the Road Board to make grants to local highway authorities, in recognition of the inability of many of them to cope unaided with the new demands imposed on them by motor vehicles.[30] The Road Board was wound up and its staff transferred to the Ministry of Transport when the latter was created in 1919.

13.78 In its early years, the Ministry had responsibility for supervising the railway companies and for administering the Road Fund. It only occasionally built roads itself.[31] This situation changed with the Trunk Roads Act 1936 under which the Ministry took over some 4,500 miles (7,200 km) of the most important national through routes, henceforth legally defined as 'trunk roads'. The main purposes of the change seem to have been to relieve local authorities of the financial burden of road improvement and maintenance and to ensure that improvements could in future be carried out nationally, 'without regard to the particular and possibly conflicting circumstances of the area through which they pass' and to consistent engineering standards.[32]

13.79 Since the 1936 Act was passed, various plans for the development of trunk roads and, later, motorways have been produced by the Ministry. The Secretary of State now has a duty under s.10(2) of the Highways Act 1980 to 'keep under review the national system of routes for through traffic in England and Wales, and if he is satisfied after taking into consideration the requirements of local and national planning, including the requirements of agriculture, that it is expedient for the purpose of extending, improving or reorganising that system' he may make an Order either to trunk or detrunk a highway. Roads may be trunked for strategic reasons (to form part of a national or international network) or in recognition of the high maintenance costs resulting from very heavy traffic flows or a high proportion of HGVs. The last major review of trunking and detrunking was in 1986.

13.80 It has been suggested that the fundamental significance of the Trunk Roads Act 1936 was that 'by giving the Ministry a direct executive responsibility for this one aspect of transport only, it inevitably resulted in an organisational commitment to inter-urban road building at the expense of an overall and balanced view of transport needs'.[33] It certainly created a body well placed to respond to the very considerable pressures for road building, both directly through its own road programme and indirectly through resource allocation and policy guidance to local authorities. The direct administration of the trunk roads programme by DOT may also have reinforced an ethos which is instinctively receptive to road-based solutions to transport problems.

13.81 The establishment of the Highways Agency prompts us to question how far a government-sponsored, national system of trunk roads is still required. Historically it may have been sensible to develop a network of national roads, financed by government, to link the country's major towns and ports with the aim of channelling long-distance traffic onto them. In the main, trunk roads end at town boundaries or form bypasses: that is consistent with the view that they are primarily for long-distance traffic. Many of the exceptions are in the former metropolitan areas (and London) and reflect a perceived inadequacy of unitary authorities to care for them.

13.82 The improvement of the motorway and trunk road network combined with the dispersal of development has, however, led to many essentially local journeys taking place on trunk roads and motorways. The M6 in the West Midlands is a particularly striking example. Yet many inter-urban roads of considerable importance and carrying heavy traffic flows are not classed as trunk roads. At the last review of trunk roads, the main criterion for trunking a road was that it would normally carry at least 300 four (or more) axled lorries a day (as a surrogate for long distance traffic). In 1993, the average length of haul of an articulated vehicle was 135 km (37 km for the heaviest rigid vehicles). The average for all lorries was 84 km.[34] At best, this test seems to be one of regional, not national importance. And only 2% of personal journeys are of 50 miles or more.[35] If a road were to be designated 'national' or 'local' in accordance with the journey distance of the majority of its existing traffic, it is likely that many stretches of trunk road would not count as national.

13.83 The government itself has acknowledged that, for the foreseeable future, its road programme will be largely restricted to widening motorways and building bypasses. There seems to be no good reason why such tasks should be carried out by the Department (which does not, after all, build airports, railways, ports or canals). We referred earlier in the report to the undesirable way in which the planning of some trunk road schemes is divorced form the other development needs of the area through which they pass. This can result in relevant factors being overlooked in both land use and road plans. It also implies that the potential of alternative modes may not be adequately considered. The problems are unlikely to be diminished by the decision to close seven of the Highways Agency's regional offices and to reduce its operating costs by 20%.

13.84 The inadequacies of the conventional justifications for the trunk road programme (6.15ff) are increasingly recognised and appear likely to be strengthened by the forthcoming report of the Standing Advisory Committee on Trunk Road Assessment (SACTRA). We have already recommended that all road schemes should be planned and decided within the land use planning system. **We further recommend a fundamental review of the definition and purpose of a separate system of trunk roads.** This review should not be about ways of delivering a road programme more efficiently, like the one which preceded the establishment of the Highways Agency. It should go to the heart of government involvement in road planning and construction. There has been much criticism that trunk road schemes are imposed from outside, without regard to the needs of an area or the development pressures they generate. If responsibility for the present trunk roads passed to local authorities, with suitable financial arrangements, this would facilitate integration of transport and land use policies and free central government from an inappropriate executive task.

Inquiry procedures

13.85 Pending the integration of national road schemes into the planning system, there is scope to improve existing practices. Procedures for public inquiries on trunk roads differ from those for other types of transport infrastructure under the Transport and Works Act 1992. The key differences concern the opportunity for Parliamentary scrutiny of infrastructure, the extent of public consultation at early stages, and the extent to which national policies may be discussed.

13.86 The lack of public discussion of the policies which are expressed in White Papers and individual road proposals, has often been criticised. The Joint Nature Conservation Committee's evidence referred to the 'lack of opportunities for debating or challenging transport policy prior to the announcement of proposals for individual schemes; and the limited input made by environmental appraisals in initial stages of consultation.' Others criticised the lack of opportunity to consider whole routes at inquiries. Instead individual schemes were promoted piecemeal, their combined effects never opened for consideration and choices effectively truncated by decisions on earlier schemes in the same route.

13.87 The government's announcement in August 1993 of measures to speed up the road programme acknowledged the force of some of these objections, stating that 'time is wasted because possible objections to a scheme are not resolved at the outset.' It is now experimenting with conferences of interested parties to discuss scheme options before public consultation takes place. This is an improvement so far as it goes and we await the results with interest. The new approach must be implemented very positively if it is to answer the criticisms cited above. We consider that the conferences must be sufficiently representative to command public confidence; must systematically explore alternatives other than road building (including the promotion of other modes of transport); work carefully through the likely interactions with land use patterns; and have available adequate environmental appraisals. At this early stage in a proposed scheme, these should be widely based (as SACTRA recommended) covering broad national and environmental issues and corridor and regional effects as appropriate. They should contain sufficient information to enable the relative merits of all options (including the use of other modes, as well as 'do nothing', and comprehensive demand management) to be compared. Even were all these conditions to be fulfilled, we doubt whether the existing institutional system would enable links to be made between road planning and wider development issues in a wholly satisfactory manner.

13.88 The public inquiry procedure for trunk road schemes has long been the subject of criticism. The Trunk Roads Act 1936 was the first time a government Minister, as distinct from a local authority, was to take powers to acquire land compulsorily and, in his autobiography, the Parliamentary draftsman reveals both the political sensitivity of the policy and how the provisions were drafted using 'modifications [to earlier Acts] of such repellent aspect that even the back-bench lobby passed them in bemused silence.'[36] He describes this as 'my worst bit of legislation by reference.' The draftsman was warmly congratulated for this work by his superiors.

13.89 Although trunk road schemes are now promoted by the Highways Agency, the Secretary of State remains responsible for the scale and content of the road programme and, following an Inspector's report, jointly with the Secretary of State for the Environment, for decisions on individual schemes. The difficulty is that, although decisions may be wholly dispassionate, such deep involvement of a single department makes it difficult to demonstrate convincingly that this is so. One of the principal objections to the procedure adopted in inquiries concerns the status of the National Road Traffic Forecasts. These are not challengeable at inquiries, as was confirmed by the House of Lords in the case of *Bushell and Brunt v Secretary of State for Transport and the Environment* (1980). The government has stated that they are not a target and that it is not desirable that they should be achieved but they are still used as evidence of the need for a road scheme. The key point is that, although road schemes can be dropped at any stage during their development, and this has happened to some high profile schemes while our report has been in progress, the impression created is that public inquiries have been largely restricted to considering the route of a proposed road and have not probed alternative strategies which would remove the need for it. The administration would defend this as a proper restriction since the role of the inquiry, in its view, is to consider public objections but not evaluate or determine policy. The proper place for the latter would be Parliament. This argument would be more convincing if there were sufficient evidence that Parliament is regularly offered the opportunity of scrutinising the road programme and if many trunk road schemes were not designed to meet regional (or even local) needs rather than national ones. In our view, there is a need both for Parliamentary scrutiny of the broad thrust of policies and for opportunities for local people to question the policies as applied in their locality. The failure to allow the latter is symptomatic of a flawed policy; it certainly creates a deep sense of frustration amongst those affected. **We therefore recommend that the rules of procedure governing inquiries into trunk road proposals and compulsory purchase orders be amended so as to permit government witnesses to answer questions about the merits of government policy and allow the inspector to take account of the interactions of the proposal with other government policies.**

The European Community

13.90 The Treaty of Rome required the adoption of a common transport policy but contained no environment chapter. That gap was filled by the Single European Act in 1987, with subsequent strengthening by the Maastricht Treaty of 1992. Nevertheless, despite the time which has passed since the passing of the Single European Act, the Community has not yet succeeded in developing a transport policy which takes adequate account of environmental imperatives and overcomes some of the defects apparent in national policies.

13.91 For the first three decades of the Community's existence, some Member States had a policy of defending their railways from competition by road (by restricting lorry movements) and therefore resisted the European Commission's attempts to establish common rules for international transport and for the conditions under which non-resident carriers may operate transport services within a Member State, both of which are called for by the Treaty. This may have had incidental environmental advantages but was seen as a failure of the common transport policy and in 1985 the European Parliament took the Council of Ministers to the European Court for failure to fulfil the requirements of the Treaty. The blockage was removed by the measures to complete the single market at the end of 1992. The European Commission itself acknowledges that 'more progress has been made in creating an integrated EC transport policy in the past five years than in the previous 30.'[37] With the notable exception of measures to reduce pollution from vehicle exhausts, the Community's legislative record of environmental achievements in the transport field is not encouraging.

13.92 The difficulty of embracing environmental concerns in the common transport policy results partly from the nature of the Community and partly from the structure of the European Commission and its method of operation. Membership of the Community constrains individual states' capacity to introduce policies unilaterally, and freedom to introduce new measures for environmental reasons is in any case limited by the requirement to observe the fundamental purposes of the Community. The Maastricht Treaty obliges Community policy on the environment to aim at a high level of protection, and provides that requirements for environmental protection must be integrated into the definition and implementation of other Community policies. The White Paper on the common transport policy published by the European Commission in 1992 seeks amongst other things to reconcile the Community's transport and environment policies under the heading of sustainable mobility but we believe that the process of reconciliation still has a long way to go.

13.93 The organisation of the European Commission in sectoral directorates-general (DG) reflects an over-compartmentalised view which lends undue weight to the aims of facilitating the movement of people and goods and takes too little account of environmental policy. Links have been created between the environment and transport DGs and at working level good relations between individuals may help to promote a less rigid adherence to sectoral thinking. The European Commission itself is a collegiate body and the draft proposals which emanate from it for discussion by the Council of Ministers have normally been agreed collectively, sometimes after vigorous debate. But despite the European Commission's growing willingness to discuss policies in open meetings, there remains an impression that some measures are not fully evaluated in the round before being put to the Council of Ministers. Further organisational improvements would be desirable, to strengthen the environmental input to transport policy developments. **We recommend that the government pursue vigorously the objective of ensuring that the consultative forum established under the EC Fifth Environment Action Programme provides a means of strengthening the environmental input to Community transport policies. We also recommend further joint Council meetings of Environment and Transport Ministers as a means of improving discussion of the environmental dimensions of transport problems.**

13.94 Techniques are needed for ensuring that a satisfactory balance is achieved in policy development. Most, if not all, the measures adopted by the Community in its first three decades were developed and agreed without any formal appraisal of their environmental effects. Since the formalising of environment policy in the Single European Act there has been a greater willingness to assess the environmental implications of new measures. The European Commission and Member States have undertaken, in proposing and implementing measures, to take full account of their environmental impact and of the principle of sustainable growth. The European Commission has for some time been considering a directive requiring the environmental assessment of policies, plans and programmes. Progress towards a formal proposal is at present blocked largely because of opposition by countries including the UK which see great practical difficulties in drafting a legislative requirement which would be equally applicable to the systems of government in all Member States. The UK has taken this view in the light of its experience in producing, in 1981, a guide for its own officials to use in assessing policies and programmes — 'Policy Appraisal and the Environment: A guide for government departments'. The government should work more positively with the European Commission and other Member States to develop a means of ensuring that policies receive thorough environmental appraisal.

13.95 Environmental assessment has an important role in the allocation of Community funds. The structural funds (which include the European Regional Development Fund) are designed to reduce the gap between the advanced and the less developed regions and promote more balanced economic development within the Community. All parts of the UK qualify for assistance under two of the Funds' objectives; and some parts qualify under additional objectives. In 1988 funding became programme-based rather than project-based with the result that adequate environmental assessment became less satisfactory. In recognition of this, the European Commission has since 1991 sought an environmental profile of Member States' policies, the priorities to be met by Community funding and their past achievements. This not only makes for better scrutiny of the environmental implications of the proposals but also in due course enables them to be audited. The European Commission is working with Member States to ensure that their programmes take account from the outset of the need for environmental protection and have been subjected to environmental assessment. We agree that Community funds must not be disbursed without adequate environmental assessment in accordance with the requirements of the Maastricht Treaty for 'sustainable and non-inflationary growth respecting the environment' and with the Fifth Environment Action Programme's aim of sustainable development.

13.96 The Maastricht Treaty established a Cohesion Fund (which is directed towards projects rather than programmes) to provide a financial contribution to certain Member States in the fields of 'environment' and 'trans-European networks'. The Community is to contribute to the establishment and development of trans-European networks for transport, telecommunications and energy infrastructures with the aim of promoting the harmonious development of the Community and the internal market.

13.97 These networks are provided for in the Maastricht Treaty, which also requires the Community to establish a series of guidelines covering the objectives, priorities and broad outlines of the measures envisaged. The principal objectives of the networks are to ensure the efficiency of the internal market, by improving the mobility of people and goods, and to reinforce economic and social cohesion.[38] Agreement has been reached on guidelines for the road, inland waterway and combined transport networks. A recent Commission proposal updates these three guidelines and includes proposals for other modes, including rail. The guidelines are indicative and do not formally commit Member States to financing them or, in the case of road, constructing them to a given standard. The Cohesion Fund may be used to finance specific transport infrastructure projects in certain Member States.

13.98 Some of the trans-European networks, for instance for combined transport, could benefit the environment. The one which arouses most concern is the Trans-European Road Network. The proposal was for a network of approximately 37,000 km (23,000 miles) of motorway or near-motorway standard roads. Of this total, about 12,000 km (7,500 miles) are motorways or high quality roads to be constructed in the next ten years, with approximately 40% being sited in the outlying countries of the Community. The European Commission described this as 'an important step forward in stressing the need to develop a traffic policy at Community level.' It considered that such a policy 'is a priority in view of the congestion on the roads, mobility trends and environmental prerogatives' and that it should 'address the challenge of the estimated 35% increase in the number of vehicles in the Community by 2010.' The only environmental measure in the original proposal for a Council Decision on the creation of the road network[39] was the requirement that the completion and functioning of the Trans-European Road Network should involve 'a common methodology for assessing the environmental impact of road projects.' This was deleted from the decision as finally agreed.[40]

13.99 The political imperatives behind the road network are perhaps understandable but the project as a whole appears to us to take far too little account of environmental considerations in themselves, of the Community's own pursuit of sustainable development and of other aspects of the common transport policy. The rapid construction of 7,500 miles of high speed road will reduce the price and increase the convenience of road transport precisely on the long journeys which might otherwise be made by rail, water or combined transport. It will encourage the growth of road traffic, not discourage it, and make it more, not less, difficult to attain environmental targets. Just as much of the traffic using the trunk road network is essentially local in character, so a similar effect applies at international level. The proportion of international traffic expected to use the Trans-European Road Network has been calculated to be of the order of 16%. Moreover the increase in the non-international traffic from 1989 to 2010 is expected to be some four times greater than the whole of present international traffic.[41] If it is true that international traffic is particularly important for trade and economic growth, then providing or reserving capacity for it would be sensible, but in fact the plan as defined at present seems to be based on the assumption that this can be done only by providing enough capacity for all the forecast local, regional and national traffic using the same roads. And there is little evidence that the construction of new roads will, in itself, promote the harmonious development of economic activities.

13.100 The European Commission is apparently aware of some of these contradictions, stating that 'the reallocation of long-distance traffic to the combined transport network and to the inland waterway network would be increased by developing systems for charging for the use of infrastructures, further integrating the external costs of transportation in terms of time lost, congestion and pollution.' These would be important and helpful measures which we would welcome but experience to date gives us no confidence that they will be agreed. Meanwhile, **in scrutinising applications for grants in support of trans-European transport networks, we recommend that the European Commission give support only to those developments which are consistent with sustainable development and respect for the environment.** It will be important to ensure that the construction of the trans-European networks, particularly the road network, does not distort the development of a more balanced common transport policy.

13.101 We expect infrastructure charging systems to play a significant role in future transport strategies within the Community but technical differences between systems in different Member States should not be allowed to impede their use by vehicles from any other part of the Community. **We therefore recommend that the government work within the European Community to develop common standards for the technology of road pricing.**

13.102 In 1992, the Community's Fifth Environment Action Programme was published.[42] It is intended as the framework for EC environment policy for the year 1993–2000 and it identifies transport as one of the target sectors for the Programme because of its suitability for Community action, its environmental effects and its role in the attempt to achieve sustainable development. The Action Programme identifies a strategy for sustainable mobility including:

improved planning of land use and economic development;

improved planning and use of transport infrastructures and the incorporation of real infrastructure and environmental costs in investment decisions and costs met by users;

the development of public transport;

technical improvement of vehicles and fuels; and

the promotion of a more environmentally rational use of the private car.

13.103 This strategy is consistent with the thinking of a discussion paper, produced in the same year, on The Impact of Transport on the Environment.[43] This was followed by a White Paper[44] on the common transport policy which set out a five pronged action programme covering: the development and integration of the Community's transport systems on the basis of the internal market; safety; environment protection; a social dimension; and external relations. The environmental measures envisaged include:

the updating and strengthening of existing legislation to control gaseous emissions, energy consumption and noise emissions from vehicles and strict application of directives on inspection and maintenance of vehicles;

the protection of areas surrounding airports from increases in noise, a ban on the establishment of new noise-sensitive activities near airports, and stricter standards for NO_x emissions from aircraft;

strategic environmental impact assessment for transport infrastructure policies, programmes and investment decisions on individual projects.

13.104 These proposals are important but are less ambitious than some of the solutions identified in the discussion paper. Whilst there are limitations in drawing up a work programme which stands a reasonable chance of success, it would not be desirable for the European Commission and Member States to neglect more difficult problems which should be addressed at the level of the Community. The European Commission attaches importance to developing a framework for charging infrastructure and other costs to users, both as a means of reducing the environmental harm caused by transport and in order to help resolve the contradictions between the common transport policy and the environmental requirements of the Single European Act and the Maastricht Treaty. Clearly such measures should be pursued at Community level so that the competitive position of Member States is not harmed.

13.105 We attach high priority to the development of an integrated Community programme to tackle vehicle pollution vigorously, partly because air pollution does not recognise national boundaries, partly because there are advantages in dealing with multinational car manufacturers, hauliers and trade associations on a Community basis. Whilst recognising that, under the principle of subsidiarity, most action to improve urban transport systems will be for local or regional authorities in conjunction with private operators within an appropriate national framework, the Community also has a role in encouraging the improvement of urban public transport systems so as to tackle the regional and global problems of air pollution by reducing the dominance of the motor vehicle.

An environmentally sustainable transport system

Chapter 14

CONCLUSIONS AND RECOMMENDATIONS

14.1 An effective transport system is vital for economic well-being and the quality of life. There is now general recognition that a continuing upward trend in road traffic would not be environmentally or socially acceptable. The need is to find transport policies for the UK and Europe which will be sustainable in the long term. Avoiding serious environmental damage, while preserving the access people want for their livelihoods and for leisure, requires a fundamentally different approach to transport policy and a radical modification of recent trends.

14.2 We endorse the general framework put forward by the government for a sustainable transport policy (1.13), which had four elements:

> To strike the right balance between the ability of transport to serve economic development and the ability to protect the environment and sustain future quality of life.
>
> To provide for the economic and social needs for access with less need for travel.
>
> To take measures which reduce the environmental impact of transport and influence the rate of traffic growth.
>
> To ensure that users pay the full social and environmental cost of their transport decisions, so improving the overall efficiency of those decisions for the economy as a whole and bringing environmental benefits.

14.3 At present pollutants from vehicles are the prime cause of poor air quality that damages human health, plants, and the fabric of buildings. Noise from vehicles and aircraft is a major source of stress and dissatisfaction, notably in towns but now intruding into many formerly tranquil areas. Construction of new roads and airports to accommodate traffic is destroying irreplaceable landscapes and features of our cultural heritage. The present generation's cavalier and constantly increasing use of non-renewable resources like oil may well foreclose the options for future generations. This is doubly irresponsible in view of the risks from global warming.

14.4 The forecasts made by DOT in 1989 showed a doubling in the overall level of road traffic by 2025. Air traffic is forecast to increase still more rapidly. Even allowing for technical improvements in vehicle design, the consequences of growth on such a scale would be unacceptable in terms of emissions, noise, resource depletion, declining physical fitness and disruption of community life. In our view, the transport system must already be regarded as unsustainable in the respects we identified at the beginning of this report (1.15), and will become progressively more so if recent trends continue. We believe this is an issue of such importance that it justifies placing significant constraints on the future evolution of the transport system.

14.5 As well as bringing many benefits, transport involves large costs, some incurred directly or indirectly by users, and some as a result of its environmental effects. Hitherto, most of the latter costs have fallen on the community rather than on the users or the builders of the transport system. Seriously misleading price signals have resulted, leading to decisions in all areas of transport which have harmed the community. Individuals and firms will continue to make such decisions until the true costs of transport become more apparent at the point of use, primarily through further increases in fuel prices but perhaps also at a later date through road pricing. To supplement and reinforce that change, the aim of future planning policies must be to reduce the need for movement (instead of stimulating ever more mobility, as has been for too long the case). This will involve a gradual shift away from lifestyles which depend on high mobility and intensive use of cars.

These changes of direction provide the essential foundation for a sustainable transport policy, and will make it possible for the economy to develop in ways which are compatible with preservation of the environment.

14.6 In our view a sustainable transport policy must have the following clear *objectives*:

A. To ensure that an effective transport policy at all levels of government is integrated with land use policy and gives priority to minimising the need for transport and increasing the proportions of trips made by environmentally less damaging modes.

B. To achieve standards of air quality that will prevent damage to human health and the environment.

C. To improve the quality of life, particularly in towns and cities, by reducing the dominance of cars and lorries and providing alternative means of access.

D. To increase the proportions of personal travel and freight transport by environmentally less damaging modes and to make the best use of existing infrastructure.

E. To halt any loss of land to transport infrastructure in areas of conservation, cultural, scenic or amenity value unless the use of the land for that purpose has been shown to be the best practicable environmental option.

F. To reduce carbon dioxide emissions from transport.

G. To reduce substantially the demands which transport infrastructure and the vehicle industry place on non-renewable materials.

H. To reduce noise nuisance from transport.

14.7 In this report we have proposed specific, and wherever possible quantified, *targets* for moving towards those objectives. We have also made recommendations about the *measures* we believe are needed in order to achieve the targets. In this concluding chapter we list targets and measures under the respective objectives. We then go on to review the overall impact of our recommendations.

Air quality

14.8 Emissions from road vehicles are the major source of outdoor exposure to air pollution (3.9). Further increases in road traffic would progressively erode the benefits of the limits recently placed on emissions from new vehicles especially in the case of particulates and nitrogen oxides (8.23–8.25). Despite uncertainties about the effects of transport pollutants, there is a clear case, especially on health grounds, for taking further action to reduce emissions. The policy objective should be:

Objective B
TO ACHIEVE STANDARDS OF AIR QUALITY THAT WILL PREVENT DAMAGE TO HUMAN HEALTH AND THE ENVIRONMENT (3.43).

14.9 We propose the following specific targets:

B1 To achieve full compliance by 2005 with World Health Organization (WHO) health-based air quality guidelines for transport-related pollutants (3.44).

B2 To establish in appropriate areas by 2005 local air quality standards based on the critical levels required to protect sensitive ecosystems (3.54).

The first target will need to be kept under review as work proceeds on the task of revising the WHO guidelines. **We endorse the recommendation of the Expert**

Recommendation 1

Panel on Air Quality Standards that for practical purposes the standard for benzene should be 5 parts per billion as an annual running average, and should be reduced to 1 part per billion at a later date (3.45).

Limits on emissions from new vehicles

14.10 An essential step in achieving full compliance with WHO guidelines for air quality is a further tightening of the limits on emissions from new vehicles. EC stage II limits will take effect in 1996–97, but decisions have yet to be taken about the stage III limits due to come into effect in 1999–2000. The European Commission and the European automobile and oil industries are reviewing the reductions in emissions needed to achieve air quality standards and how such reductions could be brought about in the most cost-effective way (8.27).

14.11 The German government has put forward proposals for stringent stage III limits (tables 8.1 and 8.3), which seem likely to be achievable without excessive cost. **We recommend that the government give support to stage III limits for all types of vehicle at the levels represented by the German proposals and press for even more stringent limits on emissions of particulates. However if alternative proposals emerge from the tripartite initiative which produce a greater net benefit for the environment, they should be preferred (8.28). We recommend that the government seek amendments to the EC test cycle for cars at the earliest possible date, so that it will adequately represent typical operating conditions (8.6).** Emissions from diesel engines are a particular priority but measures to reduce emissions of particulates could increase the amounts of nitrogen oxides produced. **We recommend that European vehicle and catalyst manufacturers pool their research on DeNOx catalysts in a programme sponsored by the European Commission (8.13).**

Recommendation 2

Recommendation 3

Recommendation 4

14.12 In order to deal with local air quality problems it may be necessary to establish special emission limits for categories of vehicle operating predominantly in urban areas or for all vehicles operating within certain defined areas. To counter high concentrations of pollutants inside vehicles, **we recommend that the government study the case for requiring filtration of the air supply to vehicle interiors (8.29).**

Recommendation 5

Reducing emissions from existing vehicles

14.13 **We recommend an immediate study to determine which categories of vehicles not designed to the latest standards justify retrofitting with catalysts or particulate traps, and whether government grants should be offered (8.67).**

Recommendation 6

14.14 There is growing recognition of the consequences of bad tuning in petrol cars and the vulnerability of catalytic converters. The present MOT tests of emissions are having little practical effect. **We recommend that more stringent standards be applied in the emissions element of the annual MOT test, and that this element become obligatory for all cars a year after registration (8.72).**

Recommendation 7

14.15 To speed up improvements in air quality in urban areas, **we recommend that vehicle excise duty on heavy vehicles be graduated according to the emission limits their engines are designed to meet, and that:**

Recommendation 8

> **a reduced rate of duty should be payable for vehicles with engines meeting planned new emission limits which are being operated before those limits become mandatory for new vehicles;**

an increased rate of duty should be payable for vehicles with engines not designed to meet the emission limits currently mandatory for new vehicles, unless they have been retrofitted with effective pollution control devices (8.68).

Fuel choice and quality

Recommendation 9

Recommendation 10

14.16 The chemical composition of fuel has an important influence on emissions. **We recommend that the differential duty which favours unleaded petrol should be retained (7.48).** However, in view of the unnecessary risk to health represented by its high aromatics content, **we recommend that the government act to end the sale of unleaded super premium petrol (8.15).**

Recommendation 11

Recommendation 12

14.17 **We recommend that the government support a reduction in the permitted benzene content of petrol to 1% (8.15).** Looking beyond that, **we recommend that the government collaborate with the oil industry and vehicle manufacturers to develop specifications for cleaner fuels which will contribute to achieving our targets for air quality (8.16).**

Recommendation 13

14.18 **We recommend that the Department of the Environment (DOE) and the Health and Safety Executive exercise their powers to require manufacturers to provide them with additional information about the combustion products of fuel additives (8.19).**

Recommendation 14

14.19 Pending the development of satisfactory electric vehicles, **we recommend that the government consider the case for incentives to operators of fleets of heavy vehicles in urban areas to use natural gas-powered vehicles (8.21).** That apart, there would not be any overall environmental advantage in widespread use in the UK of alternative fuels for internal combustion engines (8.20).

Alternative methods of propulsion

Recommendation 15

14.20 The problems of storing electric power for use by vehicles have proved difficult to overcome (8.78). Fuel cells may eventually be the preferred technology. In view of their long-term potential, **we recommend that the government increase its support for fuel cell research (8.81).**

Recommendation 16

14.21 Because of the advantages they would have in areas with particular air quality problems, **we recommend that the government encourage the development of electric power for public transport or fleet vehicles which operate in urban areas with frequent stops and for small private vehicles for neighbourhood use (8.85).**

Improved monitoring and control

Recommendation 17

Recommendation 18

14.22 **We endorse the measures recommended by the Quality of Urban Air Review Group for improving the monitoring of concentrations of air pollutants and developing modelling techniques, so as to gain a more accurate picture of exposure (3.33). We recommend that further research be carried out into the health effects both of individual transport-related pollutants and of substances in combination (3.34).**

Recommendation 19

14.23 **We recommend that local authorities be given new duties to assess ambient air quality, the sources of air pollution in their area and the risk of pollutant concentrations exceeding threshold levels to be set by the government. They should also have the duty of drawing up and implementing an air**

quality management plan if they consider this necessary in order to prevent any of the thresholds from being breached (13.30). For the major pollutants including carbon monoxide, nitrogen dioxide and PM10, the government should set thresholds at levels which enable action to be taken to prevent air quality from deteriorating to a dangerous level (13.31). We further recommend that the government explore with the Institution of Environmental Health Officers, other professional bodies and local authorities how forms of collaboration similar to the London Air Quality Network and Unit can be built on and encouraged in other conurbations (13.28).

Recommendation 20

Noise

14.24 Noise in the environment is recognised as a health hazard by WHO (4.16). This is the aspect of transport which causes the most nuisance to most people. Noise from roads in particular is very pervasive (4.5–4.7). The policy objective should be:

> **Objective H**
> **TO REDUCE NOISE NUISANCE FROM TRANSPORT (4.17).**

14.25 In pursuit of this objective, we propose the following targets (4.18):

- **H1 To reduce daytime exposure to road and rail noise to not more than 65 $dBL_{Aeq.16h}$ at the external walls of housing;**
- **H2 To reduce night-time exposure to road and rail noise to not more than 59 $dBL_{Aeq.8h}$ at the external walls of housing.**

Noise and emissions from aircraft are dealt with below (14.100–14.106).

14.26 Limits on the sound levels of new designs of road vehicles are determined by those in EC legislation, and will be tightened in 1995–96 (table 8.8). As there is only limited scope for further modifications to vehicles with internal combustion engines to make them quieter, achievement of the targets will depend to a considerable extent on traffic management, use of quieter surfacing materials and the construction of suitable screening (8.63).

14.27 **We recommend that, both in new road construction and resurfacing, porous asphalt or whisper concrete should be used at all appropriate sites; and that research and development continue in order to identify surfacing materials with an even better combination of characteristics (4.19). We recommend more extensive and innovative use of barriers to absorb and deflect sound from roads and railways, as the most cost-effective way to achieve our targets in some cases (4.20).**

Recommendation 21

Recommendation 22

14.28 Where it is not practicable to achieve our targets for external noise levels, it would be equitable for the householders affected to receive grants towards the cost of sound insulation. The liability to pay grants also adds to the pressures on the body responsible for the road or railway to find methods of reducing external noise. Grants are already available to householders affected by noise from new or substantially upgraded roads and are being extended to those affected by noise from new rail lines. **We recommend that the qualifying level for insulation grants for both road and rail noise should be reduced to 65 $dBL_{Aeq.16h}$ to match our target for daytime noise (4.22). We recommend that the government study:**

Recommendation 23

Recommendation 24

i. **how eligibility for insulation grants might be extended to house-**

holders affected by noise from existing roads or railways in cases in which it will be impracticable in the foreseeable future to achieve our targets for external noise levels;

ii. **whether eligibility under the Land Compensation Act 1973 can be extended to householders whose property has lost value since they purchased it as a result of increased traffic caused by a road improvement scheme at some distance from the property (4.24).**

Recommendation 25 **We recommend that guidance on development control should be based on the principle of preventing the exposure of new residential development or schools to noise levels which exceed our proposed targets (4.26).**

Use of materials

14.29 In addition to fuel, the transport system uses enormous quantities of materials, many of them non-renewable. These include a fifth of world steel production, a tenth of world aluminium production and a third of the aggregates used in Britain (4.63). There needs to be a clear policy objective:

Objective G

TO REDUCE SUBSTANTIALLY THE DEMANDS WHICH TRANSPORT INFRASTRUCTURE AND THE VEHICLE INDUSTRY PLACE ON NON-RENEWABLE MATERIALS (4.63).

14.30 In the case of vehicles this objective can be pursued by reducing their size and weight, using materials which are less energy-intensive or come from renewable sources, designing vehicles with a longer life, and reusing parts or recycling materials (4.70). Reductions in size and weight also promote environmental objectives by reducing fuel consumption (8.40).

Recommendation 26 14.31 **We recommend that the Department of Trade and Industry and DOE work with vehicle manufacturers and dismantlers to develop a cradle-to-grave strategy for recycling. Where necessary they should use economic instruments and other forms of regulation to implement this strategy (4.76).**

Recommendation 27 14.32 The demand for aggregates for road construction will be considerably reduced by other recommendations in this report. **If planning permission is granted for further coastal superquarries, we recommend there be a legal requirement that the quarried material must be transported by sea (4.66).** Recycling removes the need for quarrying, and often also reduces the distances over which aggregates have to be transported.

14.33 To stimulate increased recycling of materials we propose that government adopt the following targets:

G1 To increase the proportion by weight of scrapped vehicles which is recycled, or used for energy generation, from 77% at present to 85% by 2002 and 95% by 2015 (4.77);

G2 To increase the proportion of vehicle tyres recycled, or used for energy generation, from less than a third at present to 90% by 2015 (4.77);

G3 To double the proportion of recycled material used in road construction and reconstruction by 2005, and double it again by 2015 (4.69).

Changes in the Earth's atmosphere

14.34 Over two-fifths of the petroleum products used in the UK are used in road transport (table 3.1). In all, surface transport causes 21% of the carbon dioxide emissions produced by human activities in the UK, or about 24% if emissions from refining and electricity generation for transport are included. Road transport accounts for 87% of the emissions attributable to surface transport (3.58). The UK has accepted international obligations which have the aim of ultimately stabilising atmospheric concentrations of greenhouse gases and preventing dangerous climate change (3.71). On the basis of the forecast growth in road traffic, carbon dioxide emissions from the transport sector will show further substantial growth over the next 25 years (3.59, 8.54). In view of this, and the scale of reductions needed, it would not be acceptable to rely solely on reducing emissions from other sectors of the economy (3.77). Moreover policies to reduce carbon dioxide emissions from the transport sector can bring other important environmental benefits. For these reasons we believe the policy objective should be:

Objective F
TO REDUCE CARBON DIOXIDE EMISSIONS FROM TRANSPORT (3.76).

14.35 The government has set a target of reducing emissions of each greenhouse gas from all sources to the 1990 level or below by 2000; but has not adopted any target, or announced measures, to deal with the position after 2000 (3.74). The Climate Change Convention implies that the reductions made by developed countries ought to be large enough to allow scope for some growth in the energy demands of developing countries (3.78). We propose that the target for transport for the period after 2000 should be:

F1 To reduce emissions of carbon dioxide from surface transport in 2020 to no more than 80% of the 1990 level (3.78).

Emissions of greenhouse gases from aircraft have been dealt with separately (5.17–5.35).

14.36 As action to meet this target for 2020 needs to start immediately, it is sensible and logical to set an intermediate target for 2000. That can best be done by applying to the transport sector separately the target already set for the economy as a whole. In other words the target should be:

F2 To limit emissions of carbon dioxide from surface transport in 2000 to the 1990 level (3.79).

14.37 The main emphasis must be on reducing emissions of carbon dioxide from road vehicles by improving fuel efficiency. The targets we propose for such improvements are:

F3 To increase the average fuel efficiency of new cars sold in the UK by 40% between 1990 and 2005, that of new light goods vehicles by 20%, and that of new heavy duty vehicles by 10% (8.59).

14.38 An increase in the proportion of cars with diesel engines could make a valuable contribution to improving fuel efficiency. The precautionary principle prevents us from advocating such an increase until the stringent limits on emissions of pollutants recommended above (14.11) are in place (8.33–8.34).

14.39 Developments in catalyst design may enable lean-burn and two-stroke engines, which both offer high fuel efficiency, to overcome the problem of the high levels of pollutants emitted. Research on these types of engine should be given greater priority by manufacturers (8.44).

14.40 The technical potential for increased fuel efficiency will be realised only if manufacturers and purchasers expect a substantial and permanent increase in the relative price of fuel. We do not believe the present government commitment to a 5% increase in fuel duty each year is sufficient for that purpose. **We recommend that fuel duty be increased year by year so as to double the price of fuel, relative to the prices of other goods, by 2005 (7.58). We recommend that the government press for revision of the EC Directive on fuel prices so as to ensure a sustained year-by-year increase in fuel prices across the Community (7.83).**

Recommendation 28

Recommendation 29

14.41 As an additional incentive for fuel efficiency, **we recommend that the annual excise duty on cars be steeply graduated, and based on the certified fuel efficiency of a car when new (8.49). In relation to company cars we recommend further modifications to the tax rules, including abolition of the mileage thresholds and taxation of the actual value of benefits in the form of fuel, to remove the incentives for environmentally damaging behaviour (7.39). We also recommend that the government study the possibility of a pay-at-the-pump scheme (7.49).**

Recommendation 30

Recommendation 31

Recommendation 32

14.42 Better driving could reduce fuel consumption, emissions of pollutants, noise, and severity of accidents. **We recommend that the Department of Transport (DOT) conduct a major educational and advertising campaign highlighting the environmental and safety effects of driving styles (8.38).** We hope car manufacturers will increasingly emphasise safety and fuel ecconomy in their advertising, rather than speed and acceleration. **We recommend that the government explore the possibility of introducing a system of fuel efficiency labelling for cars on US lines (8.37).**

Recommendation 33

Recommendation 34

14.43 Effective enforcement of the present 60 mph and 70 mph speed limits would reduce carbon dioxide emissions from road transport by about 3%, as well as reducing noise and emissions of nitrogen oxides and improving safety (12.23–12.24). **We recommend that increased effort be devoted to enforcing speed limits, making full use of new technology (12.25). We recommend that all heavy goods vehicles over 7.5 tonnes have their speed limiters set at 56 mph as from 1996 (10.46).** Speed limits in urban areas are dealt with below (14.91).

Recommendation 35

14.44 **We recommend that the progress made towards our targets be reviewed after 2000, and that a view be taken at that stage about:**

i. the case for further increases in fuel duty after 2005 (7.59);
ii. the case for reducing the general speed limit (12.26).

Recommendation 36

14.45 Reductions in carbon dioxide emissions will be brought about in large part by people choosing more economical cars and driving them in a more economical way (appendix D). The reductions in traffic levels resulting from the increased price of fuel will therefore be on a smaller scale. More far-reaching measures, which we discuss below, are required in order to achieve the full set of objectives for a sustainable transport policy.

The impact of new construction

14.46 Significant environmental damage has been caused in recent years by the construction of transport infrastructure. Despite DOT's emphasis on widening existing roads rather than building on new alignments (6.19), there is much concern about the effects the present trunk road programme would have in damaging the landscape, causing loss of habitats or species, and damaging historic buildings and archaeological features (4.55–4.56). Providing sufficient road capacity to carry the levels of traffic predicted in the government's 1989 forecasts would require a massive programme of road building and improvement, over and above the schemes already included in the trunk road programme (6.29). If major road building programmes continue, they are likely to come into conflict with the UK's national and international commitments to protect species and habitats (4.58). Controversy has also arisen over the effects of proposals for new railways, airports and ports (4.47).

14.47 The policy objective must be:

> **Objective E**
> **TO HALT ANY LOSS OF LAND TO TRANSPORT INFRASTRUCTURE IN AREAS OF CONSERVATION, CULTURAL, SCENIC OR AMENITY VALUE UNLESS THE USE OF THE LAND FOR THAT PURPOSE HAS BEEN SHOWN TO BE THE BEST PRACTICABLE ENVIRONMENTAL OPTION (4.60).**

14.48 Statutorily designated areas should be given more effective protection against the construction of transport infrastructure than they receive at present. The aim is to give higher overall priority to environmental protection; there is not therefore any implication of a weakening of protection in relation to other areas of land. **We recommend that the following general principles should apply:** **Recommendation 37**

- **a.** **strict protection for the special areas of conservation to be designated under the EC Habitats and Species Directive;**
- **b.** **any further loss or damage to natural and semi-natural habitats or archaeological features must be reduced to the absolute minimum;**
- **c.** **where loss of a natural or semi-natural habitat cannot be avoided, the developer must be required to provide some restitution by creating an appropriate new habitat in the vicinity;**
- **d.** **where other land which has significant amenity value is used for transport infrastructure, the developer must be required to provide an equivalent area of land of equivalent amenity with equivalent access for the public;**
- **e.** **where a proposed road or railway would cause serious environmental damage, careful consideration should be given to placing it in a tunnel (4.61).**

14.49 The procedures for environmental assessment of road building and improvement schemes must be made more effective. **We recommend that all road construction proposals be subject to environmental assessment (9.60).** **Recommendation 38** **We also recommend that the role of the present Standing Advisory Committee on Trunk Road Assessment (SACTRA) be broadened to include the development of appropriate appraisal techniques to ensure consistency of evaluation methods as between the different transport modes (9.64).** **Recommendation 39**

14.50 The procedure for selecting the 'best practicable environmental option' (BPEO) was described in the Commission's Twelfth Report, in the context of controlling emissions from industrial sites. An essential part of the BPEO concept is that a wide range of alternatives should be identified and evaluated at an early stage in the decision-making process, well before plans begin to be made for specific projects. **We recommend that decision-making at all levels of transport policy be based on the identification and pursuit of the best practicable environmental option (9.56).** **Recommendation 40** To achieve this, fundamental changes will be required in the way transport policy is formed and implemented.

Creating a sustainable transport policy

14.51 An important component in any strategy to reduce the environmental effects of transport is land use planning (9.68). In the past transport and land use policies have combined to promote lifestyles which depend on high mobility and intensive use of cars, and which cannot therefore be regarded as sustainable. The central objective of a sustainable transport policy should be:

Objective A
TO ENSURE THAT AN EFFECTIVE TRANSPORT POLICY AT ALL LEVELS OF GOVERNMENT IS INTEGRATED WITH LAND USE POLICY AND GIVES PRIORITY TO MINIMISING THE NEED FOR TRANSPORT AND INCREASING THE PROPORTIONS OF TRIPS MADE BY ENVIRONMENTALLY LESS DAMAGING MODES (13.12).

Local government

14.52 Land use policy and transport policy must be integrated at the local level. The package approach now used in annual submissions of transport policies and programmes (TPPs) by local authorities to DOT is a valuable step in that direction. **We recommend that the government monitor the effectiveness of the new package approach and publish the results (13.39).** The purpose is to give local authorities greater flexibility to switch resources between different forms of transport. Flexibility is reduced by legal restrictions on the use of various types of grant and **we recommend that existing legal restrictions on the use of various types of grant be modified to improve the flexibility and effectiveness of the package approach (13.40).**

Recommendation 41

Recommendation 42

14.53 The pattern of relatively small unitary authorities, interspersed in some areas with two-tier authorities, which seems likely to emerge from the current review of local government will make it more difficult to achieve a balanced approach to transport provision. Arrangements must be made to ensure that strategic planning does not suffer. **We recommend that additional Passenger Transport Authorities (PTAs) and Executives (PTEs) should be established, including one for London, to ensure the provision of a co-ordinated and efficient system of public transport throughout their areas (13.24).** Some would be responsible for predominantly rural areas; there should be a PTA/PTE for London which should cover broadly the area within the M25.

Recommendation 43

14.54 We welcome the guidance the government has recently published on the relationship between land use and transport policies (PPG 13) and **recommend that the results of monitoring of PPG 13's effectiveness be made public, and that any shortcomings which are revealed be corrected without delay (9.43).**

Recommendation 44

14.55 Readily accessible methodologies will be needed to assess the transport implications of land use policies and location decisions as well as the land use implications of transport policies. **We recommend that the government supplement PPG 13 and other recent guidance with more specific advice on these issues.** There may also be a need for new types of training in planning education to enable these tasks to be accomplished (9.53).

Recommendation 45

14.56 Arrangements must be made to ensure that local authorities' land use and transport plans contribute to national targets for sustainable development. **We recommend that the government explore, in conjunction with the local authority associations, the potential to define and adopt targets for specific planning areas: these should be addressed in the context of local air quality plans and might relate to emissions and traffic growth as well as factors such as cycle routes built or bus lanes designated (9.45). We recommend that the government consider ways of making developers responsible for a charge reflecting the external costs of traffic induced (9.58).**

Recommendation 46

Recommendation 47

14.57 The land use planning process should be strengthened at regional level and integrated with the planning of transport infrastructure and services. **We recommend that the existing regional planning conferences should be strengthened and put on a statutory basis for each economic planning region (13.27).** These conferences should be the place in which the need for major road schemes affecting more than one local authority area is decided. **We therefore recommend that the overall need for a new or improved long-distance route, the possibilities of providing improvements by road or rail, and alternatives based on managing the level of demand or giving different priorities to short and long-distance users of the route, should be jointly assessed by the regional planning conferences and the government's newly integrated regional offices (9.48).**

Recommendation 48

Recommendation 49

14.58 Stronger mechanisms are needed to avoid undue pressure on planning authorities to grant planning permission for developments which would have undesirable effects in generating additional traffic. **We recommend that a right of appeal by third parties to the Secretary of State be introduced against decisions by local planning authorities which are contrary to up-to-date development plans (9.67).**

Recommendation 50

14.59 **All significant applications for planning permission should contain an analysis of the transport implications, including pedestrian, cycling and public transport access, and freight movements in the case of industrial developments, and we so recommend (9.65). We recommend that the Town and Country Planning (Use Classes) Order be reviewed to ensure that new uses which generate appreciably higher levels of traffic cannot take place without a fresh grant of planning permission (9.66).**

Recommendation 51

Recommendation 52

Central government

14.60 The Department of Transport was frequently criticised in evidence as being excessively dominated by roads at the expense of other elements of transport. It has recently been restructured but fundamental changes of culture are needed. **We recommend that responsibilities within the Department of Transport be restructured, from Ministerial level downwards, to reflect the fundamentally different approach which a sustainable transport policy will involve.** It will require divisions with the specific remit of bringing about the transfers of traffic to less damaging modes which are required in order to achieve environmental objectives and targets (13.76).

Recommendation 53

14.61 The emphasis in future should be on a national transport strategy covering all modes of transport. For example, the strategy should be concerned not solely with trunk roads but with trunk routes between defined points, and the full range of possible improvements in transport systems to serve those routes. **We recommend a fundamental review of the definition and purpose of a separate system of trunk roads (13.84).** If responsibility for the present trunk roads passed to local authorities, with suitable financial arrangements, transport and land use policies could be better integrated. Trunk road schemes have been criticised for being imposed without regard to the needs of an area or the ensuing development pressures. **If a system of trunk roads is retained, we recommend that all trunk road schemes be considered initially as an intrinsic part of local authority structure plans and integrated fully into the development control system (9.50). We also recommend that the rules of procedure governing inquiries into trunk road proposals and compulsory purchase orders be amended so as to permit government witnesses to answer questions about the merits of government policy and allow the inspector to take account of the interactions of the proposal with other government policies in his recommendation (13.89).**

Recommendation 54

Recommendation 55

Recommendation 56

14.62 The UK has built a high-quality interurban road network, but a side-effect has been to draw much of the life out of towns and cities. The priority in future should be to improve the accessibility of facilities and employment within urban and rural areas and improve rail and bus services. **If there continues to be a central government road programme, we recommend that, except where the construction or widening of a motorway or other trunk road represents the best practicable environmental option for meeting access needs, expenditure should be confined to maintaining existing roads and constructing new bypasses for local traffic (12.62). Road schemes for which contracts have not yet been let should be reassessed to take into account changed policies and methodology, and further progress on them should be suspended until that reassessment has been completed (12.63). Planned expenditure on motorways and other trunk roads should be reduced to about half its present level (12.65).** The resources released should be used to expand environmentally less damaging forms of transport (14.70).

Recommendation 57

Recommendation 58

14.63 Various measures can be taken to increase road capacity for interurban traffic and so avoid the need for new construction. **We recommend that, where a route designed for long-distance traffic is overloaded as a result of heavy local traffic, an assessment be made of the advantages of closing selected entry points, completely or at certain times of day, as an alternative to widening it or constructing relief roads (12.21).** Although they will take some years to develop, automated guidance systems could considerably increase the capacity of interurban roads by reducing the safe distance between vehicles. They also offer the prospect of some environmental benefits, because traffic would travel at a uniform speed. **We recommend that the government investigate the potential of automated guidance systems for increasing the capacity of motorways; and that the introduction of such systems should depend on an assessment of their environmental impact, including their impact on the total volume of road traffic (12.28).**

Recommendation 59

Recommendation 60

14.64 **If motorway tolling is introduced, we recommend that the net income, after meeting the costs of maintaining and operating the present motorway system, be used to help meet the public expenditure costs of introducing a sustainable transport policy, including the measures recommended in this report to improve public transport (12.18). In view of the potential damage if traffic transfers to environmentally more sensitive routes, we recommend that any proposal to introduce or vary tolls for a road, bridge or tunnel be the subject of an environmental assessment before it is approved (12.20).**

Recommendation 61

Recommendation 62

14.65 **Where the major purpose of a proposed bypass is environmental improvement, we recommend that:**

Recommendation 63

i. **the design standards should provide only for the traffic flows which will be diverted from the town or village bypassed, not for further large growth of traffic;**

ii. **the design, funding and implementation of comprehensive traffic measures in the area bypassed should proceed on the same timetable, and be given the same priority, as the bypass itself (12.64).**

14.66 Government departments are major employers and major users of transport and there is a standing committee of 'Green' Ministers. **We recommend that the committee of Green Ministers publish periodical reports on the results of its scrutiny of the environmental implications of government practices (13.60). We also recommend a more consistent scrutiny of the transport implications of all government departments' legal and policy responsibilities so that these can be taken fully into account and their adverse effects minimised (13.61).**

Recommendation 64

Recommendation 65

European Community

14.67 Although an environment chapter was added to the Treaty of Rome by the Single European Act in 1987, the European Community has not integrated environmental considerations adequately into the common transport policy. The government should work more positively with the European Commission and other Member States to develop a means of ensuring that policies receive thorough environmental appraisal (13.94). **We recommend that the government pursue vigorously the objective of ensuring that the consultative forum established under the Fifth Environment Action Programme provides a means of strengthening the environmental input to Community transport policies (13.93). We also recommend further joint Council meetings of Environment and Transport Ministers as a means of improving discussion of the environmental dimensions of transport problems (13.93).**

Recommendation 66

Recommendation 67

14.68 Grants are available from the Cohesion Fund for the establishment and development of trans-European networks for transport, telecommunications and energy with the aim of promoting the harmonious development of the Community and the internal market. **In scrutinising applications for grants in support of trans-European transport networks, we recommend that the European Commission give support only to those developments which are consistent with sustainable development and respect for the environment (13.100).** It will be important to ensure that the construction of the networks does not distort the development of a more balanced transport policy.

Recommendation 68

14.69 The Commission's recent Transport White Paper falls short of dealing with some critical environmental issues identified in the preceding Green Paper. Measures to increase the costs of transport to reflect environmental and social costs need to be pursued at Community level so that the competitive position of Member States is not harmed (14.40, 14.77, 14.102).

Promoting less damaging modes of transport

14.70 Trains, buses and trams are significantly more energy efficient on average than cars or lorries. The use of water to move freight is considerably more energy efficient than other modes. Increased use of the rail network can avoid the need for road building. Buses and coaches make more efficient use of existing road space. A sustainable transport policy must have as a declared objective:

> **Objective D**
> **TO INCREASE THE PROPORTIONS OF PERSONAL TRAVEL AND FREIGHT TRANSPORT BY ENVIRONMENTALLY LESS DAMAGING MODES AND TO MAKE THE BEST USE OF EXISTING INFRASTRUCTURE (10.1).**

We recommend the government ensure that the structure and operation of the privatised rail industry lead to a much greater use of rail (13.56). We recommend a substantially increased programme of investment in public transport over a ten-year period (12.68).

Recommendation 69
Recommendation 70

Personal travel

14.71 As a very substantial increase in rail, bus and coach travel is desirable, we propose the following targets for personal travel:

> **D1 To increase the proportion of passenger-kilometres carried by public transport from 12% in 1993 to 20% by 2005 and 30% by 2020 (12.54).**

The choice between public and private transport is distorted at present by differences in the way in which people make payments towards their costs (7.24).

Increases in fuel prices will go some way to mitigate the distortion but will not in themselves bring about a switch to public transport on the scale that would be desirable on environmental grounds (7.60). Public contributions to the cost of public transport can be justified on environmental grounds as well as on social grounds (7.64–7.67).

Recommendation 71

14.72 The immediate priority is to prevent any further decline in public transport use. **We recommend that the present net level of support for public transport (including fuel duty rebates for public service vehicles, the low rate of fuel duty for railways and the benefit of zero-rating for VAT) should be at least maintained in real terms until superseded by measures for achieving a major increase in the role of public transport (7.69).**

Recommendation 72

14.73 The rail system has sufficient spare capacity to make a considerable contribution to meeting our target for public transport use in 2005, as well as carrying much more freight. However, large-scale investment is necessary to utilise that capacity and improve the quality and attractiveness of services. Future public contributions to public transport costs should be in a form which gives operators incentives for efficient management (7.71). **We recommend that the criteria for the public contribution to railway costs be the environmental benefits of attracting traffic to rail from road, and whatever social criteria are considered relevant; and that existing forms of subsidy be reviewed in the light of these criteria and consolidated (12.58).**

Recommendation 73

14.74 **We recommend that no proposal be taken forward in the UK for trains running at more than 300 kph unless a comprehensive assessment has shown that the environmental benefits from transferred air traffic will outweigh the environmental costs of landtake, construction work, noise and the additional energy required to propel trains at this speed (12.49).**

Recommendation 74

Recommendation 75

14.75 **We recommend that transport and land use planning recognise the role of express coach services, and provide full facilities for them in traffic management schemes and at transport interchanges. We recommend that, in considering whether high vehicle-occupancy lanes should be designated on interurban roads, highway authorities place considerable weight on the potential benefits to express coach services (12.36).** We deal below (14.86–14.89) with further measures, particularly in urban areas, to increase the use of public transport.

Freight

14.76 An essential element in any sustainable transport policy is to move as much freight as possible by the less damaging modes. Water and pipeline are least harmful, followed by rail, road and finally air. We propose the following targets for the transfer of freight from road to other modes:

D2 To increase the proportion of tonne-kilometres carried by rail from 6.5% in 1993 to 10% by 2000 and 20% by 2010 (10.63).

D3 To increase the proportion of tonne-kilometres carried by water from 25% in 1993 to 30% by 2000, and at least maintain that share thereafter (10.64).

If pipelines retain a 5% share, achievement of these targets would reduce the proportion of freight carried by road from 63% now to 55% by 2000 and 45% by 2020.

14.77 Economic instruments should be used as well as direct regulation to reduce the adverse effects of lorries. The public sector and environmental costs imposed by the operation of heavy goods vehicles (HGVs) are substantially greater than the amounts their operators pay in road taxation (7.19). **We recommend that the government continue to press for EC legislation which will standardise the annual duty paid by operators of heavy goods vehicles at a sufficiently high level to ensure that the total amounts paid in road-specific taxes by operators in each country fully reflect external and infrastructure costs (10.12).** **Recommendation 76**

14.78 Increases in the cost of road freight must be accompanied by improvements in the facilities for sending freight by other modes. Water could be given a more important role by the use of innovative technology, such as barge-carrying ships. **We recommend that, where environmental benefits could be obtained by transferring freight from road to coastwise shipping, but the inadequacy of port facilities is preventing such a transfer, the government pay grant towards the cost of providing the facilities required (10.22).** **Recommendation 77** **In addition, for at least a transitional period, we recommend that the government make a public contribution towards the operating costs of companies carrying freight by water where that will achieve a demonstrable environmental benefit (10.22).** **Recommendation 78**

14.79 The government has relaxed the grant conditions for such facilities as private sidings, and introduced track access grants. However the amount of money available under both headings averages only £14 million a year for the next three years. **We recommend that the government keep the grant schemes for freight transport under review, and make further resources available for them if the demand exists, and if necessary modify them or adopt other measures in order to stimulate the switching of freight from road to rail or water (10.36).** **Recommendation 79**

14.80 There is a risk that freight traffic will be lost to rail as a result of the new structure of the industry and privatisation. **We recommend that the government:** **Recommendation 80**

- **i. ensure that sales of BR freight subsidiaries to the private sector take place on terms which create the most favourable prospects for increasing rail's share of the freight market;**
- **ii. authorise Railtrack to start work immediately on the modifications required to open a Channel Tunnel to Scotland route to piggy-back traffic, in advance of the general upgrading of the West Coast Main Line (10.34).**

14.81 To make transhipment of freight between modes more practical, appropriate provision will be needed in development plans. **We recommend that suitable sites for new wharves and rail depots are identified and safeguarded by planning authorities (10.37).** **Recommendation 81**

14.82 To give the necessary impetus to inter-modal transport throughout Europe and to facilitate the free operation of the market in transport services, **we recommend that the government press the European Commission to draft and introduce legislation prescribing a range of standard dimensions for containers and bodies for use in inter-modal transport and covering other matters which are essential to ensure technical compatibility (10.31).** **Recommendation 82**

14.83 There is an overwhelming case for stronger enforcement of existing laws, including the O-licence system for road hauliers. This would put an end to unfair competition by certain hauliers, improve safety, diminish emissions and noise, and reduce the infrastructure damage caused by overloading. **We recommend immediate legislation to allow impoundment of illegal operators' vehicles (10.44).** **Recommendation 83**

Reducing the dominance of motor traffic

14.84 Heavy road traffic diminishes the quality of life. Many people would like traffic in urban areas restricted to help create a more attractive environment. There is a strong case, on social and safety grounds, for reducing traffic speed and density, especially in town centres and residential districts. We consider therefore that it should be a specific objective:

> **Objective C**
> **TO IMPROVE THE QUALITY OF LIFE, PARTICULARLY IN TOWNS AND CITIES, BY REDUCING THE DOMINANCE OF CARS AND LORRIES AND PROVIDING ALTERNATIVE MEANS OF ACCESS (4.32).**

14.85 The growth in car traffic in urban areas needs to be tackled directly. The target we propose is:

> **C1 To reduce the proportion of urban journeys undertaken by car from 50% in the London area to 45% by 2000 and 35% by 2020, and from 65% in other urban areas to 60% by 2000 and 50% by 2020 (11.12).**

Achieving this target will require both a greater emphasis on restraining car traffic and the provision of attractive and appropriate alternatives.

Making public transport more attractive

14.86 People are unlikely to turn to public transport unless it is provided by high standard vehicles and regular and reliable services. **Recommendation 84** **We recommend that bus lanes should be introduced much more widely, and policed more effectively, and that buses should be given priority over other motor vehicles in urban areas, including automatic (or driver activated) priority at light-controlled junctions (11.37).** Interchanges between services should be improved and adequate information provided for customers. **Recommendation 85** **We recommend the government ensure the availability of tickets valid for all public transport modes and services in a particular area, taking full advantage of new technology. This policy should be implemented through PTAs/PTEs (11.32).** **Recommendation 86** **We recommend that the rail regulator require rail operators to make comprehensive information easily available at stations and on longer-distance trains about public transport connections at destination stations (12.40).**

14.87 New services will be needed where none is now operating and will probably require financial support for some time as people adapt to new methods of mobility. Park and ride schemes, designed to attract car users without reducing the number of people making the whole journey by public transport, should form a major element in local transport strategies.

14.88 Light rapid transit systems can improve the urban environment by reducing local noise and pollution and by improving safety. They are also one of the most energy-efficient modes. **Recommendation 87** **We recommend that the government make more resources available for new light-rail systems so that they can be built within a reasonable time in those conurbations for which they are an integral part of an overall transport strategy (11.47).**

14.89 The House of Commons Select Committee on Transport expressed disquiet about the effects of bus deregulation. We share their concerns, especially about increased congestion, the use of more polluting, older buses, and the instability of services. In London, costs have been reduced by putting services out to tender, without the environmental disadvantages of deregulation. **Recommendation 88** **We recommend that the government reform the legal framework for bus services and base it**

on the franchising of routes by PTAs/PTEs to operators who meet high standards (13.47).

Traffic restraint

14.90 Investment in improved public transport facilities should be closely linked to policies for restraining the use of private vehicles in urban areas. The measures will include traffic calming and appropriate reductions in the highway space available for private vehicles. The government should study the effects of traffic calming measures in order to give guidance on designs which produce the best combination of environmental and safety benefits.

14.91 DOT is studying the effectiveness of the 20 mph zones now being established in this country. **If they are shown to be effective, we recommend they should be used more widely (11.55). We recommend that the government should work closely with regional planning conferences, PTAs/PTEs and the private sector to develop and implement comprehensive parking strategies designed to restrain use of private cars (11.62).**

Recommendation 89
Recommendation 90

14.92 Local authorities are subject to conflicting pressures from the public and from business on the question of lorry bans. **We recommend that all urban authorities adopt a presumption against access for heavy goods vehicles over 17 tonnes and follow a uniformly strict approach to the granting of exemptions (10.50).** This is the critical measure required to encourage transhipment depots and the use of smaller and less intrusive freight vehicles. The danger that this might stimulate out-of-town developments reinforces the emphasis we place on integrating land use and transport policies. **We recommend that the government maintain the effectiveness of restrictions on access by heavy goods vehicles to Greater London (10.49). We recommend that the programme of strengthening bridges to carry 40 tonne lorries should be scaled down to the minimum necessary to provide a basic network giving such lorries access to main distribution centres (10.60).**

Recommendation 91
Recommendation 92
Recommendation 93

14.93 As technology improves, local road pricing may offer environmental advantages. **We recommend that the government work within the European Community to develop common standards for the technology of road pricing (13.101). Decisions to introduce road pricing schemes should be made locally after evaluating the environmental effects, including those on adjoining areas; and the revenue should be retained by the local authority introducing the scheme and used to finance public transport or infrastructure improvements which are not environmentally damaging (11.71).**

Recommendation 94
Recommendation 95

14.94 **We recommend that DOT:**

Recommendation 96

i. **make comparative studies of representative towns and villages before and after the completion of bypasses in order to improve understanding of their environmental and other effects;**

ii. **investigate whether some towns and villages could obtain most of the benefits of a bypass, more cost-effectively and with less environmental damage, through traffic management measures (4.34).**

Walking

14.95 Many journeys made by car are short enough to be made easily on foot or by bicycle. This would benefit the environment and promote a higher standard of physical fitness in the population. Local authorities, with government guidance and financial support, should devise the measures which will be needed to make it more attractive to make short journeys on foot, rather than by car.

Recommendation 97

We recommend that local authorities should provide networks of safe pedestrian routes, especially those which will enable children to walk to school when they live close enough to do so (11.16).

Cycling

14.96 There should be a long-term programme to encourage much greater use of cycles by both adults and children. Measures such as traffic calming will make residential roads safer for cyclists, especially children cycling to school. We propose as the target:

> **C2 To increase cycle use to 10% of all urban journeys by 2005, compared to 2.5% now, and seek further increases thereafter on the basis of targets to be set by the government (11.26).**

Since cycle use varies widely in different towns, local authorities should set targets for their own areas.

Recommendation 98

Recommendation 99

14.97 **We recommend that comprehensive networks of safe cycle routes which do not involve the use of heavily trafficked roads should be built up in all urban areas (11.22). We recommend (11.24) that:**

- **i. the government make available to local authorities the relatively modest resources required to support a 10-year programme to create high-quality cycling facilities;**
- **ii. the new appraisal method for cycling schemes be reviewed in the light of the 1995/96 TPP round.**

Recommendation 100

Recommendation 101

We recommend that British Rail and private rail operators should be required to provide adequate space for cycles on all passenger services. It will be crucial to ensure that new and substantially overhauled rolling stock is designed to meet this requirement (11.25). We also recommend that Railtrack provide secure facilities for cycle parking at all stations (11.25).

Accidents

Recommendation 102

14.98 Almost two-fifths of the accidental deaths in Britain are road accidents (4.35). **We endorse the government's target of reducing deaths and injuries from road accidents to two-thirds of the 1981–85 level by 2000 and welcome the progress already made towards meeting that target (4.41).** However, accidents to pedestrians and cyclists are still well above the level in some other countries (4.37–4.40). We propose as additional targets:

> **C3 To reduce pedestrian deaths from 2.2 per 100,000 population to not more than 1.5 per 100,000 population by 2000, and cyclist deaths from 4.1 per 100 million kilometres cycled to not more than 2 per 100 million kilometres cycled by the same date (4.41).**

DOT should monitor closely the effects of the measures we have recommended for reducing the dominance of motor traffic, to ensure that these targets are achieved as well as the others we have proposed.

Shipping

Recommendation 103

14.99 Most aspects of pollution from shipping were considered either in the Commission's Eighth Report or in its Eleventh Report. **We endorse the recommendations of the Donaldson Inquiry, particularly those relating to port state inspection (4.4).** We welcome the indications that the next significant regulation under the International Convention for the Prevention of Pollution from Ships (MARPOL) will cover emissions to air from marine diesel engines (3.8).

Aircraft

14.100 International air traffic has been increasing even more rapidly than road traffic (5.2–5.4), there are major question-marks over its environmental impact (5.25–5.26), and effective regulation must be undertaken on a global basis. We attach particular importance therefore to the recommendations we make about international measures to limit the environmental effects of air transport.

14.101 The issue of most concern is the potential effect of high altitude emissions, especially of nitrogen oxides, on the Earth's atmosphere. **We recommend that the government press for the extension of the regulatory role of the International Civil Aviation Organization to cover emissions from aircraft engines at all phases of flight, with the aim of protecting the Earth's atmosphere against irreversible or long-term changes (5.34).** **Recommendation 104**

14.102 **We recommend that the government negotiate within the EC, and more widely, for the introduction of a levy on fuel purchases by airlines that will reflect the environmental damage caused by air transport (7.75).** **Recommendation 105**

14.103 Of all forms of transport, aircraft produce the greatest intensity of noise. Although improved engine design has reduced the number of people exposed to very high levels of noise (5.12–5.13), the expansion of regional airports and growth of traffic have placed the majority of the population within earshot of aircraft (4.5). **We recommend that the government implement the Batho Committee's recommendation for a noise levy on movements at airports (7.76).** **Recommendation 106** **We recommend that the government support more stringent noise certification standards for new aircraft, if these can be met without a significant fuel penalty (5.35).** **Recommendation 107**

14.104 Deregulation of airlines in the USA in the 1980s led to increased emissions per passenger-kilometre because load factors dropped and smaller aircraft were used. **We recommend that proposals for further measures to promote competition in air services in Europe be accompanied by a full assessment of the environmental implications (5.40).** **Recommendation 108**

14.105 Government policy should be based on transferring as much traffic as possible from air to rail. **We recommend that policy on air services should be based on discouraging air travel for domestic and near-European journeys for which rail is competitive, and that the government should support the upgrading of rail links to the main international airports in order to avoid the need for development of air feeder services from regional airports (12.50).** **Recommendation 109**

14.106 **We recommend that the UK collaborate in research into the possible effects of supersonic aircraft on the stratosphere and ways of minimising those effects (including the possible imposition of route restrictions in relation to latitude and altitude), so that a comprehensive environmental assessment can be produced and considered on an international basis before decisions are taken to build and operate a new generation of commercial supersonic aircraft (5.37).** **Recommendation 110**

14.107 Our recommendations complement and reinforce each other, and must be viewed as a whole. The primary focus of this report is on the period from 2000 to 2020. In order to have a substantial effect on the situation after 2000 action must start now, and must be vigorously pursued. We have also had constantly in mind the position after 2020. The need is to identify and adopt a strategy which is likely to be sustainable for as far ahead as we can foresee, and certainly to the middle of the next century and beyond.

14.108 In the period between now and 2020 we estimate that the increases in fuel prices we have recommended might have the effect of reducing the total growth in passenger and freight transport by all modes to about 10% a decade (12.11). If so, the effect of achieving our proposed targets for modal share would be to keep road traffic nationally in 2020 to roughly the present level. After 2020 it is reasonable to expect that the cumulative long-term impact of the changes which will have taken place in land use policies, in combination with a higher level of fuel prices and a broadly constant UK population, will stabilise the level of car ownership and the overall demand for both passenger and freight transport; and that the introduction of different methods of propulsion for road vehicles will bring about further reductions in emissions of carbon dioxide and pollutants from the transport sector.

14.109 In the absence of radical changes in the approach to transport policy, if past trends were to continue unchecked, passenger and freight transport by all modes could be expected to grow by about 20% a decade between now and 2020. This is faster than the growth seen in the 1970s, but less than the growth during the 1980s. If the shares of the different modes remained as at present, the outcome would be roughly equivalent to the low forecast of road traffic made by DOT in 1989. The effect of achieving our challenging targets for increasing the proportions of traffic carried by environmentally less damaging modes, in this scenario of 20% growth a decade, would be to limit the growth in road traffic to 10% between 1992 and 2000 and a total of 35% between 1992 and 2020. Growth in road traffic on this scale would be sufficiently large to conflict with the objectives of stemming losses of land and improving the quality of life and make our objectives for air quality and carbon dioxide emissions much harder to achieve.

14.110 A sustainable transport policy can not, in our view, accommodate growth of more than about 10% a decade in overall demand for transport. At this rate of growth there are two possible forms that in theory such a policy could take. The one we are advocating, is a balanced combination of pricing and measures to promote alternatives to private road transport. We have also recommended major changes in institutions and procedures, and we believe changes on these fronts are essential to ensure that our full set of proposed targets will be achievable. The alternative would be to raise prices (for fuel and/or road use) more sharply than we have recommended, and rely entirely on that to hold down road traffic levels. This would be open to strong objections on grounds of equity. It is also likely to be less effective in achieving environmental objectives, especially in urban areas.

14.111 The combined effect of the measures we have recommended would be sufficient, we estimate, to achieve our proposed targets F1 and F2 for reducing carbon dioxide emissions. The estimates are given in appendix D.

14.112 We discussed in the chapter on economic aspects of transport (7.77–7.83) the effects on the wider economy of the increase we have recommended in the price of fuel; and also considered the general case for using public expenditure to increase the role of public transport (7.60–7.71). The total amount of public expenditure on transport will always be a matter for the government of the day to determine in the light of the state of the economy and other competing demands, but the emphasis of expenditure programmes ought to be redirected towards creating an environmentally sustainable transport system and away from the present unsustainable spiral of road building and increased traffic.

14.113 We have recommended a sharp reduction in construction of major roads (12.65) and in reconstruction of bridges to carry heavier lorries (10.60). The resources thus made available would amount to about £1.5 billion a year at 1994/95 prices. We believe these resources should be used instead for investment to enhance the quality and convenience of public transport, especially through removing barriers to achieving our proposed targets for transferring movements of people and goods to environmentally less damaging modes of transport. There will also be a need for a number of new light-rail systems in urban and suburban areas. Some part of the resources should go to improving facilities for cyclists and pedestrians.

14.114 Because of the large backlog in public transport investment and the scale of the improvements required, the resources made available through cuts in road construction are likely to be insufficient for the task. Our preliminary estimate is that the additional resources required over the next ten years could rise to £1 billion a year at 1994/95 prices. Total investment would therefore be about £2.5 billion a year, nearly double the present level.

14.115 We are aware of the government's efforts to obtain private sector funding for major capital projects, including rail projects such as the Channel Tunnel Rail Link. If these efforts are successful, the scale of public expenditure needed will be less. However, the carrying forward of a much increased programme of public transport projects should not be dependent on the success in obtaining private finance for any part of it.

14.116 We have also identified additional requirements for revenue expenditure. There is a pressing need on environmental grounds to prevent any further decline in public transport use. Achieving that may require temporary increases in public revenue contributions, maintained in real terms until public transport provides a more attractive and viable alternative to much private road use and other policies are in place which will achieve our proposed targets for modal shares. We envisage that additional expenditure during the transitional period would be not more than £2 billion a year, divided roughly equally between heavy rail and other operators. Increased public expenditure will also be needed on a more modest scale for improved enforcement of speed limits and heavy goods vehicle regulations, and to provide incentives for the adoption of alternative modes of freight transport and gas-powered heavy vehicles in urban areas.

14.117 To balance these additional requirements for expenditure, there will be a considerable increase in revenue from fuel duty. By 2000 we estimate that, after allowing for the reduced use of fuel, the additional revenue could amount to about £6.5 billion a year, an increase of 50%. Although the macroeconomic effects of the increases in fuel duty would be for the government of the day to determine (7.77–7.78), and we are not proposing that the revenue from them be hypothecated for transport purposes, we note that this additional revenue would comfortably exceed the additional requirements for public expenditure identified in this report. The timing of the additional revenue will also broadly match the phasing of the substantial injection of resources into public transport over a limited period which we have suggested will mainly be necessary.

14.118 In this comparison no allowance has been made for any additional income that may become available from road pricing. If road pricing is introduced on a significant scale, it is likely that income in the early years will be largely or entirely offset by the costs of installation. When net income becomes available from road pricing schemes in urban areas, we have recommended this should be used by the local authority for the area to meet the costs of implementing its sustainable transport policy. We also envisage local authorities using for this purpose what could be a substantial increase in the revenue from parking charges as one element in policies of traffic restraint. If charges are introduced on motorways, we have recommended that the revenue should be used, not to finance additional road building, but to meet the costs of introducing a sustainable transport policy, including improvements in public transport.

14.119 The reduction in the size of the road programme will obviously have an adverse effect on demand for the industries concerned with road building. However there will be alternative work in providing improved infrastructure for public transport, such as new light-rail systems or upgraded railway lines. Maintenance of the road network, which accounts for a substantial part of expenditure, will continue on broadly the same scale.

14.120 The present system of transport is the consequence of policy choices over a period of a century or more. Their cumulative effect has been to transform almost beyond recognition the ways in which land is settled, the ways in which people work and travel to work, and the ways in which families live. We consider this present transport system is not sustainable, because it imposes environmental costs which are so great as to compromise the choices, and the freedom, of future generations.

14.121 There is now an opportunity to take the difficult policy decisions which will lead, in the course of the coming decades, to a sustainable system of transport, mobility and access; this opportunity has been the principal subject of our report. Twenty years ago, in the period of intense concern over the environmental costs of transport in the early 1970s, policy makers failed to take these difficult decisions. It is a truly daunting prospect that, if the present opportunity were also to be lost, our successors would again be debating the increasing effects of road transport on the environment in the 2010s, following twenty more years of environmental deterioration.

14.122 We recognise that the policies we recommend will impose costs on the present generation. These policies will amount, in effect, to a decision by the present generation to take on the burden of resolving problems which were created in part by earlier generations; to reject the option of passing these problems on, in an aggravated form, to future generations.

14.123 We recognise, too, that the policies we propose could impose different burdens on different groups within the present generation. In choosing where to live and where to work, people have had expectations about the cost and availability of mobility; their choices are constrained by many different factors, and are often costly to reverse. The people who now depend on travel by private car will have made such choices on the basis of expectations which turned out to underestimate the costs of travel. Increases in costs will pose relatively heavy burdens for people who live or work in places where there is at present no satisfactory or affordable alternative to travel by private car; these places now include most rural areas and many suburban locations as well.

14.124 We have recommended that the increased cost of mobility should be imposed on the use rather than on the ownership of cars, in part because we do not consider it equitable to erect high barriers against car ownership. Increased charges for car use could likewise bear heavily on low-income households. In contrast to other taxes or charges imposed on consumption expenditure however, increases in fuel duty would not be significantly regressive, in the sense of imposing greater costs on the poor as a group than on the rich, because far fewer of the households with the lowest incomes have use of a car.[1]

14.125 One way of countering inequality of burdens would be to give low-income and rural groups direct compensation for the increased cost of private road transport through tax concessions or through increased transfer payments. That approach has been advocated by the German Council of Environmental Advisors. Our emphasis is on finding a sustainable transport policy which will maintain people's access to goods, services and participation in public life, but on the basis of lower mobility. In order to maintain such access, we believe that policies for bringing about major improvements in public transport, funded by increased public expenditure, are an essential component of any equitable and acceptable strategy for reducing private road travel by increasing its cost.

14.126 Improvements in public transport within a sustainable transport policy will bring benefits to low-income households, most of whom are entirely dependent on public transport. There will also be benefits to some rural areas in which there has hitherto been no alternative to travel by private car. Even for groups which do not benefit from the change, the outcome will be a moderately increased cost, phased in over a 10-year period, and a moderate reduction in the flexibility and convenience that could have been achieved by car travel in the absence of restraint. That must be regarded as an acceptable price to pay in the interests of future generations.

14.127 In the medium and long term, patterns of settlement, employment and consumption will change, partly as a consequence of changes in relative prices. In the short term, we consider it important to provide people with increased choice, between access by expensive private cars and by public, environmentally less damaging means of transport. The costs of the policies we recommend, over the next decades of transition to a sustainable system of transport and access, are not large in relation to the seriousness of the environmental problems. But as costs of resolving problems inherited from earlier generations, bringing benefits which will be experienced by generations in the future, they should properly be borne by the community as a whole, out of public expenditure. Society is not like a temporary 'partnership agreement in a trade of pepper and coffee', as Edmund Burke wrote in a famous passage: it is 'a partnership not only between those who are living, but between those who are living, those who are dead, and those who are to be born'.[2]

ALL OF WHICH WE HUMBLY SUBMIT FOR
YOUR MAJESTY'S GRACIOUS CONSIDERATION.

John Houghton (Chairman)
Selborne
Geoffrey Allen
Barbara Clayton
Henry Charnock
Henry Fell
John Lawton
Richard Macrory
J Gareth Morris
Donald Reeve
Emma Rothschild
Aubrey Silberston

David Lewis — Secretary

David Aspinwall — Assistant Secretaries
David Benson
Ann Booth
Julian Jones

REFERENCES

Chapter 1

1. The relevant passages are paragraphs 39 (emissions from road vehicles), 40 (emissions from aircraft) and 70–71 (transport noise), and chapter V (Global effects of atmospheric pollution).
2. The relevant passages are paragraphs 48–52 (air pollution from motor vehicles) and 150–154 (noise).
3. The Road Research Laboratory forecast illustrated in figure 6 of the Commission's First Report was for an increase of 87.5% in the number of vehicles between 1970 and 1990. The actual increase over that period was 83%. The number of vehicles recorded in 1990 differed from what had been forecast because the basis for statistics on vehicle numbers was changed in 1978 (information supplied by Department of Transport).
4. World Commission on Environment and Development (1987) (Chair: Gro Harlem Brundtland). *Our common future.* Oxford University Press; Department of the Environment (1993). *UK Strategy for Sustainable Development: consultation paper.* July 1993.
5. Page 169 in *Sustainable development: the UK strategy.* Cm 2426. HMSO. January 1994.
6. Page 169 in *Sustainable development: the UK Strategy.*

Chapter 2

1. Definitions based on those in TSGB (1993) and appendix A in NTS (1989/91); definitions of 'gross mass movement' and 'net mass movement' based on Peake, S. and Hope, C. (1993). *Measuring transport as mass movement: results for the UK, 1952–1992.* University of Cambridge Research Papers in Management Studies 1993–1994, No. 9.
2. Table 2.1 in Department of Transport (1993). *National Travel Survey 1989/91.* HMSO (cited as NTS, 1989/91).
3. Table 2.1 of NTS (1989/91) shows that the number of journeys per person per year rose by 32% between 1965 and 1989/91 and average journey length rose by 31% over the same period.
4. Figure uses data from table 9.1 in Department of Transport (1993). *Transport Statistics of Great Britain 1993.* HMSO (cited as TSGB, 1993), and information for 1993 supplied by Department of Transport Statistics Directorate.
5. Table 2A in NTS (1989/91).
6. According to table 2A in NTS (1989/91), 74% of distance travelled was by car and a further 5% by van but the classification in that case was by main mode. Table 1.1 in TSGB (1993) gives a figure of 86% for cars and vans.
7. Table 1.1 in TSGB (1993). See also page 16 of Department of Transport (1992). *Transport Statistics of Great Britain 1992.* HMSO.
8. Table 7.2b in TSGB (1993).
9. Table 1.1 in TSGB (1993).
10. Derived from tables 8.1 and 8.4 in NTS (1989/91).
11. Table 8.1 in NTS (1989/91).
12. Figure uses 1965 and 1975/76 data from Department of Transport (1979). *National Travel Survey 1975/76 Report.* HMSO (cited as NTS, 1975/76); 1985/86 data from Department of Transport (1988). *National Travel Survey 1985/86 Report. Part 1 — An analysis of personal travel.* HMSO (cited as NTS, 1985/86); and 1989/91 data from NTS (1989/91).
13. 1965 data from NTS (1975/76); 1989/91 data from NTS (1989/91).
14. Table 9.3 in TSGB (1993); 1993 information supplied by Department of Transport Statistics Directorate.

15. Information supplied by Department of Transport Statistics Directorate.

16. Table 1.12 in TSGB (1993).

17. Figure uses data from table 9.3 in TSGB (1993).

18. Figure uses GDP data from Central Statistical Office (1993). *Economic Trends Annual Supplement.* 1993 Edition. HMSO; domestic passenger traffic data from table 9.1 in TSGB (1993); freight traffic data from table 9.3 in TSGB (1993). The data on international traffic by UK airlines used to indicate international air travel by UK residents were supplied by Department of Transport Statistics Directorate for the period 1963–92.

19. Table 1.12b in TSGB (1993).

20. Table 1.12b in TSGB (1993).

21. International passenger-kilometres travelled on UK airlines to and from UK airports has been used as a proxy for international air travel by UK residents; statistics are not available before 1963 or before 1974 for non-scheduled flights.

22. Chapter 5.2 in Central Statistical Office (1994). *Social Trends 24.* HMSO (cited as Social Trends 24).

23. Table 9.5 in TSGB (1993).

24. Tables 2.4 and 3B in NTS (1989/91); see also NTS (1985/86).

25. Table 4.7 in Central Statistical Office (1993). *United Kingdom National Accounts.* 1993 Edition. HMSO. For this purpose the calculations include taxes.

26. Evidence from HM Treasury.

27. A figure of 4–8% in industrialised countries is given, without explanation, in the chapter by M. Linster on 'Background facts and figures' (page 11) in European Conference of Ministers of Transport (1990). *Transport policy and the environment: ECMT ministerial session (prepared in co-operation with OECD).* OECD, Paris.

28. Figure uses data from table 1.16 in TSGB (1993) and information supplied by Department of Transport Statistics Directorate.

29. Robinson, G. (1992). NSA: the networked society for the arts, manufacturers and commerce. *RSA Journal*, April 1992, 311.

30. Womack, J.P., Jones, D.T. and Roos, D. (1990). *The machine that changed the world.* New York.

31. Table 4.3 in Central Statistical Office (1994). *Economic Trends Annual Supplement.* 1994 Edition. HMSO.

32. Bray, J. (1992). *The rush for roads: a road programme for economic recovery*? A report by Movement Transport Consultancy for ALARM UK and Transport 2000.

33. Evidence from HM Treasury.

34. Hart, T. (1993). Transport investment and disadvantaged regions: UK and European policies since the 1950s. *Urban Studies*, **30**, No. 2, March 1993, 417–436.

35. McKinnon, A. and Woodburn, A. (1994). The consolidation of retail deliveries: its effect on CO_2 emissions. *Transport Policy*, **1**, No. 2, 125–136.

36. According to senior executives of foreign-owned companies: KPMG Peat Marwick (1994). *A survey of foreign-owned companies in the UK.*

37. Table 1.3 in Central Statistical Office (1992). *Social Trends 22.* HMSO (cited as Social Trends 22).

38. Table 1.5 in Social Trends 22.

39. Chart 3.3 in NTS (1989/91).

40. Calculated from table 3.4 in NTS (1989/91).

41. Table 4.8 in Social Trends 24.

42. Table 4.8 in Social Trends 24.

43. Table 4.12 in Social Trends 24.

44. Table 4.12 in Social Trends 24.
45. Chart 4.11 in Social Trends 24.
46. Table 8.2 in NTS (1989/91).
47. Table 3.2 in Social Trends 24.
48. Hillman, M., Adams, J. and Whitelegg, J. (1990). *One false move . . . a study of children's independent mobility.* Policy Studies Institute.
49. Table 8.1 in NTS (1989/91). Between 1981 and 1991 the population of the UK under 16 fell from 12.6 million to 11.7 million; but it is projected to rise again to 12.5 million in 2001 before declining to 11.4 million in 2031 (table 1.4 in Social Trends 24).
50. *Sustainable development: the UK strategy.* Cm 2426. January 1994. HMSO.
51. Except where otherwise indicated, data in 2.21–2.22 are drawn from: BDP Planning and Oxford Institute of Retail Management (1992). *The effects of out of town retail development: a literature review for the Department of the Environment.* HMSO; or from Institute of Grocery Distribution. *Food Retailing 1991*, quoted in National Consumer Council (1992). *Your Food, Whose Choice?* HMSO.
52. Oral evidence from Marks and Spencer.
53. Transport and Environment Studies (TEST) (1989). *Trouble in Store?* A report by TEST for the Anglo-German Foundation.
54. Headicar, P. and Bixby, B. (1992). *Concrete and tyres - local development effects of major roads: a case study of the M40.* Report commissioned by the Council for the Protection of Rural England with support from the Countryside Commission.
55. Table 8.1 in NTS (1989/91).
56. RSL survey for the supermarket group Gateway, as reported in *The Independent* and *The Times*, 10 February 1994.
57. Department of the Environment (1993). *Merry Hill impact study.* HMSO.
58. Their heart rate when walking on level ground at 3 mph exceeded 70% of the calculated maximum; reported in Health Education Authority and Sports Council (1992). *Allied Dunbar National Fitness Survey: a summary of the major findings and messages.*
59. Department of Health (1991). *Health of the Nation White Paper. A consultative document for health in England.* Cm 1523. HMSO.
60. There is a useful summary of the literature on this aspect in Tengström, E. (1992). *The use of the automobile: its implications for man, society and the environment.* TFB Report 1992: 14. Swedish Transport Research Board, Stockholm.
61. Figure uses GDP data from table 8.1 in TSGB (1993) and car ownership rates calculated from tables 8.1 and 8.3 in TSGB (1993).
62. Department of Transport (1989). *National Road Traffic Forecasts (Great Britain) 1989.* HMSO (cited as NRTF, 1989).
63. Figure uses actual data from table 9.4 in TSGB (1993) and forecast data from NRTF (1989) recalculated using the revised 1988 traffic figures reported in TSGB (1993).
64. Department of Transport, APM Division (1985). *National Road Traffic Forecasts (Great Britain).* December 1984.
65. Other factors were a higher elasticity for road freight in relation to GDP and a reduction in the assumed rates of increase in fuel prices.
66. Table 4.7 in TSGB (1993).
67. Evidence from HM Treasury.

Chapter 3

1. Table 2.1 in Department of Transport (1993). *Transport Statistics of Great Britain 1993.* HMSO (cited as TSGB, 1993).
2. Table A9 in Department of Trade and Industry (1993). *Digest of United Kingdom Energy Statistics 1993.* HMSO.

3. Tables 2.2b, 2.6b, 2.9b, 2.12b and 2.19b in Department of the Environment (1994). *Digest of Environmental Protection and Water Statistics.* No. 16. HMSO (cited as DEPWS, 1994).
4. *Marine Engineers Review*, January 1993.
5. Tables 2.6b, 2.9b, 2.12b and 2.19b in DEPWS (1994).
6. Based on estimates for 1990 emissions in Earth Resources Research (1993). *Atmospheric emissions from road transport in the United Kingdom.* Report commissioned by the Royal Commission.
7. Definitions from Quality of Urban Air Review Group (1992). *Urban Air Quality in the United Kingdom.* First Report. HMSO (cited as First QUARG Report).
8. First QUARG Report.
9. Left fuming at the kerbside. *The Times*, 12 March 1994, p21.
10. Figure 3-II redrawn from the First QUARG Report; information on House of Commons Select Committee on Transport hearing from *ENDS Report*, No. 234, July 1994, 28–29.
11. Department of Health (1991). *Ozone.* First Report of the Advisory Group on the Medical Aspects of Air Pollution Episodes. HMSO (cited as MAAPE Report on ozone).
12. Table 11.8 in DEPWS (1994).
13. *Rethinking the ozone problem in urban and regional air pollution.* National Research Council, Committee on Tropospheric Ozone Formation and Measurement. National Academy Press, Washington DC. 1991.
14. Association of London Authorities, London Boroughs Association, South East Institute of Public Health (1994). *Air quality in London.* The First Report of the London Air Quality Network. February 1994.
15. Quality of Urban Air Review Group (1993). *Diesel vehicle emissions and urban air quality.* Second Report. HMSO (cited as Second QUARG Report).
16. Second QUARG Report.
17. Evidence from Dr Simon Wolff.
18. MAAPE Report on ozone.
19. Department of Health (1992). *Sulphur dioxide, acid aerosols and particulates.* Second Report of the Advisory Group on the Medical Aspects of Air Pollution Episodes. HMSO (cited as MAAPE Report on sulphur dioxide, etc).
20. Department of Health (1993). *Oxides of nitrogen.* Third Report of the Advisory Group on the Medical Aspects of Air Pollution Episodes. HMSO (cited as MAAPE Report on oxides of nitrogen).
21. Second QUARG Report.
22. Department of the Environment (1994). *Benzene.* Report by the Expert Panel on Air Quality Standards. HMSO (cited as EPAQS Report on benzene).
23. Parliamentary Office of Science and Technology (1994). *Breathing in our cities — Urban air pollution and respiratory health.* Parliamentary Office of Science and Technology. February 1994.
24. MAAPE Report on oxides of nitrogen.
25. Devalia, J.L., Campbell, A.M. *et al.* (1993). Effects of nitrogen dioxide on synthesis of inflammatory cytokines expressed by human bronchial epithelial cells, *in vitro*. *Amer. J. Respir. Cell Mol. Biol.*, **9**, 271–278.
26. MAAPE Report on oxides of nitrogen.
27. MAAPE Report on ozone.
28. Wallace, L.A. (1989). Major sources of benzene exposure. *Environmental Health Perspectives*, **82**, 165–169.
29. *Umweltgutachten 1994.* Der Rat von Sachverständigen für Umweltfragen. Verlag Metzler Poeschel, Stuttgart. February 1994. Paragraphs 697 and 699.
30. California Environmental Protection Agency, Office of Environmental Health Hazard Assessment (1994). *Health risk assessment for diesel exhaust: preliminary draft.*

31. *Trends in applications to the Family Fund.* Social Care Research Findings 53. Joseph Rowntree Foundation, York. June 1994.

32. Rusznak, C., Devalia, J.L. and Davies, R.J. (1994). The impact of pollution on allergic disease. *Allergy*, **49**, 21–27.

33. Åberg, N. (1989). Asthma and allergic rhinitis in Swedish conscripts. *Clinical and Experimental Allergy*, **19**, 59–63; and information from Dr J. Emberlin, Pollen Research Unit, University of North London.

34. An internal report produced by a working group for the Medical Research Council Committee on Toxic Hazards in the Environment and Workplace (unpublished).

35. Greer, J.R., Abbey, D.A. and Burchette, R.J. (1993). Asthma related to occupational and ambient air pollutants in nonsmokers. *Journal of Occupational Medicine*, **35**, 909–915.

36. Seaton, A., Godden, D.J. and Brown, K. (1994). Increase in asthma: a more toxic environment or a more susceptible population? *Thorax*, **49**, 171–174.

37. Seaton *et al.* (1994).

38. von Mutius, E., Martinez, F.D., Fritzsch, C., Nicolai, T., Reitmeir, P. and Thiemann, H-H. (1994). Skin test reactivity and number of siblings. *British Medical Journal*, **308**, 692–694.

39. Strachan, D.P. (1989). Hayfever, hygiene and household size. *British Medical Journal*, **299**, 1259–1260.

40. Holgate, S. (1994). What's causing the worldwide rise in asthma? *MRC News*, No. 63, Summer 1994, 20–23.

41. Magnusson, C.G. (1986). Maternal smoking influences cord serum IgE and IgD levels and increases the risk for subsequent infant allergy. *J. Allergy Clin. Immunol.*, **78**, 898–904.

42. Unpublished MRC Report.

43. Department of the Environment (1994). *Ozone.* Report by the Expert Panel on Air Quality Standards. HMSO (cited as EPAQS Report on ozone).

44. Ayres, J.G., Noah, N.D. and Fleming, D.M. (1993). Incidence of episodes of acute asthma and acute bronchitis in general practice 1976–87. *British Journal of General Practice*, **43**, 361–364.

45. Britton, J. (1992). Asthma's changing prevalence. *British Medical Journal*, **304**, 857–858.

46. Ashmore, M.R., Saffron, L., Gunner, G., Loth, K., Panagiotopoulou, E., Riley, G. and Fitzharris, P. Effects of air pollution on the severity of hayfever symptoms in London (unpublished).

47. Ishizaki, T., Koizumi, K., Ikemori, R., Ishiyama, Y. and Kushibiki, E. (1987). Studies of prevalence of Japanese Cedar pollinosis among the residents in a densely cultivated area. *Annals of Allergy*, **58**, 265–270.

48. MAAPE Report on oxides of nitrogen.

49. Dockery, W.D., Pope, C.A. *et al.* (1993). An association between air pollution and mortality in six US cities. *The New England Journal of Medicine*, **329**, 1753–1759.

50. Bown, W. (1994). Dying from too much dust. *New Scientist*, **141**, No. 1916, 12 March 1994, 12–13.

51. Schwartz, J. and Marcus, A. (1993). Mortality and air pollution in London: a time series analysis. *American Journal of Epidemiology*, **131**, 185–194.

52. MAAPE Report on sulphur dioxide, etc.

53. Dispute over health risks from diesel emissions. *ENDS Report*, No. 230, March 1994, 6–7.

54. Bower, J. *et al.* (1994). *Air pollution in the UK: 1992/93.* Report LR 1000 (AP). Air Monitoring Group, Warren Spring Laboratory, Stevenage.

55. National Society for Clean Air. *Local Authority Air Pollution Monitoring Survey 1993.*

56. Information obtained on Commission visits.

57. Institute for Environment and Health (1994). *Air pollution and health: understanding the uncertainties.* Report of a workshop held in Leicester, 2–4 February 1994. Medical Research Council.

58. First QUARG Report.
59. Directive on air quality limit values and guide values for sulphur dioxide and suspended particulates. Directive 80/779/EEC. *Official Journal*, **L229**, 30/8/80.
60. Directive on air quality standards for nitrogen dioxide. Directive 85/203/EEC. *Official Journal*, **L087**, 27/3/85.
61. Air Quality Standards Regulations 1989 made under the Control of Pollution Act 1974. Similar Regulations came into force in Northern Ireland in 1990.
62. Directive on a limit value for lead in the air. Directive 82/884/EEC. *Official Journal*, **L378**, 31/12/82.
63. Table 2.16 in DEPWS (1994).
64. Directive on air pollution by ozone. Directive 92/72/EEC. *Official Journal*, **L297**, 13/10/92.
65. *The Ozone Monitoring and Information Regulations 1994.* Statutory Instruments 1994 No. 440. HMSO.
66. World Health Organization Regional Office for Europe (1987). *Air Quality Guidelines for Europe*. WHO Regional Publications, European Series No. 23. Copenhagen.
67. Department of the Environment, Scottish Office Environment Department, Welsh Office, Department of the Environment for Northern Ireland (1994). *Improving air quality: a discussion paper on air quality standards and management.* March 1994.
68. Tables 2.8, 2.10, 2.22, 11.4 and 11.8 in DEPWS (1994).
69. Paragraph 173 in *Environmental Health Action Plan for Europe: Fifth draft.* Second European Conference on Environment and Health (Helsinki, 20–22 June 1994).
70. Tables 2.20 and 2.21 in DEPWS (1994).
71. EPAQS Report on benzene.
72. Derwent, R.G. (1988). A better way to control pollution. *Nature*, **331**, No. 6157, 575–578.
73. Ashmore, M.R. (1984). Effects of ozone on vegetation in the UK. Pages 92–104, in: *Ozone*. P. Grennfelt (Editor).
74. Ashmore, M.R., Bell, J.N.B., Dalpra, C. and Runeckles, V.C. (1980). Visible injury of crop species by ozone in the United Kingdom. *Environmental Pollution*, **21**, 209–215.
75. Photochemical Oxidants Review Group (1994). *Ozone in the United Kingdom 1993.* Third Report. HMSO (cited as Third PORG Report).
76. Terrestrial Effects Review Group (1993). *Air pollution and tree health in the United Kingdom.* Second Report. HMSO.
77. Paragraph 2.32 and plate 5 in the Commission's Fifteenth Report.
78. Mansfield, T., Hamilton, R., Ellis, B. and Newby, P. (1991). Diesel particulate emissions and the implications for the soiling of buildings. *The Environmentalist*, **II**, 243–254.
79. EPAQS Report on ozone.
80. Critical Loads Advisory Group (1994). *Critical loads of acidity in the United Kingdom.* Report prepared at the request of the Department of the Environment. HMSO. February 1994 (cited as CLAG, 1994).
81. CLAG (1994).
82. Third PORG Report.
83. Table compiled using data from Houghton, J.T., Jenkins, G.J. and Ephraums, J.J. (Editors) (1990). *Climate change: The IPCC Scientific Assessment.* Cambridge University Press; Houghton, J.T., Callander, B.A. and Varney, S.K. (Editors) (1992). *Climate Change 1992. The Supplementary Report to the IPCC Scientific Assessment.* Cambridge University Press; Houghton, J. (1994). *Global Warming. The Complete Briefing.* Lion Publishing, Oxford; and the Third PORG Report.
84. Figure reproduced from Houghton et al (1990).
85. Tables 1.2a and 1.2b in DEPWS (1994).
86. The reference scenario was derived from Department of Trade and Industry (1992). *Energy related carbon emissions in possible future scenarios for the United Kingdom.* Energy

Paper 59. HMSO, and was used in Department of the Environment (1992). *Climate change: our national programme for CO_2 emissions — a discussion document.* The reference scenario assumes that GDP will grow on average by 2.25% a year and that fossil fuel prices in 2000 will be slightly below the 1990 level in real terms.

87. Table 2.3 in TSGB (1993).
88. Tables 1.3 and 1.4 in DEPWS (1994).
89. Dasch, J.M. (1992). Nitrous oxide emissions from vehicles. *Journal of the Air and Waste Management Association*, **42**, No. 1, 63–67.
90. Staehelin *et al.* (1993). In *Radiative forcing of climate. Draft report of the Scientific Assessment Working Group to the Intergovernmental Panel on Climate Change 1994.*
91. Hameed and Dignon (1991). In *Radiative forcing of climate. Draft report of the Scientific Assessment Working Group to the Intergovernmental Panel on Climate Change 1994.*
92. Montreal Protocol on Substances that Deplete the Ozone Layer (UNEP, 1987) and amendments to this Protocol adopted in London (1990) and Copenhagen (1992).
93. Houghton *et al.* (1990).
94. Houghton *et al.* (1992).
95. World Energy Council (1993). *Energy for tomorrow's world — the realities, the real options and the agenda for achievement.* Report of a Commission.
96. Houghton *et al.* (1992).
97. Houghton (1994).
98. Houghton (1994).
99. The Report of the Intergovernmental Negotiating Committee for a Framework Convention on Climate Change on the Work of the Second part of its Fifth Session, held at New York from 30 April to 9 May 1992.
100. *Climate Change — The UK Programme. The United Kingdom's Report under the Framework Convention on Climate Change.* Cm 2427. HMSO. January 1994.
101. Intergovernmental Panel on Climate Change (1994, forthcoming). *Radiative forcing of climate. Draft report of the Scientific Assessment Working Group.*

Chapter 4

1. Oil pollution of the sea was the subject of the Commission's Eighth Report.
2. Shell takes high ground over petrol station pollution. *ENDS Report*, No. 219, April 1993, 4–5.
3. See for example Oslo and Paris Commissions (1993). *North Sea Quality Status Report.*
4. GESAMP (1990). *The state of the marine environment.* World Meteorological Organization, Geneva (GESAMP Reports and Studies No. 39; UNEP Regional Seas Reports and Studies No. 115).
5. See for example chapters VII (Oil discharges resulting from tanker accidents) and VIII (Deliberate and accidental discharges from vessels during operations).
6. See paragraphs 6.41–6.47 (Marine transport). The report also dealt with litter from road and rail travellers (paragraphs 6.11–6.14).
7. Inquiry into the prevention of pollution from merchant shipping (Chairman: Lord Donaldson of Lymington) (1994). *Safer ships, cleaner seas.* Cm 2560. HMSO.
8. Sargent, J.W. and Fothergill, L.C. (1993). *The noise climate around our homes.* BRE Information Paper IP21/93; and tables 6.7 and 6.8 in Department of the Environment (1993). *Digest of Environmental Protection and Water Statistics.* No. 15. HMSO.
9. European Conference of Ministers of Transport (1990). *Transport policy and the environment: ECMT Ministerial Session (prepared in co-operation with OECD).* OECD, Paris.

10. Morton-Williams, J. *et al.* (1978). *Road traffic and the environment.* Social and Community Planning Research (SCPR Report P390).
11. Table 6.9 in Department of the Environment (1994). *Digest of Environmental Protection and Water Statistics.* No. 16. HMSO (cited as DEPWS, 1994).
12. Evidence from the Institution of Environmental Health Officers.
13. Evidence from Environmental Health Officers for London Boroughs.
14. Watts, G.R. (1990). *Traffic induced vibrations in buildings.* Research Report No. 246. Transport and Road Research Laboratory, Crowthorne.
15. Department of the Environment (1992). *The UK Environment.* HMSO, citing Taylor, S.M. and Wilkins, P.A. (1987). *Health effects from transportation noise reference book.* Butterworth.
16. Organisation for Economic Co-operation and Development (1991). *The state of the environment.* OECD, Paris. p157–158.
17. Jones, D.M. (1990). Noise, stress and human behaviour. *Environmental Health, Journal of the Institution of Environmental Health Officers*, **98**, No. 8, August 1990, 206–208. Cited in evidence from the Institution of Environmental Health Officers.
18. Department of Transport (1991). *Railway Noise and the Insulation of Dwellings (The Mitchell Report).* HMSO.
19. Knipschild, P., Meijer, H. and Salle, H. (1981). Aircraft noise and birth weight. *Int. Arch. Occup. Environ. Health*, **48**, 131–136.
20. Suggested by Dr Maria Heckl of Keele University in a paper to the 1993 meeting of the British Association for the Advancement of Science (reported in 'Sleeper key for cutting noise.' *New Civil Engineer*, 9 September 1993).
21. Findings of the Transport Research Laboratory reported in *Building and Civil Engineering Research Focus*, **13**, April 1993.
22. Department of the Environment (1990). *Report of the Noise Review Working Party 1990* (Chairman: W.J.S. Batho). HMSO.
23. Department of Transport (1993). *Draft Noise Insulation Regulations for New Railways and Other Guided Systems.* Circulated by DOT International Railways Division for comment in October 1993.
24. Evidence from Kent County Council.
25. World Health Organization (1980). *Noise.* Environmental Health Criteria Document No. 12. World Health Organization, Geneva.
26. Commission of the European Communities (1992). *Towards Sustainability: a European programme of policy and action in relation to the environment and sustainable development.* COM(92)23. Brussels.
27. *Second Transport Structure Plan. Part D. Government Decision. Transport in a sustainable society.* Report 20.922, No. 16. Second Chamber of the States-General, The Netherlands. Session 1989–1990.
28. United Kingdom Environmental Law Association, Noise Working Party (1992). Comments on issues raised by the EC Fifth Environment Action Programme, "Towards sustainability".
29. Report of Noise Review Working Party 1990.
30. Appleyard, D. and Lintell, M. (1972). The environmental quality of city streets: The residents' viewpoint. *American Institution of Planners Journal*, March 1972, 84–101.
31. Department of Transport (1994). *Trunk Roads in England 1994 Review.* HMSO.
32. Evidence from Transportation Planning Associates.
33. Table 7.22 in Central Statistical Office (1994). *Social Trends 24.* HMSO.
34. Information supplied by Department of Transport Statistics Directorate.
35. Table 5.31 in Department of Transport (1993). *Transport Statistics of Great Britain 1993.* HMSO (cited as TSGB, 1993).

36. Figure drawn using data from table 9.7 in TSGB (1993) and information supplied by Department of Transport Statistics Directorate.
37. Figure drawn using data from table 8.7 in TSGB (1993).
38. Table 48 in Department of Transport (1991). *Road Accidents Great Britain 1990. The Casualty Report.* HMSO; and table 48 in Department of Transport (1994). *Road Accidents Great Britain 1993. The Casualty Report.* HMSO.
39. Roberts, I.G. (1993). International trends in pedestrian injury mortality. *Archives of Disease in Childhood*, **68**, 190–192 (published by BMJ Publishing Group).
40. Table 30 in 1990 Road Accidents Casualty Report and table 30 in 1993 Road Accidents Casualty Report.
41. Casualty information supplied by Department of Transport Statistics Directorate. Information on journeys by foot made in London supplied by the London Research Centre.
42. Cyclists Touring Club (1991). *Bikes not fumes*; and Transport 2000 (1992). *Travelling cleaner: Dutch and British transport policy compared.*
43. Information included in evidence from the National Rivers Authority.
44. Road: Details of regulations are given in the main text. Rail: Goods carried by rail are subject to the Carriage of Dangerous Goods by Rail Regulations and the same classification, packaging and labelling requirements that apply to road transport. Sea: Handling etc of dangerous goods in harbours and on inland waterways are controlled by the Dangerous Substances in Harbour Areas Regulations 1987. Regulations applying to UK ports are contained in the Merchant Shipping (Dangerous Goods and Marine Pollutants) Regulations 1990. Air: ICAO requirements for the carriage of dangerous goods are implemented by the Air Navigation (Dangerous Goods) Regulations 1985.
45. Evidence from the National Rivers Authority.
46. TEST (1991). *Wrong side of the tracks? Impacts of road and rail transport on the environment: a basis for discussion.* Researched and published by TEST.
47. Oral evidence from London Regional Transport.
48. Department of the Environment (1993). *Countryside Survey 1990.* A report prepared by the Institutes of Terrestrial and Freshwater Ecology of the Natural Environment Research Council. According to this survey the built-up area of Britain is 6.9% of the total area.
49. *House of Commons Hansard*, 16 May 1991.
50. Table 9.8 in TSGB (1993).
51. Table 9.12 in TSGB (1993).
52. *Conserving England's marine heritage — a strategy, Important areas for marine wildlife around England,* and *Strategy for the sustainable use of England's estuaries,* all published English Nature reports.
53. Evidence from Joint Nature Conservation Committee, referring to a report by Davidson for the Nature Conservancy Council.
54. Adams, J.G.U. (1989). *London's green spaces — what are they worth?* Report for Friends of the Earth and the London Wildlife Trust.
55. Standing Advisory Committee on Trunk Road Assessment (1992). *Assessing the environmental impact of road schemes.* HMSO. See paragraphs 14.11–14.15.
56. Evidence from English Nature.
57. Bunce, R.G.H. and Heal, O.W. (1990). *Ecological consequences of land use change.* Report of the Institute of Terrestrial Ecology. Natural Environment Research Council, Swindon.
58. Munguira, M.L. and Thomas, J.A. (1992). The use of road verges by butterfly and burnet populations, and the effect of roads on adult dispersal and mortality. *J. appl. Ecology*, **29**, No. 2, 316–329.
59. Evidence from Scottish Natural Heritage.
60. Mader, H-J. (1984). Animal habitat isolation by roads and agricultural fields. *Biological Conservation*, **29**, 81–86.

61. Webb, N.R. (1989). Studies on the invertebrate fauna of fragmented heathland in Dorset, UK, and the implications for conservation. *Biological Conservation*, **47**, 153–165.
62. Webb, N.R. and Vermaat, A. (1990). Changes in vegetational diversity on remnant heathland fragments. *Biological Conservation*, **53**, 253–264.
63. Table 8.11 in DEPWS (1994) and information supplied by European Wildlife Division, Department of the Environment.
64. English Nature (1993). *Second Report, 1st April 1992 - 31st March 1993.* Peterborough.
65. Based on evidence from the Council for the Protection of Rural England, English Nature and the World Wide Fund for Nature and information supplied by the Department of Transport.
66. Information supplied by Department of Transport.
67. The South-East Wildlife Trusts' Transport Campaign (1990). *Head-on collision: road building and wildlife in South-East England.*
68. Information supplied by Department of Transport.
69. Council Directive 92/43/EEC of 21st May 1992 on the conservation of natural habitats and of wild fauna and flora. *Official Journal*, **L206**, Volume 35, 22/7/92. To be implemented in the UK in *The Conservation (Natural Habitats, etc) Regulations 1994.*
70. *This Common Inheritance: Britain's Environmental Strategy.* Cm 1200. HMSO. 1990.
71. Joint memorandum of evidence from the Departments of Environment and Transport.
72. Evidence from the Council for the Protection of Rural England.
73. Department of the Environment (1992). *Coastal superquarries to supply South-East England aggregate requirements.* Report by Arup Economics and Planning, London. July 1992 (unpublished).
74. Table 1.10 in TRANSNET (1990). *Energy, transport and the environment.* One estimate is that these stages account for 44% of the "pollution" produced by an average German car (Umwelt und Prognose Institut, Heidelberg (1993). *ÖKO-Bilanzen von Fahrzeugen.* 2 erweit-erte Auflage, September 1993).
75. TEST (1991) and TRANSNET (1990).
76. *Aggregates: The way ahead.* Report of the 1976 Verney Committee to the Department of the Environment.
77. 1992 Report of Arup Economics and Planning to the Department of the Environment.
78. Scottish Office (1994). *Land for mineral working.* National Planning Policy Guidance 4.
79. Stokes, G., Goodwin, P. and Kenny, F. (1992). *Trends in Transport and the Countryside — the Countryside Commission and transport policy in England.* Countryside Commission Publications, Manchester.
80. Department of the Environment (1994). *Guidelines for aggregate provision in England.* Minerals Planning Guidance Note MPG 6. HMSO.
81. Going Dutch on recycling. *Surveyor*, **178**, No. 5209, September 1992, p17.
82. Automotive Consortium on Recycling and Disposal (1992). *End of life vehicle disposal: preliminary concept*; Holman, C. (1991). *Recycling of cars in the UK.* A report for Greenpeace UK.
83. *Closing the Loop — The Car Recycling Challenge.* Euromotor Reports. 1993.
84. Five-year tyre fire burns. *Surveyor*, 27 January 1994.
85. See in particular figure 5-I and paragraphs 3.28 and 10.11 in the Commission's Seventeenth Report.
86. Dutch plant shows the way on tyre recycling. *ENDS Report*, No. 227, December 1993, p14.
87. Information supplied by Department of Trade and Industry.

Chapter 5

1. Information supplied by Department of Transport Statistics Directorate on international scheduled and non-scheduled services by UK airlines.

2. Table 7.1b in Department of Transport (1993). *Transport Statistics of Great Britain 1993.* HMSO (cited as TSGB, 1993).
3. Figure drawn using data supplied by Department of Transport Statistics Directorate.
4. Snape, D.M. and Metcalfe, M.T. *Emissions from aircraft: standards and potential for improvement.* Rolls Royce (internal report).
5. Information supplied by Department of Transport Statistics Directorate.
6. International Civil Aviation Organization (1993). *ICAO Statistical Yearbook. Civil Aviation Statistics of the World 1992* (cited as ICAO, 1993).
7. ICAO (1993).
8. International Civil Aviation Organization (no date). *Development and ICAO.*
9. Archer, L.J. (1993). *Aircraft emissions and the environment:* CO_x, SO_x, HO_x *and* NO_x. OIES Paper on Energy and the Environment No. EV17. Oxford Institute for Energy Studies (cited as Archer, 1993).
10. Science and Technology Agency (Japan) (1992). *The Fifth Technology Forecast Survey — Future Technology in Japan; "German Delphi Report" on future scientific and technological developments.* Produced by the Fraunhofer Institute for Systems Technology and Innovation Research. August 1993. Bonn.
11. Table 7.1a in TSGB (1993). 1993 data in figure 5-II supplied by Department of Transport Statistics Directorate.
12. Information supplied by Department of Transport Statistics Directorate.
13. Department of Transport (1994). *Air traffic forecasts for the United Kingdom 1994.* HMSO.
14. RUCATSE (1993). *Runway Capacity to Serve the South East.* A report by the Department of Transport's Working Group on Runway Capacity to Serve the South East (RUCATSE). HMSO.
15. Council for the Protection of Rural England (1994). *A runway too far?* Response by CPRE to the Department of Transport Working Group's Report on Runway Capacity to Serve the South East (RUCATSE).
16. London Planning Advisory Committee (1994). *Advice on Strategic Planning Guidance for London.* LPAC, Romford.
17. Information supplied by the Department of Trade and Industry.
18. SRI International (no date). *European congestion — the way out.* Summary of report for IATA entitled *A European planning strategy for air traffic to the year 2010.*
19. Grübler, A. and Nakicenovic, N. (1991). *Evolution of transport systems: past and future.* Report RR-91–8, International Institute for Applied Systems Analysis, Laxenburg, Austria.
20. Information supplied by British Airways, June 1994.
21. In *Flight International* magazine.
22. Information supplied by Department of Transport.
23. Information supplied by Department of Transport.
24. Information supplied by Department of Transport Civil Aviation Division.
25. ICAO (no date). *The Convention on International Civil Aviation: annexes 1 to 18* (cited as ICAO Convention).
26. Directive limiting aircraft noise. Directive 92/14/EEC. *Official Journal*, **L076**, 23/3/92.
27. Boeing cuts airlines sales estimate. *Flight International*, 1–7 June 1994, p8.
28. British Airways plc (1993). *1993 Annual Environmental Report.* p5.
29. Grieb, H. and Simon, B. (1990). Pollutant emissions of existing and future engines for commercial aircraft, p43–83 in: *Air traffic and the Environment — Background Tendencies and Potential Global Atmospheric Effects.* Proceedings of a DLR International Colloquium, on 15/16 November 1990, edited by U. Schumann. Lecture Notes in Engineering, No. 60. Springer-Verlag, Berlin (cited as Grieb and Simon, 1990). (Estimates were made of emissions of NO_x, HCs and CO during the LTO cycle and cruise phase of a General Electric CF6–50C engine, with a 30 minute period of cruising at a speed of Mach 0.85 and an altitude of 10.7 km).

30. Grieb and Simon (1990).

31. Report by Royal Netherlands Meteorological Institute to the Ministry of Transport (June 1994).

32. Archer (1993).

33. Barrett, M. (1991). *Aircraft pollution — Environmental impacts and future solutions.* World Wide Fund for Nature Research Paper.

34. World Meteorological Office (1991). *Scientific Assessment of Ozone Depletion, 1991.* WMO Global Ozone Research and Monitoring Project. Report No. 25.

35. Bekki, S. and Pyle, J. (1993). Potential impact of combined NO_x and SO_2 emissions from future high speed civil transport aircraft on stratospheric aerosols and ozone. *Geophysical Research Letters*, **20**, 723.

36. Stratospheric Ozone Review Group (1993). *Stratospheric Ozone 1993.* Prepared at the request of the Department of the Environment. HMSO.

37. Intergovernmental Panel on Climate Change (1994, forthcoming). *Radiative Forcing of Climate. Draft report of the Scientific Assessment Working Group.*

38. Grassl, H. (1990). Possible climatic effects of contrails and additional water vapour, p124–137 in: *Air traffic and the Environment — Background Tendencies and Potential Global Atmospheric Effects.* Proceedings of a DLR International Colloquium, on 15/16 November 1990, edited by U. Schumann. Lecture Notes in Engineering, No. 60. Springer-Verlag, Berlin.

39. Information supplied by the Department of Trade and Industry.

40. Department of Transport. Paper outlining the MOZAIC project, dated 15 June 1992.

41. ICAO Convention.

42. Schipper, L. and Meyers, S. (1992). *Energy efficiency and human activity: past trends, future prospects.* Report sponsored by the Stockholm Environment Institute. Cambridge University Press.

43. Balashov, B. and Smith, A. (1992). ICAO analyses trends in fuel consumption by world's airlines. *ICAO Journal*, August 1992, 18–21.

44. Reported in: Rochat, P. (1993). Key environmental issues range from aircraft noise to the "greenhouse" effect. *ICAO Journal*, July/August 1993, **48**, No. 6, 31–34.

45. Grieb and Simon (1990).

46. Figures quoted in International Air Transport Association (1991). *Air Transport and the Environment.* IATA, Geneva.

47. Archer (1993).

48. Rochat (1993).

49. Grieb and Simon (1990).

50. Barrett, M. (1994). *Pollution Control Strategies for Aircraft. A discussion paper.* World Wide Fund for Nature International, Gland, Switzerland.

51. Patel, T. (1993). Green designs on supersonic flight. *New Scientist*, No. 1886, 14 August 1993, 35–37.

52. Airfields Environment Trust and Airfields Environment Federation (1993). *Aviation, environmental regulation and the future: a world-wide perspective.* Proceedings of a conference held on 3 June 1993 at the Royal Society of Arts, London.

Chapter 6

1. Grübler, A. and Nakicenovic, N. (1991). *Evolution of transport systems: past and future.* Report RR-91–8, International Institute for Applied Systems Analysis, Laxenburg, Austria (cited as Grübler and Nakicenovic, 1991).

2. It has been calculated that the decline has averaged 0.9% a year over the last 180 years in the USA: pages 83–84 and figure 40 in Grübler and Nakicenovic (1991).

3. Peake, S. (1994). *Transport in transition: lessons from the history of energy.* Earthscan/Royal Institute of International Affairs (cited as Peake, 1994).
4. Peake (1994).
5. Schipper, L. and Meyers, S. (1992). *Energy efficiency and human activity: past trends, future prospects.* Report sponsored by the Stockholm Environment Institute. Cambridge University Press. p111.
6. British Road Federation (1988). *The way ahead: the cost of congestion.*
7. Mallet, V. (1993). The city now standing still. *Financial Times*, 27/28 November 1993.
8. Manning, I. (1984). *Beyond walking distance — the gains from speed in Australian urban travel.* Australian National University Urban Research Unit, Canberra.
9. Mogridge, M. (1985). *Jam yesterday, jam today, and jam tomorrow? (or how to improve traffic speeds in Central London).* Lecture given at University College London, 17 October 1985.
10. Plowden, S. (1972). *Towns against traffic.*
11. Garreau, J. (1991). *Edge City. Life on the new frontier.* Doubleday, New York.
12. Smeed, R.J. (1968). Theoretical studies and operational research on traffic congestion. *Journal of Transport Economics and Policy*, **2.**
13. Information supplied by Department of Transport Statistics Directorate.
14. Department of Transport (1989). *Roads for prosperity.* Cm 693. HMSO (cited as DOT, 1989). A more detailed and complete description of the road programme was published in February 1990 (Department of Transport, Road Programme and Resources Division (1990). *Trunk roads, England: into the 1990s.* HMSO; cited as DOT, 1990), at which time a further 20 schemes were added. England contains 76% of the surfaced roads in Great Britain and 88% of the motorways (Department of Transport (1993). *Transport Statistics of Great Britain 1993.* HMSO; cited as TSGB, 1993).
15. Paragraph 51 in DOT (1989).
16. Paragraph 4.5 in DOT (1989).
17. Paragraphs 7.2–7.3 in DOT (1990).
18. Joint memorandum of evidence from Departments of Environment and Transport.
19. Department of Transport (1994). *Trunk roads in England 1994 Review.* HMSO (cited as DOT, 1994).
20. Paragraph 4.2 in DOT (1994).
21. National Audit Office (1993). *Progress on the Department of Transport's Motorway Widening Programme.* Report by the Comptroller and Auditor General. HMSO.
22. Welsh Office (1994). *Roads in Wales: 1994 Review.* HMSO.
23. The Scottish Office (1992). *Roads, Traffic and Safety.* March 1992. HMSO.
24. HM Treasury (1991). *The Government Expenditure Plans 1991–92 to 1993–94, Northern Ireland.* Cm 1517. February 1991. HMSO.
25. Paragraph 1.6 in DOT (1990).
26. Paragraph 24 in DOT (1989).
27. McLaren, D.P. and Higman, R. (1993). *The environmental implications of congestion on the interurban network in the UK.* Paper given to 21st PTRC Summer Annual Meeting 1993, Seminar F (not included in conference proceedings).
28. Centre for Economics and Business Research (1994). *Roads and jobs: the economic impact of different levels of expenditure on the roads programme.* Study carried out for the British Road Federation.
29. Table A1 in CPRE (1993). *Driven to dig — Road-building and aggregates demand.* Council for the Protection of Rural England.
30. The findings of this study have been reported in Williams, R. and Birch, N. (1994). The longer term implications of National Road Traffic Forecasts and international network plans for local roads policy: the case of Oxfordshire. *Transport Policy*, **1**, No. 2, 95–99.

31. Goodwin, P., Hallett, S., Kenny, F. and Stokes, G. (1991). *Transport: the new realism* (Rees Jeffreys Road Fund report 624). Oxford University Transport Studies Unit.
32. Counties slam 'self-defeating' roads plan. *Surveyor*, 25 August 1993, p1. See also Association of County Councils (1991). *Towards a sustainable transport policy*. November 1991.
33. Grübler and Nakicenovic (1991).
34. Harris, R. (1993). *Monitoring Department of Transport traffic forecasts*. Paper given to 21st PTRC Summer Annual Meeting 1993, Seminar D.
35. See paragraph 26.2 in *Sustainable development: the UK strategy*. Cm 2426. HMSO. January 1994.
36. Page 169 and paragraph 26.17 in UK Sustainable Development Strategy.
37. Girardet, H. (1992). *The Gaia atlas of cities: new directions for sustainable urban living*. Gaia Books Limited. p174.
38. Wagar, W.W. (1992). *The next three futures: paradigms of things to come*. Adamantine Press (Adamantine Studies on the 21st Century).
39. Engwicht, D. (1992). *Towards an eco-city: calming the traffic*. Envirobook, Sydney. The quotations are from pages 122, 116 and 117 respectively.
40. Hughes, P. (1994). *Personal transport and the greenhouse effect. A strategy for sustainability*. Earthscan.
41. Martin, D.J. and Shock, R.A.W. (1989). *Energy Use and Energy Efficiency in UK Transport up to the Year 2010*. Department of Energy. Energy Efficiency Series 10. HMSO. September 1989. On commuting journeys, cars are assumed to have on average 1.3 occupants and on other urban journeys 2.0 occupants. The calculation was weighted to reflect the market shares of different types of cars at the time the study was carried out; it is unlikely that a calculation carried out today would give a markedly different result.
42. Information obtained during Commission visit to Zurich.
43. Derived from table 2.3 in TSGB (1993). Air transport is excluded from this calculation because most of the fuel purchased in the UK for air transport is used on international flights which are not included in UK transport statistics.
44. Net mass movement has been calculated in the way described in box 2A. The calculation of shares excludes air (for the reason explained in the previous reference) and pipelines (which are not included in the statistics of energy consumption by transport).
45. The primary source for these data is table 8.6 in TSGB (1993).
46. The primary source for these data is table 8.5 in TSGB (1993).
47. Information obtained during Commission visit to Japan.
48. Information obtained during Commission visit to Hong Kong.
49. TSGB (1993).
50. City Research Project (1993). *Meeting the transport needs of the City*. Corporation of London and London Business School.
51. Suchorzewski, W. (1993). *The effects of rising car travel in Warsaw*. Paper given to 'Travel in the City — Making it Sustainable', International Conference, Düsseldorf, 7–9 June 1993.
52. Information on Singapore obtained from reports of visits by a DOT Minister and officials; information on Norwegian cities obtained during Commission visit.
53. Information obtained during Commission visit to USA.
54. Information obtained during Commission visit to Hong Kong.
55. The Drive programme is concerned with the development of road transport telematics throughout the Community; the Adept project involves the development of electronic road pricing technology: See Harrop, P. (1993). *Charging for Road Use Worldwide; a Financial Times Management Report*.
56. Miller, L. *High-speed maglev systems*. Paper to Royal Society/Royal Academy of Engineering discussion meeting on "Passenger transport after 2000 AD", 23–24 June 1993; and O'Neill, B. (1993). Beating the bullet train. *New Scientist*, 2 October 1993, p36.

57. Pages 28–31 and figures 16 and 17 in Grübler and Nakicenovic (1991).
58. Paragraph 26.37 in UK Sustainable Development Strategy.
59. Henley Centre for Forecasting predicted in 1988 that by 1995 over 10 million full-time workers in the UK would be teleworking.
60. Kitamura, R., Nilles, J.M., Conroy, P. and Fleming, D.M. (1991). Telecommuting as a transportation planning measure: Initial results of California Pilot Scheme. *Transportation Research Record,* **1298**, 98–104; and Hamer, R.N., Kroes, E.P. and Van Ooststroom, H. (1992). *Teleworking in the Netherlands: and evaluation of changes in travel behaviour — further results.* Paper given to 20th PTRC Summer Annual Meeting 1992, Seminar C.
61. *Second Transport Structure Plan. Part D. Government Decision. Transport in a sustainable society.* Report 20.922, No. 16. Second Chamber of the States-General, The Netherlands. Session 1989–1990.
62. Information obtained during Commission visit to France; Peapod, Inc. (1992). *Booming business keeps Peapod open later.* News Release. Evanston (USA).
63. Information supplied by GPT plc.
64. World Fuel Cell Council (1991). *A practical way to a clean energy future.* September 1991.
65. Nordström, L. *The intelligent highway and the passenger.* Paper to Royal Society/Royal Academy of Engineering discussion meeting on "Passenger transport after 2000 AD", 23–24 June 1993.

Chapter 7

1. 1993/94 data given in Table 1.21 in Department of Transport (1993). *Transport Statistics Great Britain 1993.* HMSO (cited as TSGB, 1993).
2. Plowden, W. (1971). *The motor car and politics 1896–1970.* The Bodley Head. See especially chapters 3, 9 and 14.
3. Newbery, D. (1990). Pricing congestion: economic principles relevant to pricing roads. *Oxford Review of Economic Policy*, **6**, No. 2.
4. Fowkes, A.S., Nash, C.A. and Tweddle, G. (1990). *The Track and External Costs of Road Transport.* Working paper 312 for the Institute for Transport Studies, University of Leeds. September 1990.
5. Department of Transport Green Paper. *Paying for Better Motorways,* May 1993. p39.
6. Evidence of Professor David Newbery to House of Commons Select Committee on Transport. *Fifth Report: Charging for the Use of Motorways.* July 1994.
7. Organisation for Economic Co-operation and Development (1991). *The State of the Environment.* OECD, Paris. p207.
8. See for example Pearce, D. *et al.* (1994). *Blueprint 3 - measuring sustainable development.* Earthscan. This puts the external costs of congestion caused by road transport in the UK in 1991 at £13.5 billion.
9. Union International des Chemins de Fer, Paris (1994). *External effects of transport.* Interim Report, March 1994. p24. Discussions of the economics of congestion and congestion charging generally adopt the same position.
10. See Pearce *et al.* (1994). The estimated range quoted excludes the congestion costs which Pearce and his collaborators include.
11. Automobile Association (1992). *The balance sheet of motoring.* The figures used are based on a statistical average of expenditure as recorded in the 1991 Family Expenditure Survey. Costs are updated where possible and adjusted to refer to a single car.
12. Kageson, P. (1993). *Getting the prices right: A European scheme for making transport pay its full costs.* A report for the European Federation for Transport and the Environment. May 1993. p36.
13. Rothengatter, W. (1992). *Externalities of Transport.* A paper for 'Transport Economics, A state of the Art'. August 1992.

14. Information obtained during Commission visit to Hong Kong.
15. Information obtained during Commission visit to Japan.
16. Box compiled from the following sources: National Society for Clean Air (1991). *Company cars: a pre-budget briefing*; Department of Transport (1993). *National Travel Survey 1989/91.* HMSO; 1992 survey commissioned by Hertz; Kompfner, P. *et al.* (1991). *Company travel assistance in the London area.* TRL Report RR326; Potter, S. (1992). Still crude, still arbitrary and still unfair. *Town and Country Planning*, October 1992; Inland Revenue (1992). *Company cars: reform of income tax treatment.* Consultative Document; survey of executive cars in *Financial Times*, 30 June 1994, especially articles by J. Griffiths (The Revenue gnashes its teeth) and S. Daneshku (Worse than its bite).
17. See for example, Button, K. (1994). *Transport, the Environment and Economic Policy*, section 7.4, for a discussion of the combination of congestion and environmental charges in road pricing.
18. *Umweltgutachten 1994.* Der Rat von Sachverständigen für Umweltfragen. Verlag Metzler-Poeschel, Stuttgart.
19. Symons, E., Proops, J. and Gay (1994). Carbon taxes, consumer demand and carbon dioxide emissions: a simulation analysis for the UK. *Fiscal Studies*, **15**, No. 2, 19–43.
20. In April 1993, duty represented 66% of the price of leaded petrol before the addition of VAT, 61% of the price of unleaded petrol and 60% of the price of DERV (table 2.4 in TSGB (1993)).
21. United Nations (1992). *World Economic Survey 1992. Current Trends and Policies in the World Economy.* New York.
22. Evidence from Shell International Petroleum Co. Ltd.
23. *The Economist*, September 1993.
24. Evidence from the Department of Transport, March 1993.
25. Council Directive 92/82/EEC of 19 October 1992 on the approximation of the rates of excise duties on mineral oils. *Official Journal,* **L316**, 31/10/92.
26. Kageson (1993).
27. In *Umweltgutachten 1994*, paragraph 862.
28. Data supplied by Shell, reported in *The Times*, 1 December 1993.
29. Evidence from the Department of Transport.
30. Goodwin, P. (1992). A review of new demand elasticities with special reference to the short and long run effects of price changes. *Journal of Transport Economics and Policy*, May 1992, **26**, No. 2, 155–169.
31. Oum, T.H., Waters, W.G. and Jong-Say Yong (1992). Concepts of price elasticities of transport demand and recent empirical estimates. *Journal of Transport Economics and Policy*, May 1992, **26**, No. 2, 139–154.
32. Weizsacher, E.U. and Jesinghaus, J. (1992). *Ecological Tax Reform. A Policy Proposal for Sustainable Development.* Zed Books, London and Atlantic Highlands, New Jersey.
33. British Railways Board (1994). *Annual report and accounts 1993–94.* p65.
34. Information obtained during Commission visit to France.
35. In 1992, the rate was 1.35p per litre, compared with 22.9p per litre for DERV; evidence from British Rail; and table 2.4 in TSGB (1993).
36. Alamdari, F.E. and Brewer, D. (1994). Taxation policy for aircraft emissions. *Transport Policy*, **1**, No. 3, 149–159.
37. Information supplied by the Department of Transport.
38. Derived from estimates of the fuel cost and total freight charge for an average vehicle load for a median-sized (17 tonnes) heavy goods vehicle travelling 40,000 miles a year (the average for all HGVs); based on cost tables published by *Motor Transport,* April 1992.

Chapter 8

1. 10th Amendment Directive to 'Directive relating to measures to be taken against air pollution by gases from positive ignition engines of motor vehicles (70/220/EEC)'.

Directive 91/441/EEC. *Official Journal*, **L242**, 30/8/91; Directive 94/12/EEC of the European Parliament and the Council of 23 March 1994 relating to measures to be taken against air pollution by emissions from motor vehicles and amending Directive 70/220/EEC. *Official Journal,* **L100**, 19/4/94; EC Ministers agree on emissions norms and tax rules for cars, but risk fight with Parliament. *Environment Watch: Western Europe,* 17 December 1993, 12–13; US Environmental Protection Agency, Office of Mobile Sources (1993). *Exhaust emission certification standard for light-duty vehicles and trucks.* Revised 18 May 1993; California Air Resources Board (1992). *Mobile Source Emission Standards Summary,* as of June 30 1992.

2. Holman, C., Wade, J. and Fergusson, M. (1993). *Future emissions from cars 1990 to 2025: the importance of the cold start emissions penalty.* Earth Resources Research. A Report for World Wide Fund for Nature UK.
3. INRETS/TUV-Rheinland/TRRL (1992). *Actual Car Use and Operating Conditions as Emission Parameters; Derived Urban Driving Cycles.*
4. Measurements by Warren Spring Laboratory in Holman, C. (1992). *Second-stage emission limits for cars.* A Report for World Wide Fund for Nature UK. p7. Cited as Holman (1992).
5. Information from Science and Technology Section, British Embassy, Bonn.
6. California Air Resources Board, emission standards summary, 1992.
7. Page 3 in Holman (1992).
8. Bradley, M.J. (1993). *US Motor Vehicle Emissions Update.* July 1993; and information supplied by the California Air Resources Board.
9. Page 24 and table 9 in *Low-Emission Vehicle Program Costs.* Discussion paper prepared by the California Air Resources Board for the 'Low-emission Vehicle and Zero Emission Vehicle Workshop', March 25 1994.
10. Information supplied by the Japanese Environment Agency.
11. Amendment Directive to 'Directive relating to measures to be taken against air pollution by gases from positive ignition engines of motor vehicles (70/220/EEC)'. Directive 93/59/EEC. *Official Journal*, **L186**, 28/7/93.
12. Amendment Directive to 'Directive on emission of gaseous pollutants from diesel lorries and buses (88/77/EEC)'. Directive 91/542/EEC. *Official Journal*, **L295**, 25/10/91.
13. California Air Resources Board, emission standards summary, 1992.
14. Sowman, C. (1993). Going for green. *Commercial Motor*, 23–29 September 1993, 34–36.
15. Science and Technology Agency (Japan) (1992). *The Fifth Technology Forecast Survey — Future Technology in Japan*; *"German Delphi Report" on future scientific and technological developments.* Produced by the Fraunhofer Institute for Systems Technology and Innovation Research. August 1993. Bonn.
16. Directive relating to the sulphur content of certain liquid fuels. Directive 93/12/EEC. *Official Journal*, **L74**, 27/3/93; and *The Motor Fuel (Composition and Content) Regulations 1994.* Statutory Instruments 1994 No. 2295. HMSO.
17. US Auto/Oil Air Quality Research Programme, Phase 1 Final Report. May 1993.
18. Information supplied by Isuzu Motor Company.
19. Information supplied by the Department of the Environment, based on a survey carried out by Associated Octel Ltd in 1993.
20. Directive on the approximation of member states legislation on lead content of petrol, and the introduction of lead-free petrol. *Official Journal*, **L096**, 3/4/85.
21. ARCO (1993). *Choices: finding the motor fuel of the future.* Los Angeles.
22. Pages 127–128 in Centre for Exploitation of Science and Technology (1993). *The UK Environmental Foresight Project. Volume 2. Road transport and the environment. The future agenda in the UK.* HMSO (cited as CEST, 1993).
23. Cut-price supermarket petrol causes higher emissions. *ENDS Report*, No. 225, October 1993, p8.
24. California Air Resources Board, emission standards summary, 1992.
25. New aquifer threat from petrol concerns water companies. *ENDS Report*, No. 225, October 1993, 7–8.

26. 7th Amendment to Directive 67/548/EEC. Directive 92/32/EEC. *Official Journal*, **L154**, Volume 35, 30/4/92.
27. Council Regulation (EEC) No. 793/93 of 23 March 1993 on the evaluation and control of the risks of existing substances. *Official Journal*, **L84/1**, 5/4/93.
28. See results of modelling in Earth Resources Research (1993). *Atmospheric emissions from road transport in the United Kingdom.* Report commissioned by the Royal Commission (cited as ERR, 1993).
29. Broome, D. (1992). *Gasoline, diesel and the alternatives. An expanded review.* A paper written for the Royal Commission.
30. Preface by the Secretary of State for Transport to Department of Transport, Vehicle Standards and Engineering Division (1992). *Emissions from heavy duty diesel engined vehicles: the government's response to the Fifteenth Report of the Royal Commission on Environmental Pollution.* HMSO (cited as DOT, VSE Division, 1992).
31. ERR (1993).
32. The traffic levels implied by the National Road Traffic Forecasts depend on the baseline taken. ERR applied the National Road Traffic Forecasts growth rates to actual traffic levels in 1990–1992.
33. Paragraph 1 in DOT, VSE Division (1992).
34. Paragraphs 44–46 in DOT, VSE Division (1992).
35. Figure reproduced from Earth Resources Research (1989). *Atmospheric emissions from the use of transport in the United Kingdom. Volume 1: The estimation of current and future emissions.* M. Fergusson, C. Holman and M. Barrett for the World Wide Fund for Nature.
36. Britton, D. and Scheffer, J. (1992). *A comparative analysis of the total CO_2 emissions associated with the production and use of gasoline and automotive gasoil.* C389/442. Institute of Mechanical Engineers, 1992.
37. Quality of Urban Air Review Group (1993). *Diesel vehicle emissions and urban air quality.* Second Report. HMSO.
38. European Environment Bureau (1992). *Car fuel consumption.* C. Holman. p5 (cited as EEB, 1992).
39. Nieuwenhuis, P., Cope, P. and Armstrong, J. (1992). *The Green Car Guide.* Green Print.
40. Figure supplied by the Department of Transport.
41. INRETS/TUV-Rheinland/TRRL (1992).
42. Estimate of 10% from Naysmith, A. (1989). Energy conservation for car drivers, in *Energy efficiency in land transport.* Proceedings of a seminar held on 16–18 May 1988. Publication No. CD-NA-122484-EN-C. Commission of the European Communities, Luxembourg, quoted in EEB (1992). The DOE discussion document *Our National Programme for CO_2 Emissions* gives a figure of 15%.
43. For example, see surveys reported in *Local Transport Today*, 19 August 1993, p2; and *Daily Telegraph*, 6 September 1993, p13.
44. European Conference of Ministers of Transport (1993). *Reducing transport's contribution to global warming.* Conclusions of ECMT International Seminar, 30/9–1/10 1992. Document No. CEMT/CM(93).13. p7.
45. International Energy Agency/Organisation for Economic Co-operation and Development (1991). *Low consumption/low emission automobile.* Proceedings of an expert panel. OECD/IEA, Paris.
46. Evidence from the World Wide Fund for Nature.
47. White House (1993). *Historic partnership forged with auto makers.* Press Release. 29 September 1993; and information obtained during Commission visit to the White House Office of Science and Technology Policy ETC.
48 Figure updated by International Energy Studies, Lawrence Berkeley Laboratory from L. Schipper *et al.* (1993). Mind the gap (*Energy Policy,* December 1993), and Schipper and Meyers (1992). *Energy Efficiency and Human Activity* (Cambridge University Press).
49. Evidence from the Science Policy Research Unit, Sussex University.

50. ERR (1993); Wootton, J. and Poulton, M. (1993). *Reducing carbon dioxide emissions from passenger cars to 1990 levels: a discussion paper.* Transport Research Laboratory, September 1993.
51. Wade, J., Holman, C. and Fergusson, M. (1993). *Current and projected global warming potential of passenger cars in the UK.* A Report for World Wide Fund for Nature UK. June 1993.
52. Amendments to 'Directive on the approximation of laws in the Member States relating to the permissible sound level and the exhaust system of motor vehicles (70/157/EEC)'. Directive 84/424/EEC (*Official Journal*, **L238**, 6/9/84) and Directive 92/97/EEC (*Official Journal*, **L371**, 9/12/92).
53. These limits are derived from Amendment to Directive 78/1015/EEC (Directive relating to the permissible sound level and exhaust system of motorcycles). Directive 87/56/EEC. *Official Journal*, **L024**, 27/1/87.
54. COM(83)706. *Official Journal*, **C354**, 29/12/83.
55. ERR (1993).
56. Retiring old cars: Programs to save gasoline and reduce emissions. *US OTA Report Brief*, July 1992. US Congress Office of Technology Assessment.
57. Dutch go flat out for Euro 2 option. *Freight*, February 1994, p29.
58. Pages 98–100 in CEST (1993).
59. Roadside checks to sniff out dirty cars. *Daily Telegraph*, 13 June 1994; Car pollution trap short on legal firepower. *The Independent*, 21 June 1994.
60. *OTIS NEWS*, No. 2.03, 14 February 1992 quoting the *VDI Nachrichten* of 22 November 1991.
61. *Carweek*, Issue No. 17, 15 December 1993, 14–15.
62. Figure drawn from material appearing in 'Traditional propulsion systems', the introductory statement of the rapporteur to Panel 1, Dr F.F. Pischinger, in Commission of the European Communities (1993). *Written Proceedings of the European Symposium 'Auto Emissions 2000'.* Office for Official Publications of the European Communities, Luxembourg.
63. A random sample of 1,200 cars by the Environment Ministry in 1992 cited in 'Greener on the other side?' *Freight*, May 1994.
64. Pages 8–9 in Holman (1992).
65. EC Council Directive (amendment to 77/143/EEC) on the approximation of the laws of the member states relating to roadworthiness tests for motor vehicles and their trailers (exhaust emissions). Directive 92/55/EEC. *Official Journal*, **L225**, 10/8/92.
66. *Daily Telegraph*, 25 February 1994, p11.
67. *Emissions monitoring — the league table.* RAC Press Release EMLT/7494. July 1994.
68. For example, see studies reported in *New Scientist*, **187**, No. 1862, 27 February 1993, p18.
69. Page 135 in CEST (1993).
70. Brussels and the bioethanol boondoggle. *ENDS Report*, No. 212, September 1992, 22–25.
71. Information given to the Commission at the Los Angeles Motor Show, January 1994.
72. Information obtained during Commission visit to the White House Office of Science and Technology Policy.

Chapter 9

1. Owens, S. and Cope, D. (1992). *Land use planning policy and climate change*. Report to the Department of the Environment by UK Centre for Economic and Environmental Development, Cambridge. HMSO.
2. Hall, P. (1992). *Urban and regional planning.* 3rd Edition. Routledge.
3. Kellet, J.R. (1969). *The impact of railways on Victorian cities*. Studies in Social History 31. Routledge and Kegan Paul.
4. SACTRA report on generation 'could be delayed'. *Local Transport Today*, 12 May 1994, p1.

5. Paragraph 2.2.25 in Department of the Environment, Department of Transport (1993). *Reducing transport emissions through planning.* ECOTEC Research and Consulting Ltd in association with Transportation Planning Associates. HMSO (cited as ECOTEC, 1993).
6. Headicar, P. and Bixby, B. (1992). *Concrete and tyres — local development effects of major roads: a case study of the M40.* Report commissioned by the Council for the Protection of Rural England with support from the Countryside Commission.
7. Webster, F.V., Bly, P.H. and Paulley, N.J. (1988). *Urban Land-use and Transport Interaction — Policies and Models.* Report of the International Study Group on Land Use/Transport Interaction. Avebury, Aldershot.
8. KFR International Research (1993). *The M25 Development Report 1993.* Knight, Frank and Rutley.
9. The M25 Development Report 1993.
10. Croydon. *Property Week*, 9 June 1994, 36–37.
11. Goodwin, P. (1992). A review of new demand elasticities with special reference to short and long run effects of price changes. *Journal of Transport Economics and Policy*, May 1992, **26**, No. 2, 155–169.
12. Webster *et al.* (1988).
13. Dix, M.C. and Goodwin, P.B. (1981). *Understanding the effects of changing petrol prices: a synthesis of conflicting econometric and psychometric evidence.* PTRC Summer Annual Meeting 1981.
14. For evidence from existing schemes and from modelling, see Hewitt, P. (1989). *A Cleaner, Faster London: Road Pricing, Transport Policy and the Environment.* Green Paper No. 1. Institute for Public Policy Research.
15. Goldstein, G.S. and Moses, L.N. (1975). Transport controls and the spatial structure of urban areas. *American Economic Review, Papers and Proceedings*, **65**, 289–294. Cited in Button, K. (1991). *Electronic Road Pricing: Experience and Prospects.* Paper presented at the Conference on Economy and Environment in the 1990s. 26–27 August 1991, Neuchatel.
16. Owens, S. (1986). *Energy, Planning and Urban Form.* Pion, London.
17. Owens (1986) contains a review of other studies.
18. Newman, P.W.G. and Kenworthy, J.R. (1989). Gasoline consumption and cities. *Journal of the American Planning Association*, Winter 1989, 24–37.
19. ECOTEC (1993). The data on population densities in private sector developments and British cities were provided respectively by Owens, S. (1991). *Energy-Conscious Planning.* A report commissioned by the Council for the Protection of Rural England; and the Department of the Environment.
20. Goodwin, P.B. (1978). Travel choice and time budgets, in Hensher and Dalvi (Editors). *Determinants of travel choices.* Saxon House.
21. Authors arguing that density acts as an independent factor include: ECOTEC (1993); Keyes, D.L. (1982). Reducing travel and fuel use through urban planning, in Burchell, R.W. and Listoken, D. (Editors). *Energy and land use.* Centre for Urban Policy Research, New Brunswick; and Newman and Kenworthy (1989).
    Authors finding that density does not act as an independent variable include: Gordon, P. and Richardson, H. (1990). Gasoline consumption and cities: a reply. *Journal of the American Planning Association*, **55**, 342–345; Mogridge, M. (1985). Transport, land use and energy interaction. *Urban Studies*, **22**, 481–492; and Rickaby, P.A., Steadman, J.P. and Barrett, M. (1992). Patterns of land use in English towns: implications for energy use and carbon dioxide emissions, in Breheny, M.J. (Editor). *Sustainable development and urban form.* p182–196. Pion, London.
22. Hemmens, G. (1967). Experiments in urban form and structure. *Highway Research Record*, **207**, 32–41; Schneider, J. and Beck, J. (1973). *Reducing the Travel Requirements of the American City: An Investigation of Alternative Urban Spatial Structures.* Research Report 73. US Department of Transportation, Washington DC.

23. ECOTEC (1993); Fels, M.F. and Munson, M.J. (1975). Energy thrift in urban transportation: options for the future, in Williams, R.H. (Editor). *The Energy Conservation Papers: A report to the energy project of the Ford Foundation.* Ballinger, Cambridge, MA; Rickaby *et al.* (1992); Stone, P.A. (1973). *The Structure, Size and Costs of Urban Settlements.* Cambridge; and Webster *et al.* (1988).
24. Owens (1986).
25. Rickaby, P.A. (1987). Six settlement patterns compared. *Environment and Planning B, Planning and design*, **14**, 193–223.
26. Rickaby *et al.* (1992).
27. Banister, D. (1992). Energy use, transport and settlement patterns, in Breheny, M.J. (Editor). *Sustainable development and urban form.* Pion, London.
28. TEST (Transport and Environment Studies) (1991). *Changed travel — better world?*
29. Department of the Environment (1993). *Merry Hill Impact Study.* HMSO.
30. JMP Consultants. *Parking and Public Transport: the Effect on Mode Choice.* Cited in *Parking Review*, June 1993.
31. Shaw, D. (JMP Consultants Ltd) (1992). *Traffic impact study for a retail store.* Paper for the 4th Annual TRICS Conference, September 1992.
32. Merry Hill Impact Study.
33. Gossop, C. and Webb, A. (1993). Getting around: public and private transport, in Blowers, A. (Editor). *Planning for a sustainable environment. A report by the TCPA.* Earthscan.
34. Cervero, R. (1994). Transit-based housing in California: evidence on ridership impacts. *Transport Policy*, **1**, No. 3, 174–183.
35. The policy in question is known as the ABC location policy and is fully described in Amundson, C. (June 1993). *Public Policy and Public Transport,* produced with assistance from Southampton Institute of Higher Education. See also: Ministry of Housing, Physical Planning and Environment (April 1991). *The right business in the right place: Towards a location policy for businesses and services in the interests of accessibility and the environment.* The Hague.
36. Halman, G. (1994). Gummer agenda in wider perspective. *Planning*, **1063**, 8 April 1994, p10; Rhodes, J. and Henley, J. (1994). Guidance is not the end of the road out of town. *Planning*, **1063**, 8 April 1994, 22–23.
37. Earl of Caernarvon, Chairman of SERPLAN, in a letter of 20 January 1993 to the Secretary of State for Transport, quoted in *Motorway mania.* Council for the Protection of Rural England. May 1993.
38. M25 link roads inquiry postponed but minister directs mention in Surrey plan. *Local Transport Today*, 21 July 1994, p1.
39. Council Directive of 27 June 1985 on the assessment of the effects of certain public and private projects on the environment. Directive 85/337/EEC. *Official Journal*, **L175/40**, 5/7/85.
40. The Roads (Scotland) Act 1984 (sections 20A and 55A); the Roads (Northern Ireland) Order 1980 (article 39B).
41. Department of the Environment and Welsh Office (1988). *Environmental Assessment.* Circular 15/88 (DOE), Circular 23/88 (WO). July 1988. HMSO.
42. Department of Transport, The Standing Advisory Committee on Trunk Road Assessment (1992). *Assessing the environmental impact of road schemes.* HMSO (cited as DOT/SACTRA, 1992).
43. Therivel, R. and Heaney, D. *Does the sum of project EIAs equal SEA?: a transport example.* Work in progress.
44. DOT/SACTRA (1992).
45. Treweek, J.R., Thompson, S., Veitch, N. and Japp, C. (1993). Ecological assessment of proposed road development: a review of environmental statements. *Journal of Environmental Planning and Management,* **36**, 295–307.

46. Department of Transport, Scottish Office Industry Department, Welsh Office, Department of the Environment for Northern Ireland (1993). *Design Manual for Roads and Bridges. Volume 11. Environmental Assessment.* HMSO.

47. Paragraph 1.6 in Design Manual 1993.

48. Therivel and Heaney, work in progress.

49. ECOTEC (1993).

Chapter 10

1. Freight Transport Association (1991). *The transport dilemma.* p3.

2. Table 1.12 in Department of Transport (1994). *Transport Statistics of Great Britain 1994.* HMSO (cited as TSGB, 1994).

3. Allen, J. (1993). *Just-in-time transportation and the environment.* Transport Studies Group Working Paper. University of Westminster.

4. Mair, A. (1993). New growth poles? Just-in-time manufacturing and local economic development strategy. *Regional Studies*, **27**, No. 3, 207–221.

5. Nissan's cost-neutral environmental improvements. *ENDS Report*, No. 221, June 1993, p9.

6. Redrawn from figure 2 in Touche Ross Distribution and Logistics Division. *European logistics comparative costs and practice.* Prepared on behalf of the Institute of Logistics and Distribution Management and the European Logistics Association. Figure for 1991/92 from table 3 in Institute of Logistics. *Survey of distribution costs in 1991/92.*

7. Table 6.6 in Department of Transport (1993). *Transport Statistics of Great Britain 1993.* HMSO (cited as TSGB, 1993).

8. British Ports Federation and British Road Federation (1992). *Roads to the ports.*

9. EC Task Force (1990). *1992 — the environmental dimension.* Task Force Report on the Environment and the Internal Market. Economica Verlag, Bonn. Quoted in Arp, H.A. (1991). Transport and the Environment: From technical to structural solutions. *European Environment*, December 1991, 14–19.

10. Table derived from Hopfner *et al.* (1989), cited in Whitelegg, J. (1992). *Traffic congestion: is there a way out.* Leading Edge Press.

11. TEST (1991). *Wrong side of the tracks? Impacts of road and rail transport on the environment: a basis for discussion.* Researched and published by TEST. p146.

12. Teufel, D. (1989). *Die Zukunft des Autoverkehrs.* UPI Bericht 17. Umwelt und Prognose Institut, Heidelberg.

13. Table 1.12 in TSGB (1994).

14. Table 9.3 in TSGB (1993).

15. ECOTEC (1993). *The role of land use planning in encouraging waterborne freight.* Final Report of ECOTEC Research and Consulting Ltd for the Department of Transport. April 1993; and Jonathan Packer and Associates and the Stamford Research Group (1993). *Roads to water research project.* A focusing study for the Department of Transport. Final Report. March 1993.

16. BACMI economist Jerry McLaughlin quoted in *New Civil Engineer*, 8/15 April 1993.

17. Tweddle, G. and Nash, C. (1994). Barging in on the freight debate. *Surveyor*, **181**, No. 5294, 2 June 1994, 12–14.

18. Evidence from Shell UK.

19. Joint memorandum of evidence from the Departments of Environment and Transport; and oral evidence from Parcelforce.

20. Table 9.8 in TSGB (1993).

21. Transport 2000 (1992). *Travelling Cleaner. Dutch and British transport policy compared.* p12.

22. Roberts, J. *et al.* (Editors) (1992). *Travel Sickness. The need for a sustainable transport policy for Britain.* Lawrence and Wishart. p201.

23. Evidence from British Rail.

24. Re-equipping Freightliner. *Modern Railways*, May 1992, 243–245.
25. Back on the rails. *Commercial Motor*, 29 July-4 August 1993, 28–29.
26. Council Directive 91/440/EEC of 29 July 1991 on the development of the Community's railways. *Official Journal*, **L237**, 24/8/91.
27. Paragraph 3.7 in Department of the Environment and Department of Transport (1994). *Transport.* Planning Policy Guidance PPG 13. HMSO.
28. Figure drawn using data from table 4.1 in TSGB (1993).
29. McKinnon, A. and Woodburn, A. (1993). A logistical perspective on the growth of lorry traffic. *Traffic Engineering and Control*, **34**, No. 10, October 1993, 466–471.
30. Martin, D.J. and Shock, R.A.W. (1989). *Energy Use and Energy Efficiency in UK Transport up to the Year 2010.* Department of Energy. Energy Efficiency Series 10. HMSO. September 1989.
31. European Communities Committee (Sub-Committee B). Examination of witness. 24 February 1994.
32. "Rail 21 Cargo Plan", p65–66 in *Second Transport Structure Plan. Part D. Government Decision. Transport in a Sustainable Society.* Report 20.922, No. 16. Second Chamber of the States-General, the Netherlands. Session 1989–1990.

Chapter 11

1. Table 2A in Department of Transport (1993). *National Travel Survey 1989/91.* HMSO (cited as NTS, 1989/91).
2. Huntley, P. and Taylor, J. (1993). *Rural transport problems and needs.* Rural research series, 14. Rural Development Commission, Salisbury.
3. Huntley and Taylor (1993).
4. Nutley, S. (1992). Rural areas: the accessibility problem, in Hoyle, B.S. and Knowles, R.D. (1992). *Modern Transport Geography.*
5. Breheny, M., Gent, T. and Lock, D. (1993). *Alternative development patterns: new settlements.* HMSO.
6. Salveson, P. (1993). *New futures for rural rail: an agenda for action.* Transnet.
7. Surrey County Council. *A transport plan for Surrey.*
8. Lesley, L. (1993). value for money in urban transport public expenditure: the case of light rail. *Public Money and Management,* January-March 1993, 27–33.
9. Organisation for Economic Co-operation and Development (1974). *Streets for People.* OECD, Paris. See also: TEST (1988). *Quality Streets. How traditional urban centres benefit from traffic calming.* May 1988.
10. Monheim, R. (1988). Pedestrian zones in West Germany — the dynamic development of an effective instrument to enliven the city centre, in *New Life for City Centres.* C. Hass-Klau (Editor). Anglo-German Foundation.
11. Hass-Klau, C. (1993). Impact of pedestrianisation and traffic calming on retailing: a review of the evidence from Germany and the UK. *Transport Policy,* **1**, No. 1, 21–31.
12. McClintock, H. (Editor) (1992). *The Bicycle and City Traffic. Principles and Practice.* Bellhaven Press.
13. Table 2.2 in NTS (1989/91).
14. Table 2A in NTS (1989/91).
15. Bracher, T. (1989). *Policy and provision for cyclists in Europe.* Cited in McClintock (1992), p7–9.
16. Bracher, T., 'Germany', p175–189 in McClintock (1992).
17. British Medical Association (1992). *Cycling towards health and safety.* Oxford University Press.
18. *The Southampton Western Approach cycle route: cyclist flows and accidents.* Project Report 93. Transport Research Laboratory, Crowthorne.
19. *Cycling in Pedestrian Areas.* Project Report 15. Transport Research Laboratory, Crowthorne. 1993.

20. Department of Transport (17 June 1994). *Blueprint for cycling policy.* Press Notice 222.
21. Continental figures from Jones, P. (1993). *Study of Policies in Overseas Cities for Traffic and Transport.* A report by the Transport Studies Group, University of Westminster for Traffic Policy Division, Department of Transport (cited as SPOTT, 1993); English figures from Department of Transport (1993). *Transport Statistics of Great Britain 1993.* HMSO.
22. House of Commons Select Committee on Transport. *Urban Public Transport: The Light Rail Option.* HC14. 1990–91 Session.
23. SPOTT (1993).
24. Lesley, L. (1987). *Budget light rail — a product of the discount rate.* Proceedings of Electrifying Urban Public Transport Conference, Liverpool. Quoted in Lesley (1993).
25. *Financial Times,* 10 January 1994 and *New Civil Engineer,* 10 March 1994, p15.
26. SPOTT (1993).
27. AURG (1993). *L'effet tramway: Premières résultats de l'enquête ménages 1992. Evolution 1985–92.* Grenoble.
28. *Local Transport Today,* 25 November 1993, p13.
29. Devon County Council (1991). *Traffic Calming Guidelines.* Exeter.
30. Friends of the Earth (1987). *The FOE Guide to Traffic Calming in Residential Areas.*
31. Hass-Klau, C., Nold, I., Bocker, G. and Crampton, G. (1992). *Civilised Streets: a Guide to Traffic Calming, Environmental and Transport Planning.* Brighton.
32. Transport and Road Research Laboratory (1991). *Translation of Dutch 30 kph Zone Design Manual.* Crowthorne.
33. Ministry of Transport (1963). *Traffic in Towns, a study of the long-term problems of traffic in urban areas.* Reports of the Steering Group and Working Group appointed by the Minister of Transport. HMSO.
34. Paragraphs 4.4–4.11 in Department of the Environment and Department of Transport (1994). *Transport.* Planning and Policy Guidance PPG 13. HMSO.
35. Association of County Councils (1991). *Towards a sustainable transport policy.* November 1991 (cited as ACC, 1991).
36. Hughes, G. (1993). *Reducing the need to travel — National policies into local practice.* Paper given to PTRC Transport, Highways and Planning 21st Summer Annual Meeting 1993, Seminar A.
37. Triesman, M. (1993). Can Oxford survive the planners? *Oxford Magazine,* Fourth Week, Michaelmas Term 1993. See also: Buchanan, M. (1994). Can planning for Oxford survive the academics? *Oxford Magazine,* Eighth Week, Trinity Term 1994.
38. Traffic in Towns (1963).
39. ACC (1991).
40. Lex Report on Motoring (1993). *The consumer view.* Quoted in *Parking Review,* February 1993.
41. Paragraph 4.6 in PPG 13.
42. Snyder, M.C. (1993). Employees cash in at works. *The Surface Transportation Policy Project Bulletin,* **III**, No. 3, March 1993. Washington DC.
43. Button, K. (1991). *Electronic road pricing: Experience and Prospects.* Paper presented at the Conference on Economy and Environment in the 1990s, 26–27 August 1991, Neuchatel.
44. Goldstein, G.S. and Moses, L.N. (1975). Transport controls and the spatial structure of urban areas. *American Economic Review, Papers and Proceedings,* **65**, 289–294, cited in Button (1991).
45. Evidence from the Department of Transport.
46. *The Independent,* 25 March 1994.
47. Newcastle-upon-Tyne City Council (1992). *Energy and the urban environment. Strategy for a major urban centre, Newcastle-upon-Tyne, UK.*

Chapter 12

1. Tables 8A and 8B in Department of Transport (1993). *National Travel Survey 1989/91.* HMSO (cited as NTS, 1989/91); and table 3.1 in Department of Transport (1988). *National Travel Survey 1985/86 Report. Part 1 - An analysis of personal travel.* HMSO (cited as NTS, 1985/86).
2. Based on table 8A in NTS (1989/91). Analysis in terms of distance travelled for each purpose within each band of journey length (table 8B) gives a very similar picture. There are no statistics to provide an historical comparison; the pattern in 1985/86 (table 3.4 in NTS (1985/86)) seems to have been broadly similar.
3. Table 2A in NTS (1989/91).
4. For the purposes of this table all express trains are assumed to have a typical occupancy of 50%. All other energy consumption and occupancy figures from Hughes (1992) (Hughes, P. (1992). *A strategy for reducing emissions of greenhouse gases from personal travel in Britain.* Doctoral thesis submitted to the Technology Faculty of the Open University, August 1992), except that data for the French high-speed train (TGV) are from Martin and Michaelis (1992) (Martin, D.J. and Michaelis, L.A. (1992). *Research and Technology Strategy to Help Overcome the Environmental Problems in Relation to Transport (Sast Project No. 3). Global Pollution Study.* Carried out by the Energy Technology Support Unit (ETSU) for the Commission of the European Communities. Report No. EUR-14713-EN. March 1992). A range of energy consumption figures were quoted for car types and average journeys by Martin and Shock (1989) (Martin, D.J. and Shock, R.A.W. (1989). *Energy Use and Energy Efficiency in UK Transport up to the Year 2010.* Department of Energy. Energy Efficiency Series 10. HMSO): all fell within the figures for large petrol car and small diesel car given by Hughes (1992).
5. Hughes (1992).
6. Table 4.8 in Department of Transport (1993). *Transport Statistics of Great Britain 1993.* HMSO.
7. Oral evidence from the Department of Transport.
8. Table 2.5 in Department of Transport (1991). *Transport Statistics of Great Britain 1991.* HMSO.
9. Evidence from the Countryside Commission.
10. Button, K. (1994). *Transport, the Environment and Economic Policy.* Section 8.7.
11. Department of Transport, Scottish Office and Welsh Office (1993). *Paying for better motorways: issues for discussion.* Cm 2200. HMSO.
12. *Charging for the Use of Motorways.* Fifth Report of the House of Commons Select Committee on Transport. 20 July 1994. HMSO.
13. Directive on the application by Member States of taxes on certain vehicles used for the carriage of goods by road and tolls and charges for the use of certain infrastructures. Directive 93/89/EEC. *Official Journal,* **L279/32**, 12/11/93 (article 7(h)).
14. Charging for the Use of Motorways, Select Committee Report, paragraph 109.
15. Paying for better motorways (1993), paragraph 7.4.
16. Rendel Palmer and Tritton (1989). *M25 Review. Volume 1*; Keen, K.G. *et al.* (1988). *Access control by ramp metering on M6.* Paper given to PTRC Summer Annual Meeting 1988; and Hounsell, N.B. *et al.* (1992). *An investigation of flow breakdown and merge capacity on motorways.* Transport Research Laboratory (Contractor Report 338).
17. Runnymede Borough Council (1994). *M25 link roads: an alternative package.* Discussion paper. p28.
18. This point was made by staff of the California Air Resources Board. However, it is perhaps of greater significance in California, where access to a freeway is often via a relatively steep upwards ramp.
19. Department of Transport (1993). *M25 junctions 12–15 proposals for link roads: statement in response to public consultation.*
20. Government Statistical Service; Department of Transport. *Vehicle Speeds in Great Britain 1993.* Statistics Bulletin (94)30. March 1994.

21. Earth Resources Research (1993). *Atmospheric emissions from road transport in the United Kingdom.* Report commissioned by the Royal Commission (cited as ERR, 1993).
22. Coaches face third-lane ban. *Freight,* June 1994, p7.
23. ERR (1993).
24. Pages 28 and 32–35 in Runnymede Borough Council (1994).
25. Goodwin, P.B. (1992). A Review of New Demand Elasticities with Special Reference to Short and Long Run Effects of Price Changes. *Journal of Transport Economics and Policy,* **26**, No. 2, May 1992, 155–169.
26. Table 2.5 in NTS (1989/91).
27. Oum, T.H., Waters, W.G. and Jong-Say Yong (1992). Concepts of price elasticities of transport demand and recent empirical estimates. *Journal of Transport Economics and Policy,* May 1992, **26**, No. 2, 139–154.
28. Oum *et al.* (1992).
29. Goodwin *et al.* (1992).
30. Information supplied by Department of Transport Civil Aviation Division.
31. *The Guardian,* 27 May 1994.
32. Martin and Shock (1989).
33. Martin and Michaelis (1992).

Chapter 13

1. House of Commons Transport Committee (1990). *Roads for the future.* Volume 1. HMSO.
2. Department of the Environment (June 1992). *Policy Guidance to the Local Government Commission for England.* HMSO.
3. *Local Transport Today,* Issue 108, 27 May 1993.
4. Section 9(3) of The Transport Act 1968.
5. Directive on air pollution by ozone. Directive 92/72/EEC. *Official Journal,* **L297**, 13/10/92.
6. Quality of Urban Air Review Group (1992). *Urban Air Quality in the United Kingdom.* First Report. HMSO.
7. Department of Health (1993). *Oxides of nitrogen.* Third Report of the Advisory Group on the Medical Aspects of Air Pollution Episodes. HMSO.
8. Department of the Environment, Scottish Office Environment Department, Welsh Office, Department of the Environment for Northern Ireland (March 1994). *Improving Air Quality; a discussion paper on air quality standards and management.*
9. Land Use Consultants (1993). *Local moves. The funding and formulation of local transport policy.* Council for the Protection of Rural England.
10. Land Use Consultants (1993).
11. Department of the Environment (August 1993). *Local transport grants.* Circular 104/73. HMSO.
12. Department of Transport (April 1993). *Transport Policies and Programme Submissions for 1994–95.* Local Authority Circular 2/93. HMSO.
13. Cook, A.J. and Davis, A.L. (1993). *Package approach funding: a survey of English highway authorities.* Friends of the Earth.
14. Department of Transport (May 1994). *Transport policies and programme submissions for 1995–96.* Local Authority Circular 2/94. HMSO.
15. Department of Transport (May 1994). *Transport Policies and Programme Submissions for 1995/96. Supplementary guidance notes on the package approach.* HMSO.
16. House of Commons Transport Committee, Session 1992–93, Fourth Report. *The Government's proposals for the deregulation of buses in London.* HC 623-I-III. HMSO. Volumes two and three contain extensive evidence.

17. Astrop, A. (1993). *The trend in rural bus services since deregulation.* Transport Research Laboratory Project Report 21. Crowthorne.
18. House of Commons Transport Committee, Session 1992–93, Fourth Report.
19. House of Commons Transport Committee, Session 1992–93, Fourth Report. Paragraph 149.
20. House of Commons Transport Committee, Session 1992–93, Fourth Report. Paragraph 150.
21. *Local Transport Today*, Issue 120, 11 November 1993.
22. *Local Transport Today,* Issue 138, 21 July 1994.
23. National Society for Clean Air, *Clean Air*, **23**, No. 3, Autumn 1993, 118–119.
24. Department of the Environment (1991). *Policy appraisal and the environment. A guide for departments.* HMSO.
25. Highways Agency (1994). Framework document.
26. Draper, P. (1977). *Creation of the DOE.* Civil Service Studies No. 4. HMSO.
27. Council for the Protection of Rural England (1992). *Where Motor-car is Master. How the Department of Transport became bewitched by roads* (cited as CPRE, 1992).
28. CPRE (1992).
29. *Local Transport Today*, Issue 139, 4 August 1994.
30. CPRE (1992).
31. CPRE (1992).
32. CPRE (1992).
33. CPRE (1992).
34. Department of Transport (May 1994). *The transport of goods by road in Great Britain 1993.* HMSO.
35. Table 9.1 in Department of Transport (1993). *National Travel Survey 1989/91.* HMSO.
36. Kent, Sir H. (1979). *In on the Act.* Macmillan.
37. Commission of the European Communities (1993). *Transport in the 1990s*. Luxembourg.
38. Commission of the European Communities (1992). *Transport Infrastructure*. Communication from the Commission. COM(92)231 final. Brussels.
39. COM(92)231 final.
40. *Official Journal*, **L305**, 10/12/93.
41. Fournier, P. (1994). Planning a strategic motorway network for Europe. *Transport Policy*, **1**, No. 2.
42. Commission of the European Communities (1992). *Towards Sustainability: a European Community programme of policy and action in relation to the environment and sustainable development.* COM(92)23. Brussels.
43. Commission of the European Communities (1992). *Green Paper on the impact of transport on the environment. A Community strategy for 'sustainable mobility'*. Communication from the Commission. COM(92)46 final. Brussels.
44. Commission of the European Communities (1992). *The Future Development of the Common Transport Policy. A global approach to the construction of a Community framework for sustainable mobility.* COM(92)494 final. Brussels.

Chapter 14

1. In 1993, motoring took 13.7% of weekly household expenditure in the top decile, 14.6% in the fifth decile but only 5.1% in the bottom decile: tables 1.2 and 1.3 (expenditure) and 9.7 (car ownership) in Central Statistical Office (1994). *Family spending: a report on the 1993 Family Expenditure Survey.* HMSO.
2. Edmund Burke. *Reflections on the Revolution in France*. Penguin, 1984. p194–195.

Appendix A

1. Table 2.12 in Department of the Environment (1994). *Digest of Environmental Protection and Water Statistics.* No. 16. HMSO (cited as DEPWS, 1994).
2. Table 2.9 in DEPWS (1994).
3. Table 1.4 in DEPWS (1994).
4. Dasch, J.M. (1992). Nitrous oxide emissions from vehicles. *Journal of the Air and Waste Management Association*, **42**, No. 1, 63–67.
5. Intergovernmental Panel on Climate Change (1994, forthcoming). *Radiative Forcing of Climate. Draft report of the Scientific Assessment Working Group.*
6. Department of Health (1991). *Ozone.* First Report of the Advisory Group on the Medical Aspects of Air Pollution Episodes. HMSO
7. Department of Energy (1989). *An evaluation of energy-related greenhouse gas emissions and measures to ameliorate them.* Energy Paper 58. HMSO.
8. Table 2.19 in DEPWS (1994).
9. Department of the Environment (1994). *Benzene*. Report by the Expert Panel on Air Quality Standards. HMSO.
10. EPAQS Report on Benzene.
11. CONCAWE (1983). *Benzene emissions from passenger cars.* Report No. 12/83. CONCAWE, The Hague.
12. National Society for Clean Air and Environmental Pollution (NSCA) (1994). *NSCA Pollution Handbook 1994*. NSCA, Brighton. p123.
13. Table 2.6 in DEPWS (1994).
14. Quality of Urban Air Review Group (1993). *Diesel Vehicles Emissions and Urban Air Qualtiy.* Second Report. HMSO.
15. Dockery, W.D., Pope, C.A. *et al.* (1993). An association between air pollution and mortality in six US cities. *The New England Journal of Medicine*, **329**, 1753–1759.
16. Table 2.2 in DEPWS (1994).
17. Table 2.14 in DEPWS (1994).
18. Medical Research Council. *The neurophysiological effects of lead in children.* A review of research 1984–1988.
19. Collective views of a task group of the International Programme on Chemical Safety (IPCS) on inorganic lead. Draft report. February 1993.
20. Chen, W. and Morrison, G.M. (1994). Platinum in road dusts and urban river sediments. *The Science of the Total Environment*, **146/147**, 169–174.
21. Volkswagen AG (1989). Quoted in Williams, M. L. (1990). The role of motor vehicles in air pollution in the UK. *The Science of the Total Environment*, **93**, 1–8.
22. Williams (1990).

Appendix C

1. Organisation for Economic Co-operation and Development (1988). *Transport and the Environment.* OECD, Paris.
2. Kageson, P. (1993). *Getting The Prices Right: A European scheme for making transport pay its full costs.* A report for the European Federation for Transport and Environment. May 1993.
3. Grupp (1986). *Die Sozialen Kosten des Verkehrs, Grundriss zu iher Berechnung. Verkehr und Technik*, Heft 9 und 10, p359–366 and p403–407
4. Planco Consulting-GMbH (1991). *Externe Kosten des Verkehrs.*
5. UPI (1991). Umwelt und Prognose Institut, Heidelberg.
6. ECOPLAN (1992). *Externe Kosten im Agglomerationsverkehr.*
7. Infras (1992). *Internalisieren der externen Kosten des Verkehrs.* Zurich.

8. MacKenzie, J., Dower, R. and Chen, D. (1992). *The going rate; What it really costs to drive.* World Resources Institute, Washington DC.
9. CE (Centrum voor Energiebesparing en Schone Teknologie) (1991). *Brandstof heffing en inkomensverderling.* Delft.
10. UIC (1994). *External Effects of Transport.* A project for the Union International Chemins de Fer, Paris.
11. Organisation for Economic Co-operation and Development (1990). *The Social Costs of Land Transport.* Prof. Emile Quinet, Ecole Nationale des Ponts et Chaussées, Paris.
12. Quoted in TEST (1991). *Wrong Side of the Tracks? Impacts of Road and Rail Transport on the Environment.* Researched and published by TEST.
13. European Conference of Ministers of Transport (1990). *Transport policy and the environment: ECMT Ministerial Session (prepared in co-operation with OECD).* OECD, Paris. p44; OECD (1990), page 217.
14. Willeke, Weinberger and Thomassen (1990). *Kosten des Larms in der Bundersrepublik Deutschland.*
15. UIC (1994).
16. OECD (1990).
17. Fowkes, A.S., Nash, C.A. and Tweddle, G. (1990). *The Track and External Costs of Road Transport.* Working paper 312 for the Institute of Transport Studies, University of Leeds. September 1990.
18. Paragraph 3.11 in TEST (1991).
19. Figure quoted in: Fankhauser, S. (1992). *Global warming damage costs: some monetary estimates.* CSERGE working paper GEC 92–29.
20. Cline, W.R. (1992). *Economics of global warming.* Institute for International Economics, Washington DC, USA.
21. Cline (1992), also Adger, N. and Fankhauser, S. Economic analysis of the greenhouse effect: optimal abatement level and strategies for mitigation. *International Journal of Environment and Pollution*, **3**.
22. Houghton, J. (1994). *Global Warming. The Complete Briefing.* Lion Publishing, Oxford. p111.
23. For the purposes of this calculation, the combined GDP of the Developed Countries included the national income of countries in North America, southern and western Europe (excluding the countries of the former East European bloc, Cyprus, Malta and the former Yugoslavia and Albania), Australia, Japan, New Zealand and South Africa. (See United Nations *World Economic Survey 1992*, page xiii).

Appendix D

1. Peake, S. (1994). *Transport in transition: lessons from the history of energy.* Earthscan/Royal Institute of International Affairs.

Appendix E

1. Department of Transport, Vehicle Standards and Engineering Division (1992). *Emissions from heavy duty diesel engined vehicles: the government's response to the Fifteenth Report of the Royal Commission on Environmental Pollution.* HMSO.

Appendix F

1. Department of the Environment (1992). *Climate change: our national programme for* CO_2 *emissions — a discussion document.*

Appendix A

POLLUTANT EMISSIONS FROM TRANSPORT

Carbon monoxide

A.1 Transport is responsible for about 90% of all UK carbon monoxide emissions; emissions from road transport have increased by over 30% in the last ten years.[1]

A.2 Under normal atmospheric conditions, carbon monoxide is converted in the atmosphere to carbon dioxide as a result of reaction with hydroxyl radicals. High concentrations of carbon monoxide can lead to local depletion of hydroxyl radicals (since the latter react with carbon monoxide more readily than with other pollutants such as tropospheric ozone and methane) and therefore to a build-up of these two pollutants, both of which are greenhouse gases.

A.3 Carbon monoxide is toxic to vertebrates (and some invertebrates) because it combines with haemoglobin in the blood to form the stable complex carboxyhaemoglobin, thereby reducing the blood's oxygen-carrying capacity. Exposure to high concentrations results in loss of consciousness and death. At lower concentrations, carbon monoxide affects the functioning of the central nervous system, causing impairment of vision and slowing down reflexes and mental functions; it can also cause headaches and drowsiness.

A.4 WHO air quality guidelines (see table 3.4) relate to the levels of carboxyhaemoglobin circulating in the blood for which symptoms have been observed. The guideline concentrations are most likely to be exceeded in urban areas during the winter.

Nitrogen oxides

A.5 Road vehicles are responsible for nearly 90% of emissions of nitrogen oxides from the transport sector, and for over 50% of all emissions of nitrogen oxides emissions in the UK.[2] Nearly all of the initial nitrogen oxides from vehicles is released as nitric oxide but this is rapidly oxidised under normal daytime atmospheric conditions (less rapidly at night) to the secondary pollutant, nitrogen dioxide. Nitrogen dioxide is involved in the formation of another secondary pollutant, tropospheric ozone.

A.6 Nitrogen oxides are also involved in reactions leading to the formation of nitrous and nitric acid, which are removed from the atmosphere by dry or wet deposition and contribute to eutrophication or acidification of the environment.

A.7 At high concentrations, nitrogen dioxide can cause respiratory irritation. It damages and inflames the epithelium of the airways, predisposing individuals to the development of respiratory infections and bronchitis. Traffic is not the only source of nitrogen dioxide to which individuals may be exposed; the pollutant can occur at high concentrations inside buildings where gas is used for heating or cooking.

Nitrous oxide

A.8 Road transport is a minor but significant source (giving rise to 7% of all UK emissions) of nitrous oxide[3] which is formed, along with other oxides of nitrogen, during all atmospheric combustion processes. Under normal atmospheric conditions, nitrous oxide is rapidly oxidised to nitric oxide. Like nitric oxide, nitrous oxide is effectively non-toxic to humans. However, it differs from other oxides of nitrogen in being a greenhouse gas.

A.9 Emissions of nitrous oxide in the exhaust gases of vehicles with catalytic converters are an order of magnitude higher than those from cars without.[4] The increasing use of catalytic converters has led to an increase in nitrous oxide emissions from cars[5] but, although transport is consequently likely to become a major anthropogenic source, natural sources of nitrous oxide account for more than 70% of the total.

Tropospheric ozone

A.10 Ozone, which occurs naturally in the troposphere at very low concentrations, is a secondary pollutant, formed in a series of chemical and photochemical reactions involving oxygen, nitrogen dioxide and radicals derived from hydrocarbons. Its formation is greatly enhanced in air polluted by nitrogen oxides and hydrocarbons (see A.16). Ozone is involved in many atmospheric oxidation processes, including those that lead to the formation of acid rain.

A.11 There is evidence that high concentrations of ozone increase susceptibility to infections, irritate mucous membranes and reduce lung function, resulting in temporary respiratory difficulties in sensitive individuals and in those taking vigorous exercise.[6] In studies in the USA, it was found that exposure to ozone concentrations of 160–300 $\mu g/m^3$ for periods of an hour reduced lung function in adults and children taking vigorous exercise. Studies in the USA have found a wide variation in sensitivity to ozone but, although 5–10% of the population is sensitive, asthmatics appear to be no more or less so than others.

A.12 WHO guideline limits for ozone (see table 3.4) are based on its effects on the respiratory system. The upper guideline limit of an hourly mean concentration of 100 ppb of ozone (200 $\mu g/m^3$) has been shown to irritate eyes, nose and throat and cause coughing, chest pain and nausea. Background levels in the UK are between 30 and 40 ppb but the WHO limits are regularly exceeded in the summer months.

Methane

A.13 Vehicle exhaust contains small amounts of methane, which is a greenhouse gas; motor vehicles contribute about 1% of total UK emissions.[7] Some methane also escapes during the production and distribution of petroleum products and natural gas. Vehicles running on natural gas are a potential source of emissions to the atmosphere.

Volatile organic compounds

A.14 Volatile organic compounds (VOCs) associated with transport include hydrocarbons and their derivatives formed during combustion, hydrocarbons from the evaporation of fuel and halogenated compounds used in the manufacture and maintenance of vehicles and aircraft. Many VOCs are also derived from natural sources. In the UK about 38% of all VOC emissions come from vehicle exhausts and from fuel evaporation. VOC emissions from road vehicles, including evaporative emissions, increased by about 8% between 1982 and 1992.[8]

A.15 VOCs are absorbed by the lung and gut. Their breakdown in the body can give rise to carcinogenic metabolites. Many hydrocarbons are themselves carcinogenic or are suspected carcinogens (eg benzene and 1,3-butadiene). Other VOCs, for example some aldehydes, are toxic. Combustion of alternative fuels such as methanol or ethanol produces more aldehydes, particularly formaldehyde and acetaldehyde, than the combustion of petrol.

A.16 Both hydrocarbons and aldehydes can cause irritation of skin and mucous membranes and may lead to breathing difficulties; long-term exposure to hydrocarbons has been shown to lead to impairment of lung function. Hydrocarbons are also involved (with nitrogen dioxide) in the formation of tropospheric ozone and photochemical smogs, which in turn may cause respiratory problems.

A.17 There are few quantitative data on the effects of exposure to VOCs in exhaust emissions, nor is it known to what extent the individual components of exhaust may contribute to health effects. There is a WHO guideline value for formaldehyde (see table 3.4) but, on the basis that there are no completely safe levels of exposure to carcinogenic compounds, WHO does not recommend guidelines for benzene, which is perhaps the most important VOC arising from transport.

A.18 Unleaded petrol generally contains more aromatic compounds (which serve as functional replacements for lead compounds) than leaded petrol. These aromatics include benzene, a stable compound of which there are no significant natural sources other than petroleum.[9] About 78% of atmospheric benzene in the UK comes from petrol engine exhausts, a further 9% from diesel engine emissions and 10% from fuel evaporation and petrol refining and distribution.[10] However, most of the benzene in vehicle exhausts is formed during combustion, as a result of the thermal de-alkylation of other aromatics and the trimerisation of ethylene; only about 35–40% is there as a result of the benzene in the fuel itself.[11]

A.19 Catalytic converters are capable of reducing hydrocarbon emissions, including those of benzene and of the suspected carcinogen 1,3-butadiene (formed by thermal degradation of olefins in the fuel) by 90%.[12]

Particulates

A.20 Black smoke from vehicle emissions includes particles derived from fuel combustion products and from unburnt fuel. Emissions of black smoke from road transport increased by about 85% between 1982 and 1992.[13] Diesel vehicles are a major source of black smoke; they are responsible for about 40% of all emissions in the UK. The situation is more extreme in urban areas since a greater proportion of total black smoke emissions there is from traffic; for example, diesel vehicles are responsible for 87% of black smoke emissions in London.[14] The very fine particles of carbon can act as nuclei for condensation of other substances, including hydrocarbons.

A.21 Particulate material from vehicle exhausts can irritate mucous membranes lining the respiratory tract and may give rise to breathing difficulties. Some constituents (eg polyaromatic hydrocarbons, derived from hydrocarbons in fuel) may be carcinogenic.

A.22 The size of particles has an important bearing on their respiratory effects. Particles with an aerodynamic diameter of 10 μm or less (PM10) are inhalable, that is, small enough to be breathed in; thoracic particles are defined as those which penetrate beyond the larynx; and those with a diameter of less than 2.5 μm (referred to as respirable particles) are small enough to penetrate to the deep lung where they are retained. Particulates may be amongst the most harmful components of vehicle exhaust; a recent American study indicates a link between mortality rate and airborne particles.[15]

A.23 WHO guidelines for particulates are based on the minimum concentrations of sulphur dioxide and particulates that cause respiratory effects. EC air quality standards for smoke and sulphur dioxide are intended to protect both human health and the environment, with limit values based on the results of epidemiological studies.

Sulphur dioxide

A.24 The direct contribution of transport to total national emissions is comparatively small; the principal source is the combustion of sulphur-containing fossil fuels, especially coal (power stations were responsible for 69% of emissions in 1992 and refineries for 4%). Road transport was responsible for 2% of UK emissions in 1992 (a proportion which is likely to decline as sulphur levels in diesel fuel are reduced) and coastal shipping for 2%.[16] Emissions from transport can be significant in certain localities, for example in harbour areas, or in streets where traffic is congested and includes a high proportion of diesel-fuelled vehicles. Sulphur dioxide can give rise to respiratory problems.

Lead

A.25 Additives in petrol are still a major source of airborne lead (other sources being combustion processes and smelting works) but, despite substantial increases in petrol consumption, emissions of lead from UK road transport have fallen dramatically (they are now about 20% of those recorded in the 1970s[17]) as a consequence of reductions (under the terms of an EC Directive) in the permitted maximum amount of lead in petrol in 1981 and 1985, and the introduction of unleaded petrol in the UK in 1986.

A.26 Leaded petrol cannot be used in cars fitted with catalytic converters; emissions are therefore expected to continue to decline as the proportion of cars with catalysts increases.

A.27 Most airborne lead is in the form of fine particles. Inhaled particles of lead may be deposited in the lungs or absorbed by the gut. Absorption rates are highest in children and may be enhanced by poor diet. Like many other metals, lead is a cumulative poison. It is accumulated in the liver, kidneys, brain, bone and nervous tissue. Long-term exposure to high doses can affect many organ systems. The most serious effects are on the nervous system, haemoglobin synthesis and haemopoiesis.

A.28 A review of studies (undertaken between 1984 and 1988) on the neuro-psychological effects of low-level lead exposure in children found that the limitations of epidemiological studies were such that it was not possible to conclude that exposure to lead at urban levels then current were harmful but suggested that it would be prudent to continue to reduce environmental levels.[18] A more recent review by a task group of the International Programme on Chemical Safety (IPCS) confirmed these findings for the relationship between blood lead and IQ. The IPCS group also reported that there was conflicting epidemiological evidence of a dose-dependent association between blood lead and foetal growth.[19]

Other airborne vehicle pollutants

A.29 Increased use of catalytic converters containing *precious metals* (platinum, palladium and rhodium) has raised the question of whether these metals are being released to the environment in significant quantities. A recent paper[20] reports that platinum levels in all size fractions of road dusts in an area of Göteborg, Sweden, have increased since 1984. For example, for the less than 63 μm size fraction, platinum levels increased from 3.0 ng/g in 1984 to 8.9 ng/g in 1991. Concentrations in urban river sediments were found to reflect those of the road dusts and, in 1991, ranged from less than 0.5 to 2.2 ng/g. The significance of these relatively low concentrations for human health is not known.

A.30 *Asbestos*, from the brake linings of road and rail vehicles, is released into the atmosphere in small amounts. The amounts of asbestos entering the environment from this source are probably too low to represent a health hazard.

A.31 *Ammonia* is present in very small amounts in exhaust gases from engines without catalytic converters. There is some evidence that emissions of ammonia from vehicles fitted with three-way catalytic converters are very much higher than from non-catalyst vehicles (a German study[21] found that average emissions were 85.9 mg/km from cars fitted with catalysts, compared with 2.2 mg/km from non-catalyst cars). On this basis, if all petrol cars were fitted with catalysts, they could contribute as much as 10% of total UK emissions of ammonia.[22]

Odour

A.32 Odours associated with transport arise as a result of the volatile or gaseous nature of fuels and their combustion products. There is, as with noise pollution, a subjective element in the perception of odour as a nuisance. Some object to the smell of kerosene, petrol or diesel fuel itself, others do not. With improvements in engine technology, diesel vehicles are generally less smelly than they were 10 or 20 years ago and many people are now more concerned about the smell of hydrogen sulphide which is released from cars with new three-way catalysts under certain driving conditions. The smell of rancid cooking oil or vinegar, from the exhaust fumes of buses running on bio-fuels, are, at least for some people, a negative feature of these alternatives to petroleum-derived fuels.

Appendix B

PROJECTIONS OF EMISSIONS FROM ROAD TRANSPORT

Produced for the Royal Commission by Earth Resources Research

B.1 The Commission asked Earth Resources Research (ERR) to carry out a modelling study in order to assess the effects of the forecast increase in traffic on the emissions from road transport. Two alternative assumptions were made about the proportion of new cars which will be diesel-powered: that the proportion will increase to 16% in 2000 and 20% in 2005 and then remain constant, and that it will grow linearly from 20% in 2005 to 40% in 2025.

B.2 The assumptions made about emission limits for new vehicles are summarised at 8.22(i). It was also assumed that by the turn of the century the catalytic converters fitted to new cars will perform adequately from cold start. The key points which emerge from the study are summarised in chapter 8 (8.23–8.25).

B.3 This appendix contains the projections of emissions of airborne pollutants for high and low forecasts of traffic growth and the lower assumption on diesel market share. It also illustrates the results for the main vehicle classes for the low forecast of traffic growth. It covers the following pollutants:

carbon monoxide: (table B.1 and figure B-I);
volatile organic compounds: (table B.2 and figure B-II);
nitrogen oxides: (table B.3 and figure B-III);
sulphur dioxide: (table B.4 and figure B-IV);
particulates: (table B.5).

B.4 In the case of particulates the results for the higher assumption on diesel market share are illustrated:

figure B-V is based on the low forecast of traffic growth;
figure B-VI is based on the high forecast of traffic growth;
figure B-VII, also based on the high forecast of traffic growth, shows the split between urban and non-urban roads.

B.5 Projections for lead, not reported in detail here, confirmed that emissions will decline rapidly and remain at a low level.

B.6 The modelling study also covered carbon dioxide emissions and the effects of measures to reduce them. The baseline projections of carbon dioxide emissions are shown in chapter 8 (figures 8-IV and 8-V) and the results discussed at 8.54 and 8.56.

Table B.1
Projected emissions of carbon monoxide

in kilotonnes

	Vehicle Classes								
Low Forecast	Cars	LGVs	MCycles	HGVR	HGVA	PSVs	Urban	Ex-Urban	Total
1990	6326.7	521.1	101.7	145.4	61.4	59.7	4436.9	2779.1	7216.0
1995	5824.2	497.0	88.2	124.8	59.4	55.0	3840.3	2808.2	6648.5
2000	4692.1	387.2	62.5	72.1	41.3	38.8	2680.3	2613.6	5293.9
2005	2565.2	194.7	39.5	31.4	27.6	21.0	973.8	1905.6	2879.4
2010	1739.8	127.8	32.2	30.7	30.3	10.7	452.3	1519.2	1971.5
2015	1681.3	128.2	31.6	31.2	33.7	10.6	447.9	1468.7	1916.5
2020	1766.7	140.0	31.6	31.7	37.5	10.6	471.2	1547.0	2018.2
2025	1874.9	153.7	31.6	32.3	41.6	10.6	499.3	1645.5	2144.8
High Forecast									
1990	6326.7	521.1	101.7	145.4	61.4	59.7	4436.9	2779.1	7216.0
1995	6000.9	521.5	88.2	125.6	63.0	55.0	3952.5	2901.6	6854.1
2000	5103.5	432.5	62.5	73.2	47.0	38.8	2908.1	2849.4	5757.5
2005	2898.8	231.4	39.5	32.3	33.6	21.0	1097.4	2159.2	3256.6
2010	2023.9	161.1	32.2	32.0	39.3	10.7	523.8	1775.4	2299.2
2015	1989.4	171.8	31.6	33.1	46.6	10.6	529.2	1753.8	2283.1
2020	2126.8	199.5	31.6	34.3	55.2	10.6	568.9	1889.1	2458.0
2025	2276.4	232.6	31.6	35.8	65.2	10.6	612.0	2040.3	2652.3

Figure B-I
Projected emissions of carbon monoxide: low forecast of traffic growth

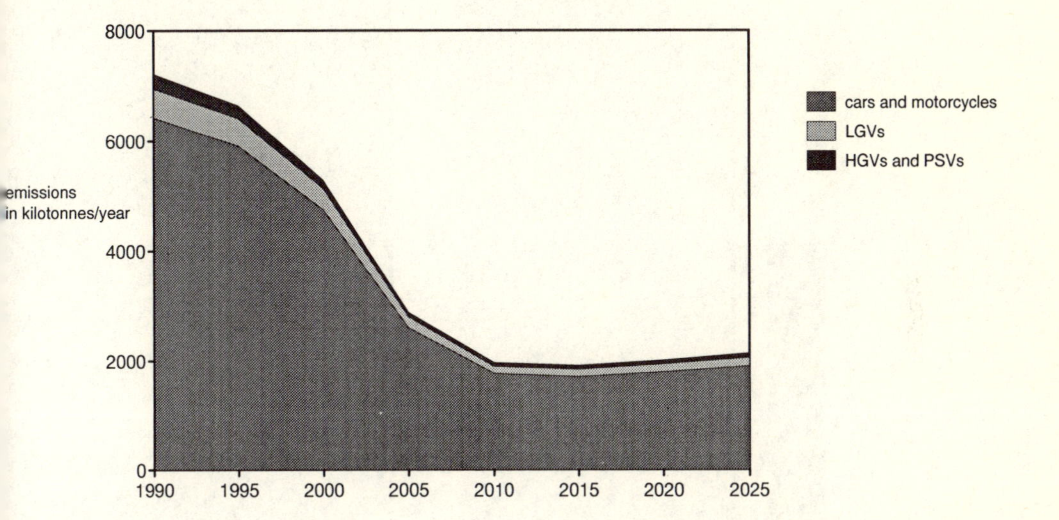

Table B.2
Projected emissions of volatile organic compounds

in kilotonnes

	Vehicle Classes								
Low Forecast	Cars	LGVs	MCycles	HGVR	HGVA	PSVs	Urban	Ex-Urban	Total
1990	802.0	68.4	24.0	26.9	27.5	11.3	581.4	378.8	960.2
1995	709.7	64.2	20.9	25.6	28.5	11.8	510.4	350.2	860.7
2000	528.8	47.7	14.5	21.9	25.9	10.0	367.1	281.7	648.8
2005	232.3	20.4	8.6	18.9	24.2	8.2	151.9	160.7	312.6
2010	131.4	11.2	6.7	19.0	26.8	7.0	84.6	117.5	202.1
2015	127.2	11.0	6.6	19.3	29.8	7.1	84.1	116.8	200.9
2020	133.8	12.0	6.6	19.6	33.1	7.0	88.3	123.9	212.2
2025	142.0	13.2	6.6	20.0	36.8	7.0	93.1	132.5	225.6
High Forecast									
1990	802.0	68.4	24.0	26.9	27.5	11.3	581.4	378.8	960.2
1995	730.5	67.4	20.9	25.7	30.2	11.8	525.2	361.4	886.5
2000	574.5	53.3	14.5	22.3	29.4	10.0	397.5	306.6	704.0
2005	262.3	24.3	8.6	19.4	29.4	8.2	170.0	182.3	352.3
2010	152.9	14.1	6.7	19.8	34.7	7.0	96.8	138.4	235.2
2015	150.5	14.8	6.6	20.4	41.2	7.1	98.4	142.1	240.6
2020	161.1	17.1	6.6	21.2	48.8	7.0	105.9	156.0	261.9
2025	172.4	20.0	6.6	22.1	57.7	7.0	114.1	171.7	285.8

Figure B-II
Projected emissions of volatile organic compounds: low forecast of traffic growth

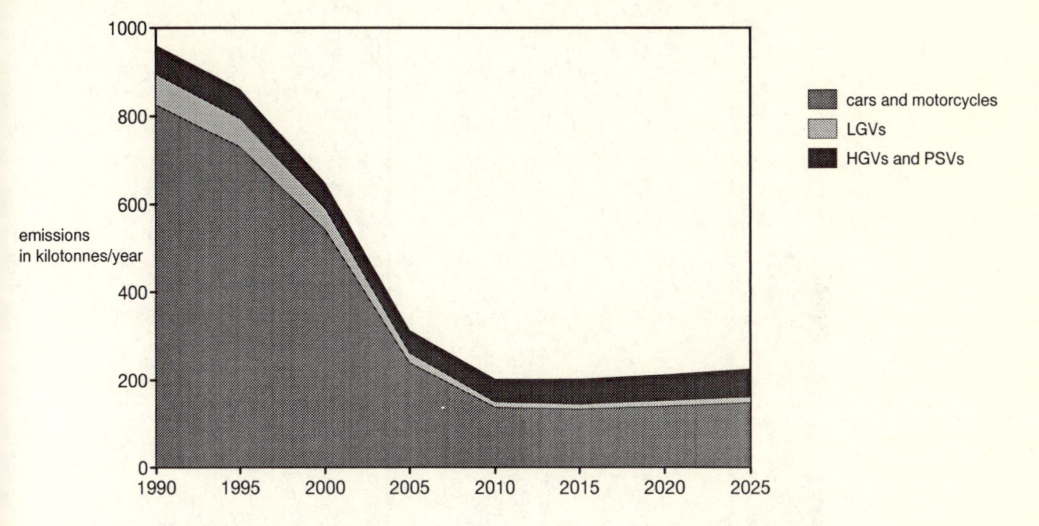

Table B.3
Projected emissions of nitrogen oxides

in kilotonnes

	Vehicle Classes								
Low Forecast	Cars	LGVs	MCycles	HGVR	HGVA	PSVs	Urban	Ex-Urban	Total
1990	1055.4	95.1	0.8	134.8	139.4	37.4	473.5	989.5	1463.0
1995	928.6	90.5	0.7	132.3	146.7	43.5	431.4	910.9	1342.3
2000	685.6	70.6	0.7	125.3	132.5	41.1	333.6	722.2	1055.8
2005	315.2	37.7	0.6	119.4	120.5	37.1	190.1	440.5	630.5
2010	207.3	28.0	0.6	120.4	132.0	34.3	151.6	371.0	522.6
2015	216.8	29.4	0.6	122.4	147.0	34.6	158.2	392.7	550.8
2020	230.5	32.2	0.6	124.6	163.5	34.3	165.9	419.7	585.6
2025	244.6	35.3	0.6	127.0	181.5	34.2	174.0	449.3	623.3
High Forecast									
1990	1055.4	95.1	0.8	134.8	139.4	37.4	473.5	989.5	1463.0
1995	955.6	95.0	0.7	133.1	155.4	43.5	443.2	940.2	1383.4
2000	744.7	78.8	0.7	127.4	150.6	41.1	358.3	785.0	1143.2
2005	355.8	44.8	0.6	122.8	146.5	37.1	209.0	498.7	707.7
2010	241.2	35.3	0.6	125.6	171.3	34.3	170.4	438.0	608.4
2015	256.6	39.4	0.6	129.8	203.3	34.6	182.4	481.8	664.3
2020	277.5	45.8	0.6	134.8	240.7	34.3	197.0	536.7	733.7
2025	297.0	53.4	0.6	140.6	284.4	34.2	212.5	597.8	810.2

Figure B-III
Projected emissions of nitrogen oxides: low forecast of traffic growth

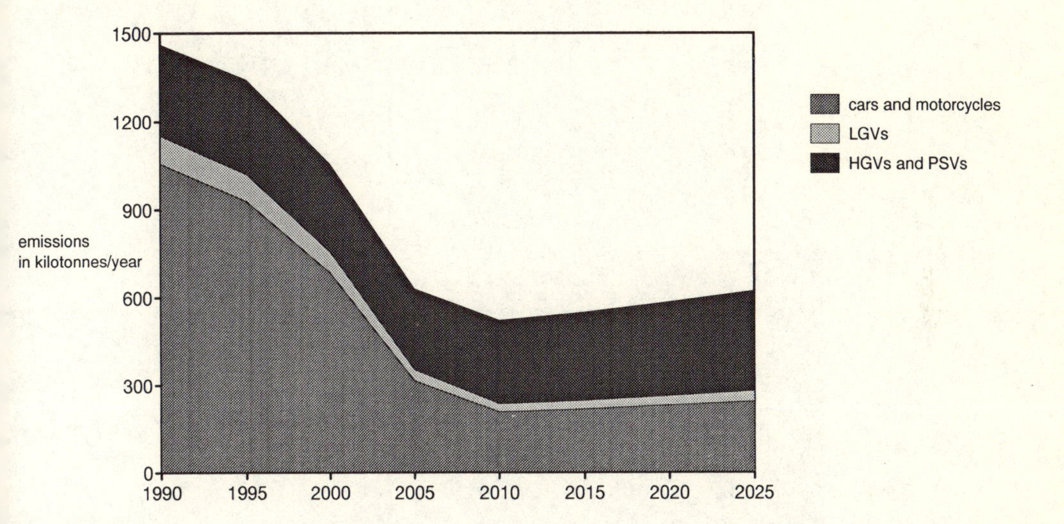

Table B.4
Projected emissions of sulphur dioxide

in kilotonnes

	Vehicle Classes								
Low Forecast	Cars	LGVs	MCycles	HGVR	HGVA	PSVs	Urban	Ex-Urban	Total
1990	23.22	5.96	0.22	15.94	14.00	3.80	24.05	39.08	63.13
1995	23.61	5.90	0.18	12.54	12.17	3.57	22.43	35.54	57.97
2000	21.83	3.22	0.17	3.17	3.43	0.92	13.67	19.07	32.74
2005	21.47	3.29	0.16	3.22	3.87	0.91	13.61	19.32	32.93
2010	21.90	3.51	0.15	3.27	4.34	0.91	13.97	20.11	34.08
2015	23.04	3.82	0.15	3.32	4.83	0.92	14.71	21.38	36.10
2020	24.37	4.18	0.15	3.38	5.38	0.91	15.55	22.82	38.38
2025	25.86	4.59	0.15	3.45	5.97	0.91	16.51	24.43	40.93
High Forecast									
1990	23.22	5.96	0.22	15.94	14.00	3.80	24.05	39.08	63.13
1995	24.39	6.20	0.18	12.62	12.89	3.57	23.05	36.80	59.86
2000	23.78	3.59	0.17	3.23	3.90	0.92	14.80	20.79	35.58
2005	24.26	3.91	0.16	3.31	4.71	0.91	15.28	21.97	37.26
2010	25.47	4.42	0.15	3.41	5.63	0.91	16.21	23.80	40.00
2015	27.27	5.12	0.15	3.52	6.69	0.92	17.50	26.17	43.67
2020	29.33	5.95	0.15	3.66	7.92	0.91	19.00	28.93	47.93
2025	31.40	6.94	0.15	3.81	9.36	0.91	20.61	31.97	52.58

Figure B-IV
Projected emissions of sulphur dioxide: low forecast of traffic growth

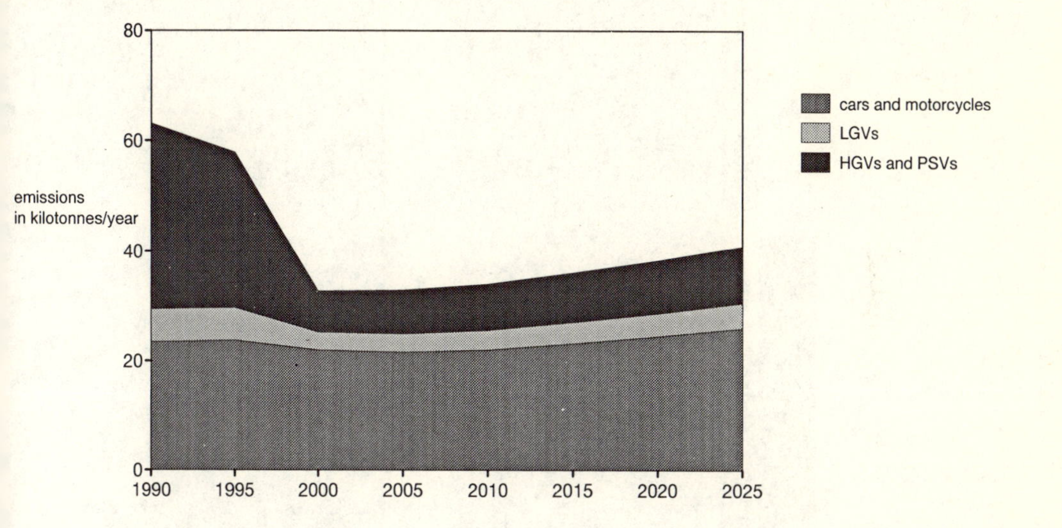

Table B.5
Projected emissions of particulates

in kilotonnes

	Vehicle Classes								
Low Forecast	**Cars**	**LGVs**	**MCycles**	**HGVR**	**HGVA**	**PSVs**	**Urban**	**Ex-Urban**	**Total**
1990	2.55	3.21	0.00	21.30	13.04	4.54	12.56	32.09	44.64
1995	3.57	3.97	0.00	19.40	13.47	5.17	13.50	32.07	45.57
2000	5.71	4.43	0.00	11.46	8.54	4.14	11.53	22.76	34.28
2005	6.89	3.68	0.00	3.23	2.57	2.37	7.68	11.06	18.74
2010	7.43	2.96	0.00	1.91	1.68	0.92	6.16	8.73	14.90
2015	7.76	2.76	0.00	1.80	1.76	0.65	6.02	8.71	14.73
2020	8.46	3.00	0.00	1.83	1.96	0.65	6.48	9.43	15.91
2025	8.98	3.30	0.00	1.87	2.18	0.65	6.89	10.08	16.97
High Forecast									
1990	2.55	3.21	0.00	21.30	13.04	4.54	12.56	32.09	44.64
1995	3.70	4.17	0.00	19.52	14.27	5.17	13.80	33.03	46.83
2000	6.22	4.95	0.00	11.65	9.70	4.14	12.21	24.45	36.66
2005	7.78	4.36	0.00	3.32	3.12	2.37	8.50	12.46	20.95
2010	8.64	3.73	0.00	1.99	2.18	0.92	7.16	10.31	17.47
2015	9.18	3.70	0.00	1.91	2.44	0.65	7.22	10.66	17.87
2020	10.19	4.28	0.00	1.98	2.89	0.65	8.02	11.96	19.98
2025	10.91	4.99	0.00	2.07	3.41	0.65	8.77	13.25	22.02

Figure B-V
Projected emissions of particulates: low forecast of traffic growth, high share of diesel cars

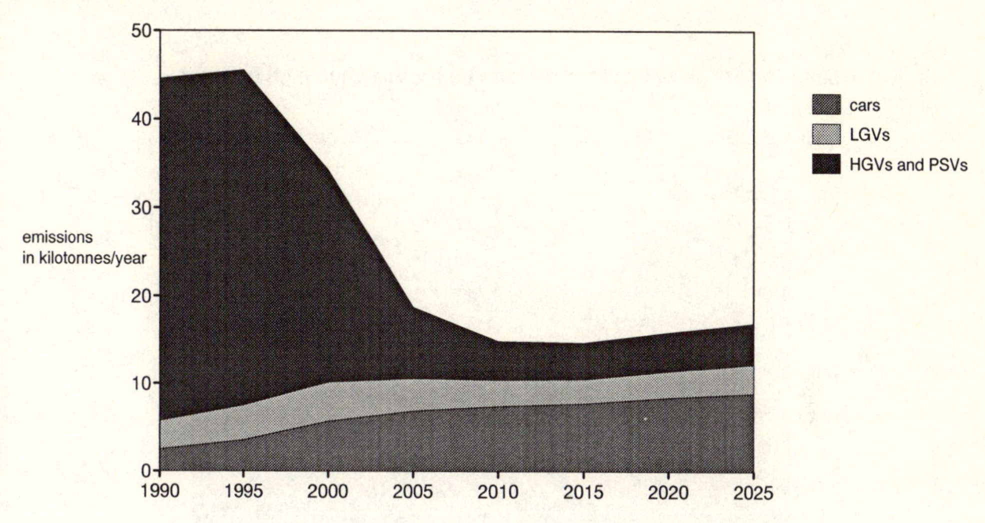

Figure B-VI
Projected emissions of particulates: high forecast of traffic growth, high share of diesel cars

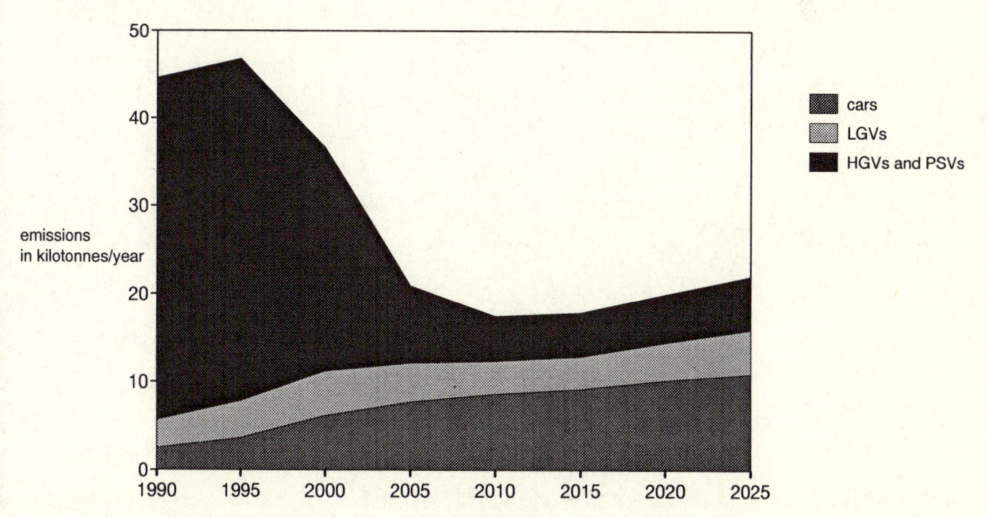

Figure B-VII
Projected emissions of particulates, urban and non-urban split: high forecast of traffic growth, high share of diesel cars

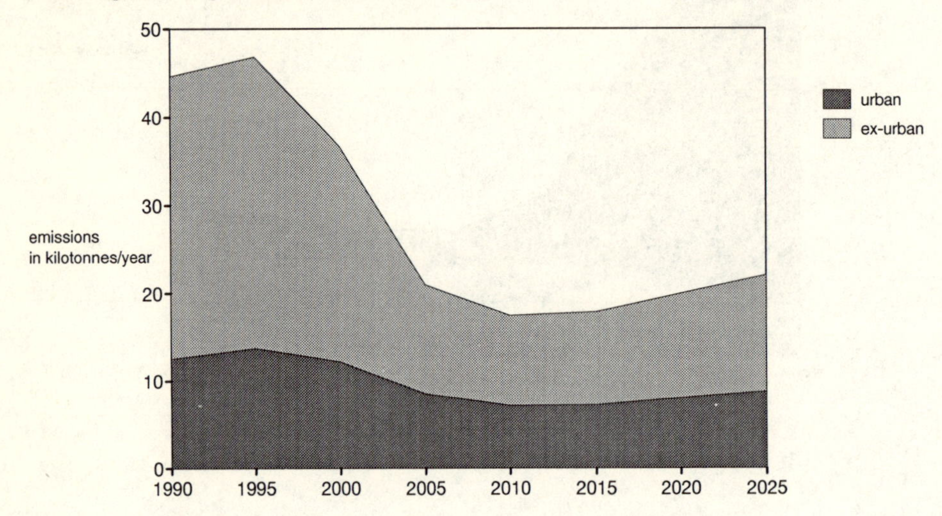

Appendix C

EXTERNAL COSTS OF TRANSPORT

C.1 Evidence on the costs of the damage caused to the environment is both limited and fragmentary. This reflects the difficulties of estimating the nature of environmental effects, the extent of damage caused, and the cost to the community of such damage in money terms. Nonetheless, a number of studies have been made of the impact of transport on the environment, and these have yielded a range of estimates of the costs of damage, or of the implicit valuations behind remedial or avoidance expenditures. This appendix looks at those environmental effects for which some estimates have been made of the cost of damage to the community, with such costs expressed in money terms. The techniques used to produce such estimates are described in box C. Those environmental effects for which no credible estimates of damage or other costs in money terms are available are discussed in chapter 7.

C.2 Most of the studies referred to were carried out abroad, so for comparative purposes the cost estimates they produce are discussed here in terms of a percentage of the Gross Domestic Product (GDP), or national income, of the country concerned, or of the UK or Great Britain as appropriate. The applicability of such studies to the UK must be considered in the light of differences between the respective countries, but within the limits of the methods and information employed they provide a useful indication of the broad order of magnitude of the costs incurred. On the basis of these studies, broad order of magnitude estimates have been made (in the form of a wide range of values) of the costs of two categories of environmental effects that are at present quantifiable in money terms, namely *air pollution* and *noise* from transport in Britain. These estimates are included in table 7.2 in chapter 7. For air pollution, the costs relate in part to damage caused to other countries by pollutants originating in Britain.

C.3 Another category of environmental cost included in table 7.2 is an estimate of the cost attributable to the UK for the *climate change* or *global warming* effects of carbon dioxide emissions from transport. This is an estimate of the cost that Britain's transport imposes on the world; climate change is a global pollution problem, so it is appropriate to cost the damage that Britain's pollution imposes globally, rather

BOX C — **TECHNIQUES FOR PUTTING A MONEY VALUE ON ENVIRONMENTAL DAMAGE**

Preventive expenditure: the amount paid to prevent or ameliorate unwanted effects. An example is expenditure on insulation and double-glazing to keep out noise. Sometimes a community valuation can be inferred, as when governments provide grants towards such expenditure.

Replacement/restoration cost: the amount public bodies or individuals spend, for example, to restore damaged buildings or landscapes.

Property valuation: differences in the market value of similar properties which reflect differences in the local environment, for example, the amount by which the price of a house is lower because it is next to a busy road.

Loss of earnings: loss of productive output through injury or ill health.

Changes in productivity: the money value of a reduction in crop, forestry or fishery yield due to environmental damage.

Contingent valuation: the amount people say they would be willing to pay to avoid unwanted effects. This may differ from the amount they would be willing (or able) to pay in practice.

than to look at the effects on this country. Apart from the difficulties of limited information that apply also to air pollution and noise effects, there is an additional uncertainty in these estimates due to the fact that the extent of potential damage from global warming is both uncertain and to an extent speculative. Moreover, most of the costs will be incurred in the longer term, mainly by future generations, and so do not impinge directly on present welfare. Present emissions contribute to future social costs, and should be taken into account in estimates of the costs of environmental damage today, but to try to do so raises problems of estimation and interpretation. A fourth element of the social costs of transport included in table 7.2 is the cost of *transport accidents*, in so far as these are borne by the whole community; these are also in a sense environmental consequences of transport.

C.4 In addition to these four categories of social costs which are quantifiable to a limited extent at present, other environmental effects of transport remain, which are not amenable to quantification or to costing in money terms at present. These are discussed in chapter 7. They include the disruption of life by transport in cities; from the evidence the Commission received it is clear that people are concerned at the disturbance in cities caused by high levels of traffic. Severance, reduced social interaction, and other loss of amenity are major costs of the disruption caused by high levels of road traffic in both urban and rural areas, and to a lesser extent by aircraft and trains. These costs are over and above the costs of damage by air pollution and noise which are quantified in this appendix. Other environmental effects not taken into account in the environmental costs in table 7.2 include the contamination of groundwater by leaching and spillages of pollutants from transport, and damage to fauna and flora.

C.5 There is also a high social cost from the use of land for transport infrastructure, and from the disruption road and rail track causes. These effects are discussed in chapter 7, but they are not considered to be quantifiable (and they are offset to some extent by the social benefits provided by the increased access to the countryside afforded by growing transport), so no allowance for the social cost of the use of land for transport is included in table 7.2.

C.6 The Commission has relied on the results of a number of studies that have been made in the UK and abroad of the costs of environmental damage. It is considered that, taken together, these studies provide useful evidence that the costs of the environmental damage caused by transport are substantial, and indicate the order of magnitude of the costs that arise from the limited categories of environmental effects covered.

The cost of air pollution in the UK

C.7 Air pollution damages health and comfort. Road vehicles are the main influence on air quality over the large areas of the UK in which there are no significant industrial emissions (3.9). There remain many uncertainties about the effects of vehicle emissions on health, but individuals suffering from respiratory disorders may experience a worsening of symptoms when there are high concentrations of nitrogen dioxide and associated pollutants, especially particulates (3.24), and high ozone levels cause breathing difficulties in susceptible individuals (3.20). Other possible health effects are still under investigation but have caused much concern. In addition to the medical costs of ill health and the costs of resulting loss of production, there are the probably much higher costs of suffering by those affected and by their relatives.

C.8 Air pollution from transport dirties and damages the fabric of buildings, causing crumbling and the erosion of surfaces (3.49). Air pollutants also damage agriculture, forestry and water and soil (3.46–3.48). For example, high ozone levels affect crop yields, acidification of surface waters damages fisheries (and may also affect the quality of drinking water) and air pollution contributes to tree damage.

C.9 Recent estimates of the overall cost of air pollution in the UK tend to extrapolate from studies in other countries. The 1988 OECD report on transport and the environment[1] quoted estimates of air pollution damage attributable to transport as a few per cent of GDP in its member countries. A number of studies reported by Kageson[2] provide estimates of the costs of damage from air pollution from transport to health, buildings and materials, wildlife and habitats, which he summarised in terms of an equivalent percentage of the GDP of the country concerned. Limiting coverage to emissions of nitrogen oxides and Volatile Organic Compounds (VOCs), most studies indicated a cost of between 0.4 and 0.7% of GDP. The results are summarised below (the UPI study was considered by Kageson 'to be based on rather extreme assumptions and some double counting'):

External costs of air pollution from transport as % of GDP

Study reported		*Pollution cost (% GDP)*
Grupp[3]	(1986)	0.41%
Planco[4]	(1990)	0.67%
UPI[5]	(1991)	1.39%
ECOPLAN[6]	(1992)	0.68%
Infras[7]	(1992)	0.72%
MacKenzie[8]	(1992)	0.71%
CE[9]	(1988)	0.27–0.38%

C.10 Kageson concluded that the inclusion of other pollutants, in particular sulphur dioxide, would not materially alter the estimates of the total costs caused by transport; no account was taken of the cost of damage from particulates from diesels. The estimates reported also exclude the potential costs of global warming from emissions of carbon dioxide.

C.11 This range of values is supported by a major study of the external effects of transport in European countries for the UIC.[10] Air pollution was limited to emissions from transport of nitrogen oxides and VOCs; the results included detailed estimates for the UK (based on unit costs estimated for Sweden and UK emission levels) that indicated that the cost for the UK in 1993 might be between 0.4% and 1.1% of GDP. The inclusion of further pollutants was regarded as unlikely to increase significantly the total external costs of air pollution from transport.

C.12 The broad order of magnitude of air pollution costs is supported by other research. Estimates from a number of studies reviewed for the OECD[11], pointed to costs of air pollution from transport in the range of 0.1% to 1.4% of GDP; most were in the range 0.2% to 0.5% of GDP. Studies in 1986, 1987 and 1990 by different German authors[12] of the costs of air pollution from motor vehicles gave estimates which range between 0.7 and 1.5% of West German GDP. They covered health effects, the general dirt and soiling of buildings from local pollution, and wider regional effects like acid rain; however, they also included the contribution to global warming.

C.13 Excluding any contribution from emission of greenhouse gases to global warming, the studies reviewed suggest that the total cost of air pollution from transport in Britain is likely to be within the range of about 0.4% to 1.0% of GDP. This range excludes some outliers but most of the estimates were within this range. The lower value is higher than those reported by some studies, although none of the studies included all types of damage. Similarly, some studies have reported higher costs than those implied by the upper end of this range, but those may reflect unrealistic assumptions.

The cost of noise and vibration

C.14 Road traffic is the most pervasive source of noise in the environment and the most common source of outside noise heard in the home (4.5–4.7). There have been efforts in several countries to assign a monetary value to the cost of noise, normally from land-based transport. This is complex given the wide range of sensitivity to noise shown by different people. The effects of traffic noise have been variously evaluated in terms of productivity losses, decreases in asset values (normally house values) and costs of abatement measures (eg improved engine technology).

C.15 Differences in findings between countries reflect not only different ways of calculating the cost but also differences in exposure to noise. A number of the estimates made relate to Germany. Comparative studies suggest that the number of people exposed to seriously disturbing noise from road traffic (over 65 dB(A)) is broadly similar in the UK and West Germany (6–8 million) and that significantly more people in the UK are exposed to noise levels of 55–65 dB(A). However, railway noise appears to be more extensive in Germany.[13]

C.16 Vibration is a rather neglected topic and there has been no useful investigation of costs. Land-based transport is the prime source of ground-borne vibration; from tyre and road contact in the case of HGVs, buses and coaches and from rail/wheel contact in the case of freight trains. Airborne vibration (very low frequency noise) comes from diesel-powered engines and exhaust systems and although no physically damaging level has been established it causes unpleasant symptoms.

C.17 Estimates of the total cost of transport noise made in a number of European countries and in the USA mainly range between 0.06 and 1% of GDP. Studies reported by Kageson estimated the costs of noise to be equal to 0.1 to 0.9% of the GDP of the country concerned. These are summarised below; the reservation about the UPI estimate (C.9) applies also to noise. Comparison with the estimates of the costs of air pollution shows that the estimates of the cost of noise pollution are lower than the estimated cost of air pollution.

External costs of noise from transport as % of GDP

Study reported		*Noise cost (% GDP)*
Grupp	(1986)	0.08%
Planco	(1990)	0.09%
UPI	(1991)	1.98%
ECOPLAN	(1992)	0.52%
Infas	(1992)	0.71%
MacKenzie	(1992)	0.17%
CE	(1988)	0.06%
Willeke[14]	(1990)	0.90%
CETUR	(1984)	0.35%

C.18 The study of the external effects of transport in European countries for the UIC[15] yielded a single estimate of 0.64% of GDP for the cost of noise pollution from transport in the UK (based on unit costs estimated for Sweden and UK exposure levels). Estimates from a number of studies reviewed for the OECD[16] pointed to costs of noise from transport in the range of 0.02% to 0.9% of GDP.

C.19 In view of the higher level of exposure to noise in Britain, and its higher level of population density, the cost of noise and vibration is likely to be higher than in many of the countries covered in these studies. A small allowance for the cost of vibration damage and disturbance also has to be added. A range of 0.25 to 1.0% of GDP has therefore been taken as a likely range of costs for noise.

The cost of accidents

C.20 Assigning monetary costs to transport accidents is problematic; no price can be put on life, yet society needs criteria by which to make spending decisions that affect safety. So, despite the difficulty, valuations are made and used (to a much greater extent than for air or noise pollution). They play an important part in deciding where limited resources may best be employed to improve transport safety.

C.21 In assessing the cost of road casualties, DOT includes public sector payments, which include NHS costs for treating accident casualties, and police, ambulance and public legal costs; and the estimated cost of the lost output of those off work because of injury. The casualty valuations also include a notional allowance for the human cost of loss of life and suffering, based on estimates of willingness of individuals to pay to reduce their own level of risk of death or injury. For fatal accidents, these human costs form the major part of the costs estimated by DOT. The resultant estimates of casualty and accident costs are used to take account of expected changes in the number or severity of accidents arising from road schemes and to help in the assessment of safety proposals. Applying the latest DOT values to the 1993 casualty figures yields the figures in table C.1:

Table C.I
Cost of accident casualties (1993)

	cost per casualty	number of casualties	total
fatal	£744,060	3,814	£2.84 billion
serious	£84,260	45,009	£3.79 billion
slight	£6,540	257,197	£1.68 billion
total cost			£8.31 billion

C.22 Damage to vehicles and other property from all road accidents adds £2–2.5 billion per year, bringing the total estimated cost of accidents to about £10.5 billion. Not all these costs fall on the community, most of the costs of material damage are borne by road users either directly or indirectly through accident and health insurance premiums, as are some of the costs of injury and deaths. The University of Leeds Institute for Transport Studies suggests that the share of accident costs not borne by users is close to 50% of total accident costs.[17] This is equal to nearly 1% of GDP; but the real value may be higher because of under-reporting of accidents — Adams reported that 30% of traffic accident casualties and 70% of cyclist casualties were not recorded in police statistics.[18]

C.23 An alternative estimate of the total cost of road accidents that is borne by the whole community rather than by transport users, might include the whole casualty costs of fatal accidents, the cost of all injuries to pedestrians, cyclists and children, and part of the costs associated with the higher involvement of heavy goods vehicles in serious accidents. Adding in the relatively smaller cost of air, rail and water transport casualties, the total costs of accidents in Britain not borne by users, including all social costs, would amount to some £5.5 billion.

The cost of climate change

C.24 There is great uncertainty about both the extent of future climate change and its effects. Effects would vary significantly from country to country — for some the damage would be serious while for others there might even be net gains. Whatever the effects on this country, it can be argued on moral grounds that policy should be based primarily on accepting the implications of Britain's share of responsibility for climatic change effects. In consequence, in estimating a broad order of magnitude of costs, attention is focused here on the worldwide effects of emissions from transport in Britain on global warming, without giving separate attention to the effects of such pollution on this country. It is assumed that Britain should bear a responsibility for worldwide emissions of carbon dioxide in proportion to its share of world emissions; and that, for transport, cost responsibility should similarly be proportional to emissions of carbon dioxide.

C.25 A number of attempts have been made to estimate in very broad terms the possible cost worldwide of climate change, or global warming. These recognise the very large degree of uncertainty about both the magnitude of the potential global warming effect and, more importantly, about its effects and the costs of the consequent damage and loss, or of amelioration and prevention expenditures. Damage and other costs to the economy of developed countries has been calculated as likely to be between 0.25 and 2% of their GDP.[19] Cline[20] estimated the cost to the USA as 1% of its GDP. Studies[21] suggest that on average the cost of damage to developed countries will be within a range of 1% to 1.5% of GDP. Some 25% of the contribution to this effect is ascribed to transport worldwide, which is a similar position to that in the UK. The cost of damage in developing countries is expected to be more serious; estimates suggest costs at between 2% and 6% of their GDP.[22] These estimates do not include any estimate of the 'human' or 'social' costs associated with deaths, disease and suffering which might be caused by global warming.

C.26 Estimates of the costs of uncertain future damage must inevitably be extremely tentative. However, an appropriate allowance for the contribution of transport in the UK to climate change effects each year can be calculated in very broad terms. In doing so it has to be acknowledged that there is a further problem of interpretation: the global warming effect is expected to have relatively little impact in the short term, but to give rise to increasing costs in the long term; yet present emissions will be contributing to that growing future effect. So, while the global warming effect is described here in terms of a percentage of world GDP averaged over a long period, there is no close relationship between emissions in a particular year and the global warming effect in future years. Despite the difficulties of interpretation this causes, the average cost to world GDP each year over the period to 2050 arguably provides an appropriate basis on which to estimate the cost responsibility of the damage caused by carbon dioxide emissions from transport in Britain.

C.27 This provides the basis on which the cost of global warming attributed to emissions from transport in Britain was calculated. The stages in this calculation were:

(1) it has been estimated that, for all sources of carbon dioxide emissions in 1990, Britain was responsible for 2.8% of world emissions, and that transport was responsible for 24% of these emissions, or 0.7% of world emissions;

(2) the damage from global warming caused by emissions each year is assumed to be equivalent to between 1% and 1.5% of the GDP of developed countries[23] and between 2% and 6% of the GDP of other countries in the year in which the emissions occur;

(3) on the basis of 1994 levels of world GDP, these assumptions imply a cost responsibility for UK transport of £1.8 billion to £3.6 billion in 1994/95. The degree of uncertainty attached to this estimate is large.

These costs are shown in table 7.2.

Appendix D

CARBON DIOXIDE EMISSIONS FROM SURFACE TRANSPORT: EFFECTS OF COMMISSION'S RECOMMENDED MEASURES

D.1 A number of measures recommended in the report are intended to reduce carbon dioxide emissions from road transport. The combined effect of these measures on emissions in 2000 and 2020 from surface transport is summarised in table D.1. This is based on the low forecast of the National Road Traffic Forecasts published by the Department of Transport in 1989. The final column of the table shows emissions in 2000 and 2020 as a percentage of emissions in 1990. It shows that the recommended measures would achieve the Commission's proposed targets limiting carbon dioxide emissions from surface transport in 2000 to the 1990 level and reducing them to 80% of 1990 levels by 2020.

D.2 The effects of a particular measure on its own would not be the same as its effects when combined with other measures. In addition, the reduction produced by a given measure will vary in amount, although not as a percentage of the preceding total of emissions, according to the order in which the contributions of the measures are assessed.

D.3 The measures included in the table are:

a doubling of fuel prices by 2005. Increases in fuel duty are recommended to double the price of fuel in real terms over 10 years (7.58);

increased effort devoted to enforcement of 60 mph and 70 mph speed limits on single and dual carriageway roads (12.25);

40% improvement in average fuel efficiency of new cars sold in the UK by 2005, together with a 20% improvement for new light goods vehicles and a 10% improvement for new heavy duty vehicles (8.59). Part of the increase in vehicle fuel efficiency will result from the increase in the price of fuel; the table shows the additional effect if targets are achieved in full;

halving the growth of car traffic in urban areas. It has been assumed that, spread over all built-up areas, the effect of achieving target C1 for reducing the proportion of journeys made by car will be to reduce the growth in car traffic to half the overall rate indicated in the low forecast made in 1989, after adjusting for the effects of the other measures listed.

D.4 The estimates in the table allow for additional carbon dioxide emissions from public transport as a result of the transfer of passenger and freight traffic. These include the indirect transport-related emissions of carbon dioxide from power stations due to increased running of electric trains. Indirect transport-related emissions from oil refineries are not covered by the table.

D.5 Other measures recommended in the report will influence carbon dioxide emissions but it is difficult to separate out their effects. In most cases these are likely to be on a smaller scale than the effects shown in the table and would not add significantly to overall reductions.

D.6 A recently published study identifies a group of measures which would reduce energy use from transport over the next few decades at the long-term rate at which the primary energy intensity of the economy as a whole has been falling.[1] They overlap with the measures recommended in this report, but the Commission's proposal for transport over this period is for a faster reduction in energy use.

Table D.1
Combined effect of recommended measures on carbon dioxide emissions from surface transport

million tonnes carbon/year

		cars	road	total	% of 1990
	1990	19.7	30.5	32.3	100
baseline	2000	22.4	34.7	36.4	113
doubling of fuel prices by 2005		20.2	31.3	33.2	103
enforcement of 60 mph and 70 mph speed limits		19.1	30.1	32.0	99
40% improvement in fuel efficiency of new cars by 2005		18.4	29.1	31.1	97
halving of growth of car traffic in urban areas		18.2	28.9	30.9	96
baseline	2020	26.9	42.9	44.7	138
doubling of fuel prices by 2005		18.2	29.5	32.0	99
enforcement of 60 mph and 70 mph speed limits		17.1	28.4	30.9	96
40% improvement in fuel effiency of new cars by 2005		13.1	23.8	26.3	82
halving of growth of car traffic in urban areas		12.6	23.2	25.7	80

These calculations are based on the low forecast of traffic growth in the National Road Traffic Forecasts. The baseline for carbon dioxide emissions is taken from the modelling study carried out for the Commission by ERR.

Appendix E

COMMISSION'S FIFTEENTH REPORT: LETTER FROM THE CHAIRMAN REPLYING TO THE GOVERNMENT'S RESPONSE

16 December 1992

Rt Hon John MacGregor OBE MP
Secretary of State for Transport

Emissions from heavy duty diesel-engined vehicles

The Royal Commission welcomes the detailed response published by the Government to its Fifteenth Report.[1] We are encouraged to note that many of the recommendations made in that Report are being carried forward. We are disappointed however that some of the recommendations have not been taken up, and wish to re-emphasise our concern regarding the following crucial points.

Replacement of engines in old buses (para 40 of the Government's response)

Buses are major contributors to air pollution blackspots in urban centres. This is in large measure due to the age of their engines, some of which are very old. Emissions standards for new engines, even when accompanied by in-service smoke tests, will not deal with this problem, especially with the levels of nitrogen oxides emitted. We also doubt whether the availability of capital allowances for new engines will be sufficient to induce the large-scale scrapping of old engines. We continue to believe therefore that the Government ought to provide direct incentives for the fitting of new or rebuilt engines to old buses.

Economic instruments (para 32–35)

We agree major gains for environmental quality can come from the use of economic instruments that work with the grain of the market. There are nevertheless circumstances where that particular type of mechanism will not be effective and a more positive approach is required, involving specific grants or tax concessions. We are disturbed by the implication at several points in the response that there is an overriding objective of tax neutrality in Government policy which takes no account of the economic distortions caused by pollution, and we do not believe this position is consistent with the Environment White Papers. The replacement of engines in existing buses is one context where tax concessions could be used in a cost-effective manner. Limiting the benefit to buses should not be an insuperable problem.

Fuel additives (para 65)

We must again express our concern that sufficient attention is not given to the possibility of harmful effects from fuel additives. Without adequate testing we doubt that advice given by the Government to manufacturers can be fully informed, or that there will be a basis for the Government to take decisions about the use of its powers under the Environmental Protection Act. New procedures are required to ensure there is adequate testing and discussion before new additives are brought into widespread use.

In accordance with the Commission's practice I am making this letter available to the press and public.

SIR JOHN HOUGHTON

Appendix F

STATEMENT BY THE COMMISSION ON COUNTERING CLIMATE CHANGE

The Royal Commission on Environmental Pollution welcomes the present national debate about the UK programme for limiting carbon dioxide emissions. The Government's discussion document[1] contains valuable information about the sources of, and trends in, such emissions, and possible measures to reduce them.

Action to counter climate change will need the widest possible public support. But relying solely on voluntary action would not produce commitments to cut carbon dioxide emissions which can be monitored and guaranteed. The Commission envisages Government action, employing economic instruments and other forms of regulation, as playing a major role in achieving specific targets for reductions.

The Government's discussion document in effect deals only with the position up to 2000. That reflects the terms of the obligation which the UN Framework Convention on Climate Change placed on the UK and other countries. It is important however to start formulating policy for the period after 2000, when reductions in emissions will be necessary for environmental reasons in the face of likely strong pressures towards increases.

Vigorous efforts will be required to achieve much greater efficiency in energy use. The main problem area is transport, and predominantly the car. The tables in the discussion document show that:

- total UK carbon dioxide emissions fell by 10% between 1970 and 1990, but emissions from transport increased by 65%
- transport accounts for the whole of the net increase projected in UK carbon dioxide emissions between 1970 and 2020 (an increase of 39 million tonnes of carbon a year)
- two-thirds of that projected increase in carbon dioxide emissions is accounted for by private cars.

The Commission is carrying out a study of transport and the environment. The main focus of its report, due to be published next year, will be on the period after 2000. The intention is to provide guidance which will help overcome the challenge of reconciling people's needs and desires for mobility with an overall plan for the UK to achieve sustainable development. In the course of its study the Commission is examining, not only the environmental impact of transport, but the relevance of land use policies, technology, economic issues and the institutional framework.

The evidence submitted and the available data already show that, in view of society's present dependence on energy-intensive road transport, fundamental changes will be needed to create a sustainable transport system. There are other important effects on the environment, besides those from energy use. Transport is the main source of levels of air pollution which are causing concern on health grounds. Projected increases in numbers of vehicles could cancel out the benefits from cleaner engines and catalytic converters. A response is also needed to the widespread public concerns about noise from transport and about damage to protected areas from road building.

Among the crucial issues are:

- the tendency of energy use for transport to grow much more rapidly than the economy — the reverse of what has happened with other energy uses
- the attitudes of the public, industry and government towards car performance
- ways of ensuring that decisions by car users reflect the costs their vehicles impose on the environment

the feasibility of patterns of land use which, rather than encouraging maximum use of the car, reduce the need for journeys

the availability of public transport where it represents the best way of limiting energy use and pollution, and the attractiveness of fares and services.

The difficult longer-term questions about policies after 2000 will soon have to be confronted. Transport policies will have to be devised which are both environmentally sustainable and socially acceptable. As an essential first step, the Commission urges the Government to accept that specific targets must be set for reducing carbon dioxide emissions from the transport sector, and realistic programmes put in place for achieving those targets. This will be an important contribution to ensuring that all major decisions about transport and land use policies are taken in the light of their long-term consequences for the environment.

11 March 1993

Appendix G

INVITATION TO SUBMIT EVIDENCE

On 14 May 1992 the Commission issued a press release in the following terms:

TRANSPORT AND THE ENVIRONMENT

The Royal Commission on Environmental Pollution has today invited evidence for a new study on transport and the environment.

The study will explore the options for developing transport strategies aimed at reconciling the necessary movement of people and goods with the need to protect the environment. It will look at the scope for technological, regulatory, fiscal and other measures which could contribute, in the medium to long term, to the development of an environmentally sustainable transport policy.

Among the issues the Commission intends to investigate are the contribution of transport to emissions of greenhouse gases and atmospheric pollutants, and the effects of traffic and transport infrastructure on natural habitats and the manmade environment. The Commission will take into account the relationship between transport, land use and planning policies.

The Commission will focus primarily on the United Kingdom but developments in Europe and elsewhere will be considered. It intends to give particular attention to the following:

the institutional framework for transport policy;

the nature, scale and costs of the environmental effects of transport;

land use and planning policy considerations, both urban and rural;

issues relating to the movement of freight;

the potential contribution of developments in technology to the alleviation of transport-related environmental problems;

an economic analysis of these problems and of measures to mitigate them.

The Commission would welcome evidence from any organisations or individuals with an interest in these issues.

Appendix H

ORGANISATIONS AND INDIVIDUALS CONTRIBUTING TO THIS STUDY

Government bodies

Department of the Environment
Department of Health
Department of Trade and Industry
Department of Transport
Foreign and Commonwealth Office
HM Treasury
Scottish Office
Welsh Office

Commission of the European Communities

Other organisations

Agricultural and Food Research Council
Air Operators Association
Air Transport Users Committee
Airfields Environment Federation
Associated Octel
Association of Chief Police Officers
Association of County Councils
Association of District Councils
Association of Local Bus Company Managers
Association of London Authorities
Association of Metropolitan Authorities
Automobile Association
Baltic Exchange
Berkeley Hanover Consulting
Berkshire County Council
Birmingham City Council
Blackburn, Hyndburn and Ribble Valley PHA
BMW (GB) Limited
Bristol Civic Society
British Airways
British Medical Association
British Ports Federation
British Rail Network SouthEast
British Railways Board
British Road Federation
British Telecom
British Waterways Board
Bus and Coach Council
Cambridgeshire County Council
Cardiff Business School
Central Transport Consultative Committee
Centre for Exploitation of Science and Technology
Centre for Local Economic Strategies
Chamber of Shipping
Chartered Institute of Transport

Chelmsford Borough Council
Chemical Industries Association
Child Accident Prevention Trust
City of York
Confederation of British Industry
Construction Industry Council
Convention of Scottish Local Authorities
Council for the Protection of Rural England
Council of Science and Technology Institutes
Countryside Commission
Countryside Council for Wales
Cranfield Centre for Logistics and Transportation
Cyclists' Touring Club
Derbyshire County Council
Earth Resources Research
Economic and Social Research Council
Energy Technology Support Unit
English Heritage
English Nature
English Tourist Board
Environmental Health Officers from London Boroughs
European Conference of Ministers of Transport
European Environmental Bureau
European Federation for Transport and the Environment
European Foundation for the Improvement of Living and Working Conditions
European Ports Working Group
Ford Motors
Freight Transport Association
Friends of the Earth
General Motors
Greater Manchester Transport Action Group
Greenpeace
Hamworthy Pumps and Compressors Limited
Health and Safety Executive
Hong Kong Government
Inland Waterways Association
Institute for European Environmental Policy
Institute of Logistics and Distribution Management
Institution of Civil Engineers
Institution of Environmental Health Officers
Institution of Highways and Transportation
International Institute for Environment and Development
International Maritime Organization
International Tanker Owners Federation
Joint Nature Conservation Committee
Kent County Council
Law Society
Local Government Commission
London Amenity and Transport Association
London Borough of Tower Hamlets
London Boroughs' Association
London Ecology Unit
London Planning Advisory Committee
London Regional Transport
London Rivers Association
London Transport Planning
Marks and Spencer
Manchester Diocese Board for Social Responsibility

Meteorological Office
Midlands Amenity Societies Association
National Economic Development Office
National Farmers' Union
National Federation of Bus Users
National Federation of Women's Institutes
National Rivers Authority
National Society for Clean Air
National Trust
National Voluntary Council for Children's Play
Natural Environment Research Council
Natural Gas Vehicles Association
New Economics Foundation
Newcastle upon Tyne City Council
Northumberland County Council
Nottinghamshire County Council
Open University
Oxford City Council
Oxfordshire County Council
Oxford University Transport Studies Unit
Parcelforce
Parliamentary Advisory Council for Transport Policy
Pedestrians' Association
Perkins Technology Ltd
Port of London Authority
Public Enterprise Group
Public Health Alliance
Railstore Ltd
Railtrack
Railway Development Society
Railway Path and Cycle Route Construction Co (Sustrans)
Road Haulage Association
Rolls-Royce
Royal Academy of Engineering
RAC Foundation for Motoring and the Environment
Royal Institution of Chartered Surveyors
Royal Mail
Royal Society for the Prevention of Accidents
Royal Society for the Protection of Birds
Royal Society of Arts
Royal Town Planning Institute
Ryedale District Council
Save Our Severnside
Save Waste and Prosper
Science and Engineering Research Council
Scottish Natural Heritage
Shell UK
Shell International
Slough Borough Council
Society of Motor Manufacturers and Traders
South Coast Air Quality Management District (California)
South East Regional Planning Conference
South Yorkshire Passenger Transport Executive
South Yorkshire Transport Ltd
Southampton University (Institute of Sound and Vibration Research)
Stafford Borough Council
Standing Advisory Committee on Trunk Road Assessment
Strathclyde Regional Council

Town and Country Planning Association
Transport Development Group (Bedford)
Transport and General Workers Union
Transport and Health Study Group
Transport Director for London
Transport 2000
Transportation Planning Associates
Tyne and Wear Passenger Transport Executive
United Kingdom Petroleum Industry Association
University College London Transport Studies Group
University of Leeds Institute for Transport Studies
Vauxhall Motors Limited
Volkswagen Audi (UK)
Volvo Car Corporation
Warwickshire Nature Conservation Trust
West Midlands Passenger Transport Executive (Centro)
World Wide Fund for Nature

Individuals

Dr J G U Adams, University College London
Mr N Allonby
Mr C Amundson
Professor R Anderson, St George's Hospital
Dr L J Archer, Oxford Institute of Energy Studies
Dr M R Ashmore, Imperial College of Science, Technology and Medicine
Dr J G Ayres, Birmingham Heartlands Hospital
Dr D J Banister, University College London
Dr T Barker, Cambridge Econometrics
Dr M Barrett, Pollen
Mr T Bendixson
Mr T Benjamin, International Drivers' Behaviour Research Association
Dr S A Boehmer-Christiansen, Science Policy Research Unit, University of Sussex
Dr P Brimblecombe, University of East Anglia
Mr D Broome
Professor K J Button, Loughborough University of Technology
Dr C L Cheeseman, MAFF Central Science Laboratory
Professor M D I Chisholm, University of Cambridge
Professor R U Cooke, University College London
Professor R J Davies, St Bartholomew's Hospital
Dr R G Derwent, Meteorological Office
Dr J L Devalia, St Bartholomew's Hospital
Dr C Eaglen
Dr J Emberlin, Pollen Research Unit, University of North London
Professor H-J Ewers, University of Münster
Mr S Fankhauser, CSERGE, University College London
Mr J Farman, European Ozone Research Co-ordinating Unit, Cambridge
Mr M Fergusson, Earth Resources Research
Mr W Fransen, Royal Netherlands Meteorological Institute
Mr A Frogley
Dr P B Goodwin, University of Oxford Transport Studies Unit
Professor H Grassl, University of Hamburg
Dr P W Greig-Smith, MAFF Central Science Laboratory
Mr N Haigh, Institute for European Environmental Policy
Professor P G Hall, University College London
Dr R S Hamilton, Middlesex University
Professor R Harriman, Scottish Office Agriculture and Fisheries Department, Freshwater Fisheries Laboratory, Pitlochry

Mr T Hart, University of Glasgow
Dr D Hill, Ecoscope
Dr M Hillman, Policy Studies Institute?
Dr C Holman
Dr M Hornung, NERC Institute of Terrestrial Ecology
Dr P S Hughes
Dr C M Jefferson, University of the West of England, Bristol
Dr C E Johnson, AEA Harwell
Dr H C Longuet-Higgins, University of Sussex
Mr D Lowe, European Re-Engining Consortium
Dr A C McKinnon, Heriot-Watt University
Mr R McQuillan
Mr C G B Mitchell, Transport Research Laboratory
Professor C A Nash, University of Leeds Institute for Transport Studies
Professor D E Newland, University of Cambridge
Professor P Nijkamp, Free University of Amsterdam
Dr S E Owens, University of Cambridge
Professor D W Pearce, University College London
Mr F Pearce, New Scientist
Mr S Plowden
Mr M Quinn
Mr E Relton
Professor T M Ridley, Imperial College of Science, Technology and Medicine
Professor W Rothengatter, University of Karlsruhe
Professor M F Russell, Institute of Sound and Vibration Research, University of Southampton
Dr L Schipper, International Energy Studies, Lawrence Berkeley Laboratory
Sir Clive Sinclair
Dr S Smith, Institute of Fiscal Studies
Dr H Somerville, British Airways
Mr S Sorrell, Science Policy Research Unit, University of Sussex
Mr D C Statham, National Park Officer, North Yorks Moors National Park
Professor D H Tidmarsh, Sheffield Hallam University
Dr M Treisman
Mr D Turner, Traffic Director for London
Mr A Webb, National Power
Mr A Wenban-Smith
Dr J Whitelegg, University of Lancaster
Dr K G Willis, University of Newcastle upon Tyne
Dr S P Wolff, University College and Middlesex School of Medicine
Dr J Wootton, Transport Research Laboratory
Mr F J Worsford, University of Westminster Transport Studies Group

Visits

Members of the Commission made visits in order to meet representatives of the organisations listed below.

30 June to 2 July 1993 Yorkshire, Tyneside and Strathclyde

Department of Transport Regional Office Newcastle upon Tyne
Department of Environment Regional Office Newcastle upon Tyne
Ryedale District Council
York City Council
North Yorkshire County Council
Newcastle upon Tyne City Council
Tyne and Wear Passenger Transport Executive
University of Northumbria, Department of Politics
University of Newcastle, Department of Engineering
Strathclyde City Council

1–3 July 1992 The Netherlands

—The Hague
Ministry for Housing, Physical Planning and Environment
Commission for Long Term Environmental Policy
National Council for Environmental Protection
Strichtung Natuur en Milieu

—Utrecht
Commission for Environmental Impact Assessment
Netherlands Railways

—Delft
City Development Office

10 December 1993 Hong Kong
Planning, Environment and Lands Branch
Environmental Protection Department

13–16 December 1993 Japan
Environment Agency
National Police Agency
Ministry of Transport
Ministry of International Trade and Industry
Ministry of Construction
Metropolitan Expressway Public Corporation
National Land Agency
Tokyo Metropolitan Government Environment Protection Agency
Railway Technical Research Institute
Ministry of Transport, Traffic Safety and Nuisance Research Institute
Isuzo Motors Ltd
Tokyo Metropolitan Government Research Institute for Environment Protection

11–18 January 1994 USA

—Los Angeles
California Air Resources Board
South Coast Air Quality Management District
Southern California Association of Governments
Los Angeles County Metropolitan Transportation Authority
Warner Center Transportation Management Organisation

—Portland, Oregon
City Planning Bureau
City Energy Office
City Office of Transportation
Metropolitan Service District
Tri-County Metropolitan Transportation District
Policy Adviser to the State Governor
Oregon Environmental Council
1000 Friends of Oregon

—Washington DC
White House Office of Environmental Policy
White House Office of Science and Technology Policy
National Science Foundation
National Academy of Engineering
Department of Commerce
Environmental Protection Agency
Washington Metropolitan Area Transit Authority
American Planning Association

11–12 January 1994 Sweden
  Ministry of Environment, Stockholm
  Nature Conservation Society, Stockholm
  Swedish National Chemicals Inspectorate, Solna

13 January 1994 Norway
  Public Roads Administration
  Oslo City Council, Environment and Transport
  Department of Environment, Regional Air Pollution and Transport
  Centre for International Social Science Research

24–26 January 1994 Switzerland
  Zurich City Planning Authority
  Transport Planning, Zurich City
  Regional Rapid Transit Authority

14–16 March 1994 France
  Ministry of Transport, Paris
  Centre d'Etudes des Transports Urbains
  Le Syndicat des Transports Parisiens
  L'Agence de l'Environement et de la Maîtrise de l'Energie
  Société Nationale des Chemins de Fer Français
  Fédération Nationale des Usagers de Transport
  Régie Autonome des Transports Parisiens
  Conseil Régional de l'Ile-de-France
  Syndicat Mixte des Transports en Commun de l'Agglomération Grenobloise
  Conseil Général de la Savoie
  Société Lyonnaise des Transports en Commun

Appendix J

MEMBERS OF THE ROYAL COMMISSION

Chairman

SIR JOHN HOUGHTON CBE FRS

Member of the Government Panel on Sustainable Development
Co-Chairman of the Scientific Assessment Working Group of the Intergovernmental Panel on Climate Change
Chief Executive (previously Director-General) of the Meteorological Office 1983–91
Formerly Deputy Director of the Rutherford-Appleton Laboratory, Science and Engineering Research Council
Professor of Atmospheric Physics, Oxford University 1976–83
President, Royal Meteorological Society 1976–78
Vice-President of World Meteorological Organization 1987–91

Members

SIR GEOFFREY ALLEN PhD FRS FEng FIC FIM FRSC FInstP

Executive Adviser to Kobe Steel Ltd
President of the Institute of Materials
Chairman of Enviromed plc
Chancellor of the University of East Anglia

PROFESSOR G S BOULTON FRS FRSE†

Regius Professor of Geology, University of Edinburgh
Extraordinary Professor, University of Amsterdam 1980–86
Member, Natural Environment Research Council
Chairman, Natural Environment Research Council, Earth Science and Technology Board

PROFESSOR H CHARNOCK CBE FRS¶

Visiting Fellow, Department of Oceanography, University of Southampton
Senior Visiting Fellow, Institute of Oceanographic Sciences
Formerly Director, Institute of Oceanographic Sciences
President, International Union of Geodesy and Geophysics 1971–75
President, Royal Meteorological Society 1982–84

PROFESSOR C E D CHILVERS BSc(Econ) MSc Hon MFPHM†

Professor of Epidemiology, University of Nottingham and Vice Dean of the Faculty of Medicine and Health Sciences
Non-executive Director of Nottingham Community Health NHS Trust
Member, Committee on Carcinogenicity of Chemicals in Food, Consumer Products and the Environment
Director, Trent Institute for Health Services Research

PROFESSOR DAME BARBARA CLAYTON DBE MD PhD Hon DSc FRCP FRCPE FRCPath

Honorary Research Professor in Metabolism, University of Southampton
Past-President, Royal College of Pathologists

Chairman, Medical Research Council Committee on Toxic Hazards in the Environment and the Workplace
Chairman, Health of the Nation Nutrition Task Force
Chairman, Standing Committee on Postgraduate Medical and Dental Education
Chairman, Medical and Scientific Panel of the Leukaemia Research Fund

DR PETER DOYLE CBE FRSE#

Executive Director, Zeneca Group plc
Member, Medical Research Council
Member of Council, Centre for the Exploitation of Science and Technology
Member, Advisory Council on Science and Technology 1989–93
Member of Department of Health's Research and Development Committee's Standing Group on Health Technology Assessment
Member of Council, University College London
Member of Council, Royal Society of Chemistry

H R FELL FRAgS NDA MRAC

Managing Director, H R Fell and Sons Ltd
Council Member, Royal Agricultural Society of England
Member, Minister of Agriculture's Advisory Council on Agriculture and Horticulture 1972–81
Commissioner, Meat and Livestock Commission 1969–78
Past-Chairman, Tenant Farmers Association

SIR MARTIN HOLDGATE†

President of the Zoological Society of London
Chairman of the Government's Energy Advisory Panel
Chief Scientist and Deputy Secretary, Department of the Environment, 1976–88
Director General of the International Union for Conservation of Nature and Natural Resources, 1988–94

P R A JACQUES CBE BSc¶

Special Adviser, Health and Safety, Trades Union Congress
Member of the Health and Safety Commission

PROFESSOR J H LAWTON BSc PhD DSc FRS

Director, Natural Environment Research Council Centre for Population Biology
Professor of Community Ecology, Imperial College of Science, Technology and Medicine
Honorary Research Fellow, Natural History Museum, London
Adjunct Scientist, Institute of Ecosystem Studies, New York
Chairman, Royal Society for the Protection of Birds

PROFESSOR R MACRORY Barrister MA(Oxon)

Director-designate, Environmental Change Unit, Oxford University
Denton Hall Professor of Environmental Law, Imperial College of Science, Technology and Medicine
Associate Director, Imperial College Centre for Environmental Technology
First Chairman of the UK Environmental Law Association 1986–88
Editor–in–Chief of the Journal of Environmental Law
Honorary Chairman, Merchant Ivory Productions Ltd

PROFESSOR J G MORRIS CBE DPhil FIBiol FRS

Professor of Microbiology, University of Wales, Aberystwyth
Chairman, Biological Sciences Committee of the Science and Engineering Research Council 1978–81
Chairman, Biological Sciences Committee of the University Grants Committee 1981–86

D A D REEVE CBE BSc FEng FICE FIWEM*

Deputy Chairman and Chief Executive, Severn Trent Water Authority 1983–85
Past-President, Institute of Water Pollution Control
Past-President, Institution of Civil Engineers
Member, Advisory Council on Research and Development, Department of Trade and Industry
Additional Member, Monopolies and Mergers Commission

EMMA ROTHSCHILD MA¶

Director, Centre for History and Economics and Fellow, King's College, Cambridge
Research Fellow, Sloan School of Management, Massachusetts Institute of Technology (MIT)
Associate Professor of Science, Technology and Society, MIT 1978–88
Member, OECD Group of Experts on Science and Technology in the New Socio-Economic Context 1976–80
Board Member, British Council

THE EARL OF SELBORNE KBE FRS

Managing Director, The Blackmoor Estate Ltd
Chairman, Joint Nature Conservation Committee
Chairman, House of Lords Select Committee on Science and Technology
Chairman, Agricultural and Food Research Council 1983–89

PROFESSOR Z A SILBERSTON CBE MA

Senior Research Fellow, Management School, Imperial College of Science, Technology and Medicine
Professor Emeritus of Economics, University of London
Director, Brussels Office, London Economics
Member, Biotechnology Advisory Commission, Stockholm Environment Institute
Vice-President, Royal Economic Society
Secretary-General, Royal Economic Society (1979–92)
Member, Restrictive Practices Court (1986–92)
President, Confederation of European Economic Associations (1987–89)
Formerly Specialist Adviser, House of Lords Select Committee on the European Communities

* Mr Reeve died on 28 June 1994
\# Dr Doyle joined the Commission on 12 May 1994
† Joined Commission on 1 September 1994
¶ Left Commission on completion of the Transport Study

GENERAL INDEX